T0255653

Project Management

Project Management: Planning and Scheduling Techniques is a highly readable guide to the essentials of project planning, scheduling, and control aimed at readers looking for an introduction to the core concepts of planning and scheduling, including the 'Critical Path Method', but also the 'Precedence Diagramming Method', the 'Line of Balance' technique, and the 'Programme Evaluation and Review Technique'. This book explains the theory behind the methods and makes effective use of learning outcomes, exercises, diagrams, and examples to provide clear and actionable knowledge for students and project managers.

The book can be used as a classroom textbook or as a self-study guide for project managers taking their professional qualifications, and it includes examples from a wide range of project management scenarios. It is suitable for planning and scheduling courses in the fields of industrial, civil, and mechanical engineering, construction, and management.

Vijay Kumar Bansal is a faculty member of the Department of Civil Engineering at the National Institute of Technology (NIT) Hamirpur, India. He has more than two decades of experience in teaching, research, and fieldwork in project management: he is a resource person for many international e-learning hubs, has guided many Ph.D. and Master's Programme theses, and has completed many research projects. The author's research interests include 4D scheduling, repetitive scheduling, location-based planning scheduling, etc. He has contributed many research papers to national and international journals, conferences, and magazines, published mainly by Taylor & Francis, American Society of Civil Engineers, Elsevier, Springer, etc. Apart from teaching and research, the author has worked in relevant positions such as executive engineer, associate dean, etc.

Project Management
Planning and Scheduling Techniques

Vijay Kumar Bansal

LONDON AND NEW YORK

Designed cover image: © Twenty47studio/Getty Images

First published 2024
by Routledge
4 Park Square, Milton Park, Abingdon, Oxon OX14 4RN

and by Routledge
605 Third Avenue, New York, NY 10158

Routledge is an imprint of the Taylor & Francis Group, an informa business

British Library Cataloguing-in-Publication Data
A catalogue record for this book is available from the British Library

ISBN: 978-1-032-55091-6 (hbk)
ISBN: 978-1-032-54937-8 (pbk)
ISBN: 978-1-003-42899-2 (ebk)

DOI: 10.1201/9781003428992

Typeset in Times New Roman
by Deanta Global Publishing Services, Chennai, India

Do not let others do your planning for you. Develop your plan, schedule it, and execute it. Make your own decisions. Your plan is within you; when your creativity is focused, the plan that will come is executable….

Contents

Preface

Project management is the process of leading a project team to achieve project objectives, or the process of leading a project team to complete a project within the agreed time duration, allocated budget, and quality. Project management focuses on the delivery of project deliverables within the constraints of time duration, cost, and acceptable quality. The need for effective project management is ever-increasing with regard to developing an executable project plan, achieving large profits, and lowering expenditures. Knowledge of project management helps individuals or organizations stay alive in a cut-throat market.

Project management is a multidisciplinary field that has emerged to cover the management of all the stages of a project, from start to finish. The author, in his teaching and professional career, has felt a need for a textbook that can be used as self-study material for students and project management professionals. The author has always felt a need for a textbook that explains all the key concepts of project management clearly and without any shortcuts, aimed at students and project management professionals. Further, the author has felt a need for a simple textbook which addresses the needs of the average student. Many of the available project management books are focused on the *critical path method*, in which only activity-on-arrow networks are used, so the author felt that the *precedence diagramming method* needed to be documented in the easiest way possible to make readers and project management professionals familiar with it. With these requirements in mind, the idea of writing that textbook developed.

We may provide an outline for this textbook. It starts with an introductory chapter that covers the basic concepts of project management. For effective project management, a project is divided into small manageable parts called activities. The logical process of obtaining a complete list of project activities is developing the *work breakdown structure* of a project. Chapter 2 covers the development of a project's work breakdown structure. The inter-dependencies between the project's activities are defined to develop its execution sequence. Defining inter-dependencies is the process of defining preceding and subsequent activities for each activity in a project. Chapter 3 covers how an execution sequence is developed and represented in the form of a *bar chart*. Chapter 4 covers how an execution sequence is developed and represented in the form of a *network diagram*.

Chapters 5 and 6 cover the *critical path method*, a widely used technique for project planning and scheduling. It is used in the planning and scheduling of projects in which a planner is familiar with the project in question. Additional resources are sometimes used to reduce the time duration of a project in a process called *project crashing*, discussed in Chapter 7. The objective of project crashing is to determine the optimum time duration of a project, that at which the total project cost is minimized. Projects are sometimes scheduled in such a way that the requirements of various types of resources are kept almost uniform throughout the time duration of a project, because variations in resources requirements are difficult to manage in practice. Thus, the objective is to avoid peaks and troughs in resources usage profiles; this process is called the *levelling of project resources*, and is covered in Chapter 8. A project rarely goes according to the plan and schedule; deviations from the original plan and schedule may occur. When the execution of a project begins, the actual progress achieved at a given point in time is compared with the planned progress, and corrective measures are decided upon. The original plan and schedule are modified or updated to incorporate the corrective measures in a process called *network updating*, which is covered in Chapter 9.

The *programme evaluation and review technique* is used for the planning and scheduling of projects of a non-routine nature, or for projects in which the time duration estimates for their various activities lack a fair degree of accuracy. Such projects have a large amount of uncertainty in their estimated project activity time durations. Therefore, a probability value is associated with the time duration of the project. This probability depends upon the uncertainty involved in the estimated time durations of the various activities in the project. Chapters 10 and 11 cover the program evaluation and review technique.

The networks used in the critical path method are based on a relationship in which an activity starts only after the activities preceding it are completed, which is called a finish-to-start relationship. In the *precedence diagramming method* three additional relationships (finish-to-finish, start-to-start, and start-to-finish) are used, as discussed in Chapter 12. This chapter also describes the preparation of a calendar date schedule. For scheduling repetitive projects, *line of balance* is the widely used technique that ensures continuous crew engagement. Chapter 12 covers the *line of balance* technique in detail.

During the last few years, knowledge of project management and understanding of project management has matured; almost every company is using project management in one or another form. This textbook addresses readers who aspire to understand project planning and scheduling techniques. These techniques are simple tools for project management. The contents covered in the book are fundamental in nature. Each chapter starts with learning objectives and ends with a chapter conclusion. Illustrated examples are provided for every technique covered. These examples have been developed by the author and solved manually for the sake of learning and practice, however they may be solved with the help of software tools. The textbook also includes an exercise at the end of each chapter with descriptive answers. The book is also an easy self-study tool for non–project managers wishing to learn project planning and scheduling techniques. The author has illustrated each project planning and scheduling technique with theory, diagrams, and examples to help develop clear, understandable, and actionable content for project managers.

The book is for both under-graduate and graduate courses in engineering, business administration, management science, and information technology. The book addresses students wishing to improve their project management skills and also those managers and executives who wish to provide continuous support to projects. The author hopes he has succeeded in providing this, but he is sure that students, teachers, and project management professionals will have suggestions, corrections, and criticisms of the text presented in the book. The author encourages readers to send their valuable suggestions, corrections, and criticisms so that he can include these necessary changes in future editions. Please mail your comments to:

Vijay Kumar Bansal
Website: https://nith.ac.in/

1 Project Management

1.1 Learning Objectives

After the completion of this chapter, readers will be able to understand:

- Project management and the different steps involved in project management,
- The difference between project planning and project scheduling, and
- The widely used project planning and project scheduling techniques.

1.2 Introduction

The past is history; by tomorrow, today will also have become history. This history is never neglected; rather it is documented, and it teaches us to manage tomorrow in a better way, using our experiences of today and of the past. Some incidents today are out of our control, but equally, some are in our control and are manageable. The incidents which are in our control and are manageable are managed based on our skills, knowledge, and experience in handling such incidents. *Management* is a process that comprises the *planning*, *organizing*, *directing*, *executing*, and *controlling* of incidents, businesses, organizations, or resources. Plans are developed for managing future activities based on our experience of the present and the past. Planning is the process of developing plans for the future. *Planning* is concerned with the future and is done for the future, managing future activities or incidents to make them better for mankind. Planning reduces the risk of things getting worse. Risk can never be completely eliminated, however, but planning may reduce risk to some extent. *Organizing* is the establishment of effective authority relationships among people to allow them to work together to execute the plan. These relationships are represented in the form of an organizational structure for effective management. Assigning responsibilities to the people in an organizational structure is called *directing*. *Execution* is the implementation of a plan by bringing together those concerned to convert the plan into reality. Finally, *controlling* involves a comparison of actual progress with planned progress, to keep everything on track.

Project management is a particular field of management that is concerned with managing projects. In general a project is a one-time activity which has to be completed within a limited time duration, and has a well-defined outcome. Project management is the management of the project only; management, on the other hand, is a continuous process with no visible beginning or end.

1.3 Project Management

Project management is the process of leading the project team to achieve project objectives or complete project deliverables within the agreed time duration allocated budget, and quality. Project management involves the application of procedures, techniques, knowledge, and

DOI: 10.1201/9781003428992-1

experience to the management of a project to achieve the project objectives. The team involved in project management needs a wide range of technical and management skills, experience, and business aptitude. Project management involves dividing a project into manageable units, estimating the time duration for each unit, developing an execution sequence for different units, scheduling different units, executing, and controlling to meet project objectives. In other words, it is the art of managing all aspects of a project from start to end. Project management focuses on the management of a project and all work related to it; management, on the other hand, is an ongoing process that focuses on the general activities of a business.

1.4 Project Management Objectives

The main objective of project management is to complete the project within the constraints of the allocated time duration, available budget, and prescribed quality. However, to maintain a reputation for providing good quality deliverables or services, project management must involve developing a flexible organizational structure, continuously motivating the people engaged to get the best out of them, and providing a safe and satisfactory working environment for the people involved. This is graphically represented by the well-known project management triangle as shown in Figure 1.1.

1.5 Basic Steps in Project Management

Project management mainly involves *planning*, *scheduling*, *execution*, and *controlling*. A project starts with project planning, comprising the formulation of a number of alternative means for achieving the project objectives and finalizing the best-suited alternative, keeping in view the project constraints. The plan is made flexible so it can accommodate any unexpected changes required during the project execution. Project scheduling is generally done when the project plan is finalized. Project execution starts with the finalization of the project schedule. Once a project starts, actual progress made during the execution of a project may be as per the plan, ahead of the plan, or behind the plan. To keep the project progressing as per the plan requires a certain level of control. Therefore, in project management, project execution and controlling run parallel. Project planning and scheduling are not static processes; a project is planned and scheduled, and if the plan and schedule are not satisfactory or acceptable, both are revised as per the project requirements and constraints. Planning and scheduling are performed repeatedly until the results are satisfactory. Similarly, as the project execution starts, execution, controlling, and schedule refinements become cyclic, as shown in Figure 1.2. The discussion of project

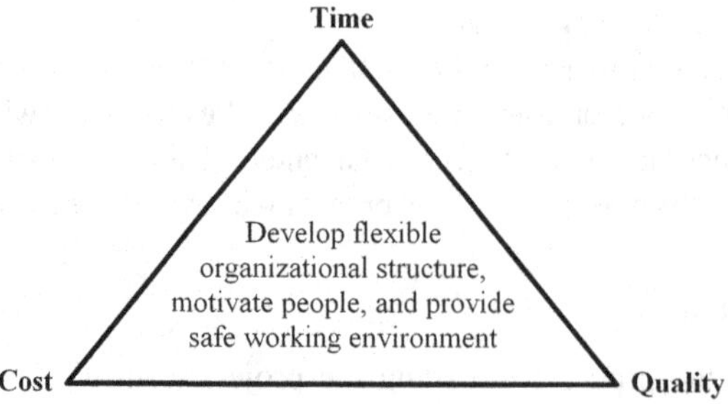

Figure 1.1 Project management triangle

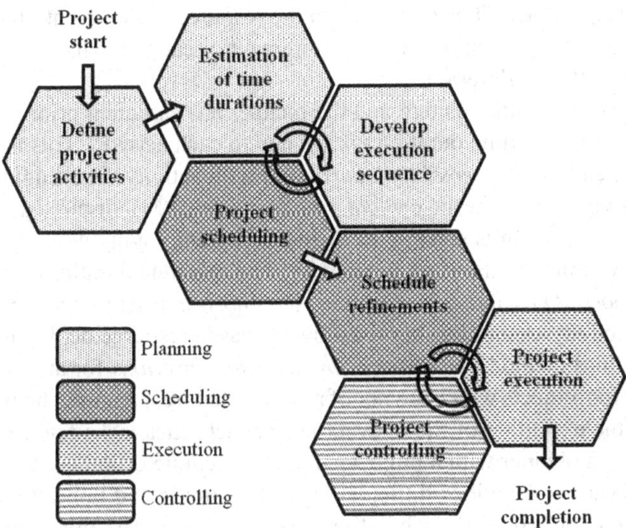

Figure 1.2 Basic steps of project management

management under this heading has been divided into seven sub-headings. Each sub-heading covers the topic in brief, however each topic contains a lot of details that cannot be covered in the present chapter. Hence, each sub-heading has been discussed in detail in subsequent chapters.

1.5.1 Defining Activities: Work Breakdown Structure

A project is expressed in terms of its objectives and scope. To manage a project effectively, it is divided into small manageable units called activities. A project consists of hundreds of activities; identifying all the activities involved in a project without the help of any logical process is difficult. Developing a project's work breakdown structure (WBS) is the logical process of obtaining a complete list of project activities. A WBS involves the decomposition of a project into smaller, more manageable parts which eventually describe the project activities. The lowest level of a WBS yields project activities that become too simple for estimation of their time durations, required resources, and costs. The activities found at the lowest level of a WBS fulfill the objectives or sub-objectives of a project. Chapter 2 covers the process of developing a WBS in detail.

1.5.2 Estimation of Time Durations and Resource Requirements

Time durations and resource requirements are estimated for the activities mentioned under the last sub-heading. For timely completion, each activity in the list requires resources. With more assigned resources the time duration may decrease; on the other hand, with fewer assigned resources, the time duration may increase. Further, the use of better technology may shorten the time duration and improve the quality of work obtained. The accuracy of project activities' time durations depends upon the prior experience of the planner in handling such projects and the familiarity of a planner with the resources to be employed. The use of resource output charts is sometimes helpful in cases where expert knowledge is not available. It is assumed that preceding and subsequent activities do not affect the resource requirement of the activity under consideration. Further, the normal resource requirement of each activity is considered in order to determine its normal time duration.

When a project is of a routine nature and the planner has sufficient prior experience in handling such projects, a single time duration is assigned to each activity. This is the first method; the time duration is called the *normal time duration* for an activity. The single estimate of time duration for an activity is used in the *critical path method* (CPM), discussed in Chapters 5 and 6. However, the time duration is sometimes shortened by employing more resources. The second method uses two time durations for an activity. The two time durations are the *normal time duration* and the *shortest possible time duration*. The *shortest possible time duration* is the time beyond which an activity cannot be shortened by increased resource deployment, as discussed in Chapter 7. The third method uses *three estimated time durations* for an activity. Three time durations are used for projects which are not of a routine nature, meaning the time duration estimate required for the execution of the relevant project activities lacks fair degree of accuracy. The unfamiliarity of a planner with a project can make accurate estimates of time durations for the completion of various activities difficult. Therefore, lower and upper time duration limits are decided for the completion of each activity in the project. The third time duration is the estimated duration of an activity under normal conditions; its value lies between the lower and upper limits. Three estimated time durations are used in the *programme evaluation and review technique* (PERT), discussed in Chapters 10 and 11.

1.5.3 Development of an Execution Sequence

The development of an execution sequence for the different activities mentioned under the first sub-heading involves defining inter-dependencies among them. Defining inter-dependencies means defining the preceding and subsequent activities of each activity in the list. This requires input from personnel who have previously handled such projects. Project constraints are kept in view while deciding on inter-dependencies. Once the inter-dependencies among the various activities are defined, the execution sequence for the various activities of a project is depicted in the form of a bar chart, network, or line of balance (LOB) graph. Chapter 3 describes how an execution sequence is developed and represented in the form of a bar chart. Chapter 4 describes how an execution sequence is developed and represented in the form of a network. Networks are based on the premise that a subsequent activity starts only after its preceding activities are completed; this is called a finish-to-start relationship. Chapter 12 covers three additional relationships (finish-to-finish, start-to-start, and start-to-finish) used in networks to describe inter-dependencies more practically, and in which overlaps between various activities are also modeled. LOB graphs are widely used for representing the execution sequence of repetitive projects, and are covered in Chapter 13.

1.5.4 Project Scheduling

The execution sequence is represented in the form of a bar chart, network, or LOB graph. The next step is the development of a timetable for the execution of project activities. The development of a timetable means the development of a schedule. The execution sequence, along with the time durations of various activities of a project, are used as inputs in project scheduling. Project scheduling involves calculating a time duration for the completion of a project, and the start and finish times for each activity in a project. The calendar date scheduling of a project is also involved. When the project completion date falls before the allowed date, the schedule developed can be finalized. However, when the project completion date falls beyond the allowed date, the project plan and schedule are revised as shown in Figure 1.2. Project planning and scheduling are not static processes; Chapters 5, 6, 10, 11, 12, and 13 cover these topics in detail.

1.5.5 Schedule Refinements

The schedule developed under the last sub-heading is refined or improved to make it more practical or executable. Sometimes additional resources are used to shorten the time duration of a project. The reason for project shortening is to deliver a project earlier than the planned duration. The shortening of a project's time duration by adding more resources is called project crashing. Project crashing uses two time durations for each activity: the normal time duration and the shortest possible time duration. The shortest possible time duration is the time beyond which an activity cannot be shortened by increasing resources. The purpose of project crashing is to speed up project delivery. Sometimes the objective of project crashing is to determine the optimum time duration of a project at which project cost is minimized, as discussed in Chapter 7. The schedule finalized under the last sub-heading, therefore, is refined through the process of project crashing.

Resources are assumed to be either unlimited or limited in the planning stage. When a project is planned with the assumption that resources are unlimited, the previously decided start and finish times of different activities are changed in such a way that the daily requirement of each type of resource is almost uniform, however the previously decided time duration of the project is not changed. This method is the topic of Chapter 8. In fact, available resources are not unlimited and are not available as and when they are required. Practically, available resources are considered limited, therefore projects are scheduled keeping in view the availability of the various types of resources required. In this case the time duration of the project is changed, generally exceeding the allowed date; thus, the schedule finalized under the last sub-heading is refined again.

1.5.6 Project Execution and Related Arrangements

No matter how effective a project plan is, a project cannot be successful unless the plan is effectively implemented. Project execution needs a project team, authority and responsibility relationships among team members, and other required resources. Therefore, project execution has been covered under the following three sub-parts.

Organizing: This involves deciding on the required number of project team members and setting up authority and responsibility relationships among them to help execute a project smoothly and efficiently. The relationships between the project team members are set up and represented in the form of an organizational structure which is depicted with the help of a chart. Each member of a project team is made aware of his or her authority and responsibility.

Staffing and Directing: Staffing is appointing or recruiting the right people based on their skills, knowledge, and experience in the different positions created within an organizational structure. Sometimes training or introductory programs are also arranged for the newly appointed people. Directing is the direction of appointed people to help them get their work done. It is the process in which appointed people are instructed, guided, and motivated to bring out their maximum performance in achieving project objectives.

Project Execution: This is the implementation of the project plan through the bringing together of the members of a project team. The project is executed by putting everything in the project plan into action. Project execution is the longest and most complex stage of a project's lifecycle and consumes a lot of energy and resources. The primary objective of project execution is the construction of project deliverables and the consistent monitoring of project progress to deliver the project deliverables within the agreed limits of time, budget, risk, and project-specific constraints. The execution of a project involves many decisions; decisions which are not of routine nature are made by higher-level management, and decisions which are of routine nature are made by lower-level management.

1.5.7 Project Controlling

The actual project progress achieved at a point of time may be as per the plan, ahead of the plan, or behind the plan. To keep project progress as per the plan, a project requires certain level of control. Therefore, suitable control is necessary during the execution of a project in order to complete the project within the allocated time duration. Project control is also necessary to ensure the effective and efficient work of project team members. It involves a constant comparison of planned and actual performance, so as to rectify deviations by deciding on appropriate corrective measures. This requires an accurate information flow about actual work done. This is the topic of Chapter 9. Project management also requires very good coordination among project team members. Regular meetings among different management levels or among members of the same management level are essential for good coordination. Regular meetings are also the best platform to discuss deviations and decide on remedies. Further, the planning of the inflow and outflow of funds over the time duration of a project and their correct monitoring are two of the most important tasks of project management. Planned expenditures are compared with the actual expenditures that occurred during the execution of a project, S-curves being widely used, as discussed in detail in Chapter 9.

1.6 Project Planning

Planning in general is the process of thinking and documenting the essential actions/steps needed to achieve a specific goal. It involves the visualization of the course of actions for their documentation in the form of a plan. The documentation of all the essential actions/steps is necessary to convert the plan into reality after the planning process. Planning is a habit of thinking of intellectual human beings, allowing for the conversion of their ideas into reality. It is an intellectual process for deciding on the various courses of action by which a goal can be achieved. Planning is used in many ways and in different contexts; thus, it is not a professional activity but, in general, it is an activity of everyday life. However, it is an essential part of every professional job, particularly in business and projects.

Planning is also essential in the case of a project; in the case of a project, planning is called *project planning*. Project planning is generally done in two phases: the first is the *preliminary planning* phase, and the second is the *detailed planning* phase. The preliminary project planning phase involves deciding beforehand what to do, why to do it, how to do it, who will do it, and where to do it, keeping in view the project objectives to be achieved. These questions give a general idea about the project in hand.

Detailed project planning involves critically examining a project, formulating a number of alternatives for achieving the project objectives, and finalizing the best-suited alternative for the allocated time duration, budget, available resources, location, and other project constraints imposed. Once an alternative is finalized, the project is divided into small manageable parts, their time durations are estimated, and all parts are arranged in an execution sequence to complete a project within the constraints imposed. A detailed plan includes a list of all manageable parts and their time durations, the required resources, the allocated budget, the execution sequence, and the schedule.

The finalization of the project plan requires many professionals to make decisions. The finalized plan must be flexible to accommodate any unexpected changes that may occur during the execution of a project. As situations change, a plan is modified or abandoned. A plan exists as long as it helps in accomplishing the project objectives; otherwise, it is modified or abandoned. Once a plan is developed it is possible to compare the planned and actual project progress, to achieve effective project control, and decide on corrective measures accordingly.

1.7 Project Scheduling

Every intellectual human being makes a timetable to organize his or her routine activities. The timetable is a plan of the time when routine activities are to take place. For example, the start time of one's journey to the workplace, the time for shopping, and the time for going for a walk. Every intellectual human is also a human resource for others; an individual cannot be present at two or more places at the same time, hence the necessity of his or her routine timetable. Sometimes an individual requires a particular resource on a particular date and time, but the resource may not be available or free on the date and time decided. The required resource needs to be ordered or booked in advance as per the previously decided timetable. An individual also needs to communicate his or her timetable to others, so that his or her requirements can be fulfilled as per previously decided dates and times. In simple words, the development of a timetable means the development of a schedule.

The development of the schedule for a project means much more than the development of a timetable for the execution of various activities. A project schedule also includes the schedules of different resources involved in the project execution, depending upon the project's scope. Project resource schedules include finance schedules, manpower schedules, materials schedules, equipment schedules, etc. Scheduling is developing the timetable for performing all parts of a project. It is the fitting of a detailed plan onto a time duration or onto calendar dates. It shows the start and finish times, the time durations, and the orders of execution of all the manageable parts of a project. Planning and scheduling are two terms which are often thought of as synonymous. However, they are different: scheduling only deals with the *when to do it* aspect. In addition to the timetable, the project schedule provides the following information to a project team.

The Time Duration of a Project: The calculation of the time duration of a project is of prime concern in project management; scheduling calculations provide the time duration of a project. This helps team members plan their tasks to meet project deadlines.

Start and Finish Times of Activities: Each activity in a project has a time duration. To finish an activity in the allocated time duration different types of resources are required; the start and finish times of each activity are decided accordingly in the process of project scheduling.

Resource Planning: A schedule helps in financial planning by calculating funding requirements across the time duration of a project. The requirement of other types of resources across the time duration of a project is also generated from the project schedule.

Project Control: To control a project, planned and actual progress is compared at an instant of time. The schedule provides planned progress with regards to time, and the actual progress is the progress made during the execution of a project.

Coordination Between Team Members: When the start and finish times of each activity are known, it helps in sharing resources and coordination among project team members. Some activities may have more resources and others may have less; depending upon the available time duration, the sharing of resources may be planned.

When a project is not scheduled properly, the project may go off track, the schedule no longer being followed. This results in delays to the completion of a project, more expenditure than the budget allocated, and disputes between the executing and funding agencies. Thus, experienced planners are employed for project planning and scheduling. The schedule developed by experienced planners may help to avoid many problems. However, outside the experience of planners, there are widely used project planning and scheduling techniques discussed in the subsequent section. These techniques help project planners make their plans and schedules more efficient and executable.

1.8 Planning and Scheduling Techniques

There are a variety of planning and scheduling techniques documented in the literature, however this book covers the techniques which are widely used in practice. This section provides an overview of the following techniques.

Work Breakdown Structure: This is a simple technique used for breaking down a project into small manageable parts. The division of a project into small manageable parts is called work breakdown. It is the splitting up of a project into small divisions, sub-divisions, and further sub-divisions. WBS is the hierarchical representation of all divisions and sub-divisions of a project. It is easy to assign time durations to these small manageable parts. The accurate estimation of the resources required to complete these manageable parts within the assigned time durations is also possible. The creation of a WBS is the first step toward the development of an execution schedule for a project. Chapter 2 covers WBS in detail.

Bar Chart: This is a simple graphical chart used to represent the plan and schedule of a project. A bar chart does not require special technical knowledge to be understood. A bar chart lists activities to be performed on the y-axis and the time duration of a project on the x-axis. The activities in a bar chart are represented using bars. The lengths of the bars in a bar chart are equal to their time durations. The start of a bar depicts the start time and the end of a bar depicts the finish time of the corresponding activity. The inter-dependencies among various activities are also represented in a bar chart. Chapter 3 covers the bar chart in detail. Bar charts are good for the planning and scheduling of simple and small projects but are not suitable for complex and large-scale projects. For the planning and scheduling of complex and large-scale projects, network-based techniques are widely used.

Critical Path Method: This is a network-based technique to facilitate the planning and scheduling of complex and/or large-scale projects. It also eliminates the limitations of bar charts. The various activities of a project and the inter-dependencies among them are accurately and easily represented with the help of networks in CPM. Chapter 4 covers the development of networks. The plan of a project is represented in the form of a network and network calculations are used for the scheduling of a project. Planning and scheduling are done separately in CPM, however planning and scheduling are done together in the case of bar charts. Chapters 5 and 6 cover CPM in detail. Additional resources are used to reduce the time duration of a project; this is called project crashing, and is discussed in Chapter 7. Projects are also scheduled in such a way that requirements of various types of resources are almost uniform throughout the time duration of a project because variations in the requirement of resources are difficult to manage. This process is called the leveling of project resources, covered in

Chapter 8. A project rarely goes as per the plan and schedule; deviation from the original plan and schedule always takes place. When the execution of a project starts, actual progress made at an instant of time is compared with the planned progress and corrective measures are decided on. The original plan and schedule are modified or updated to incorporate the corrective measures. This process is called network updating; Chapter 9 covers the network updating process in detail.

Program Evaluation and Review Technique: CPM is used when a planner is thoroughly familiar with a project. CPM is used in projects which are of routine nature or where a planner has sufficient prior experience in handling such projects. The *Program Evaluation and Review Technique* (PERT), on the other hand, is used for the planning and scheduling of projects of a non-routine nature or for projects in which the time duration estimates of various activities have a low degree of accuracy. Such projects include a large amount of uncertainty in the estimated time durations of project activities. Therefore in PERT, a probability number is associated with the occurrence of any project event or project time duration. This probability depends upon the uncertainty involved in the estimated time durations of various activities in a project. Similarly to CPM, PERT also uses a network to represent the project plan and to calculate the time duration of a project and the time of occurrence of any event. Chapters 10 and 11 cover PERT in detail.

Precedence Diagramming Method: The networks used in CPM are based on a relationship in which an activity starts only after its preceding activities are completed, called a finish-to-start relationship. In the *precedence diagramming method* (PDM) three additional relationships (finish-to-finish, start-to-start, and start-to-finish), discussed in Chapter 12, are also used in a project network. Further, lags or leads to time durations between the preceding and subsequent activities in a network are also modeled. A lag is a temporal delay to the start of an activity after the end of the preceding activity. A lead, on the other hand, is when an activity starts before the end of the preceding activity. It is an overlap of time durations between subsequent and preceding activities. PDM represents practical situations among various activities in a better way than CPM; hence, PDM is better suited for the planning and scheduling of complex projects.

Line of Balance: Network-based techniques are commonly used in the planning and scheduling of large and complex projects. However, network-based techniques have limitations when applied to repetitive projects. Repetitive projects are projects where repetitions of various activities take place. The main limitation of network-based techniques is that continuous crew engagement is not ensured. For scheduling repetitive projects, *line of balance* (LOB) is a widely used technique that ensures continuous crew engagement. Repetitive activities are treated as a single activity in LOB. Thus, the entire repetitive task is represented by a single line and is considered a single activity. An execution schedule in an LOB graph has two axes: time duration, and distance or number of units. In general, the time duration of a project is plotted along the x-axis. Distance or number of units is plotted along the y-axis. Activities are represented by lines with constant or sometimes changing slopes. The slope of a line represents the progress rate of the activity. Chapter 12 covers LOB in detail.

1.9 Conclusion

Project management includes project planning, scheduling, execution, and controlling. A project starts with project planning; scheduling is the timetable made when a plan is finalized. The schedule is putting the developed plan into a time duration. Once a project starts, actual progress made during execution may be as per the plan, ahead of the plan, or behind the plan.

To keep project progress as per the plan requires certain level of control called project controlling. Planning and scheduling are done repeatedly until the results are satisfactory. Similarly, as project execution starts, execution, controlling, and schedule refinements become cyclic. There are a variety of planning and scheduling techniques used in project management like WBS, bar charts, CPM, PERT, PDM, and LOB. Each technique has its benefits and limitations. This book covers all these techniques in detail in subsequent chapters.

Exercises

Question 1.1: What makes general planning different from project planning?
Question 1.2: Differentiate between project planning and project scheduling.
Question 1.3: Why is project schedule refinement recommended in project management?
Question 1.4: What are the different project scheduling techniques; under what circumstances they are used?
Question 1.5: What are single time duration estimates, two time duration estimates, and three time duration estimates? Under what circumstances are these used?
Question 1.6: What is project control? Why is it important?

2 Work Breakdown Structure

2.1 Learning Objectives

After the completion of this chapter, readers will be able to:

- Understand work breakdown structures and their development,
- Explore applications of work breakdown structures, and
- Use work breakdown structures as the first step in project management.

2.2 Introduction

As discussed in the previous chapter, the first step in project management is the identification of project activities. It is difficult to prepare a complete list of project activities without any logical process because large-scale projects contain hundreds of activities. Further, different management levels need project information with different levels of detail. For example, upper management levels are responsible for decision-making that does not require operational level details, however the personnel involved at the operational level require operational level details on a project. It is difficult and time-consuming to prepare the information required for the different levels of management. Work breakdown structures (WBS) are a very simple tool used widely for the logical identification of project activities. The United States Department of Defense and NASA initially developed WBS. A WBS helps provide project information of the necessary depth to every team member or management level. It helps provide the information required from the lowest to the uppermost management level. WBS is the hierarchical and deliverable-oriented division of a project into manageable parts.

2.3 Project Work Breakdown Structures

The first step in project management is breaking down a project into small manageable parts. These small manageable parts are easy to assign time durations to. The accurate estimation of the resources required to complete these manageable parts within their assigned time durations is also possible. The decomposition of a project into small manageable parts is called work breakdown. This is the splitting up of a project into small divisions, sub-divisions, and further sub-divisions. A WBS is the hierarchical representation of all the divisions and sub-divisions of a project. The hierarchical representation of all the divisions and sub-divisions of a project facilitates its management and completion in the allocated time duration using the assigned resources.

DOI: 10.1201/9781003428992-2

The divisions and sub-divisions in a WBS are used to foresee the project time duration, project cost, coordination among divisions and sub-divisions, and controlling of a project. A WBS describes project objectives and the hierarchical decomposition of the project into smaller or more manageable deliverables. A WBS helps project team members to understand a project in a much better way. This helps in assigning a task to an individual or a team and helps in tracking the project's progress.

There are different ways to achieve project objectives. These different ways are thought of as different alternatives for achieving project objectives. These different alternatives are examined, the best two or three out of those are evaluated in detail, and a detailed project report is prepared. The one best out of the two or three is selected, keeping in view the project's constraints. A WBS is developed only for the best alternative selected. Further, there may be different alternatives for achieving each objective of a project. A WBS is developed only for the best alternative selected; a WBS does not include all possible alternatives in its structure.

2.3.1 Hierarchical Levels

The development of a WBS starts, as shown in Figure 2.1, at the highest level and moves toward lower levels with the identification of all possible units at all levels. The divisions and sub-divisions of a WBS are continued from the highest level to the lower levels, but the lower levels are less complex than the higher levels. The management of the lower levels is easier than that of the upper levels. Thus, the number of levels is decided in such a way that a WBS reaches the lowest level where it becomes manageable for planning, scheduling, execution, and controlling. Therefore, the lowest level is generally represented in terms of project activities.

The work breakdown levels in a WBS may be level-I (*project level*), level-II (*sub-project level*), level-III (*task level*), level-IV (*work package level*), and level-V (*activity level*). A WBS works top to bottom; it is like beginning at the top of a pyramid and expanding downward. The lower levels of a WBS have more detailed descriptions and lower complexity than preceding levels. The number of levels in a WBS depends upon the size, complexity, and scope of a project or work. The number of levels also depends upon the experience of a planner in handling such projects. If a planner has more experience, the number of levels chosen may be lower, however, if a planner has less experience, the number of levels may be higher. A planner may restrict a WBS to reaching only the task level, while others may go to the work package or activity level. As a planner goes down to lower levels, estimates of time durations and costs become more

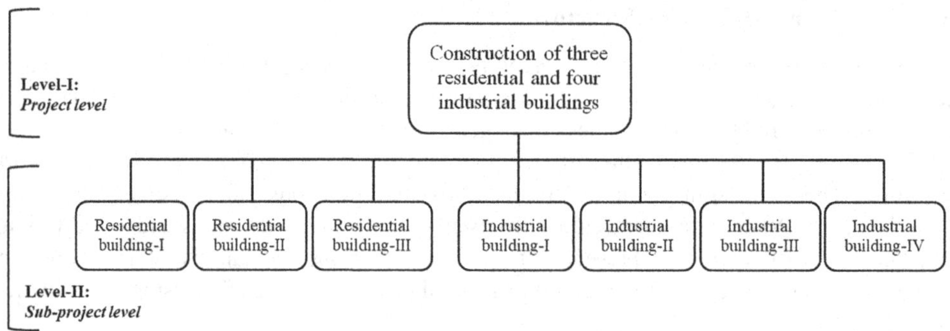

Figure 2.1 Two levels of a work breakdown structure

accurate, whereas at upper levels, the accuracy of estimations of time durations and costs is lower.

The project level is the top level of a WBS. It describes a project and the final deliverable required; a WBS starts from this level. It involves deciding on project objectives and providing project descriptions. The next level of a WBS is the sub-project level. It includes major parts or units which together constitute a project; all units need to be completed to deliver the project. The major parts or units at the sub-project level are part deliverables that include a large volume of work. The third level is the task level; it involves breaking down the part deliverables at the sub-project level into smaller and more manageable identifiable parts or units, and deliverables containing one or more *work packages*. The fourth level is the work package level; it involves breaking the part deliverables of the task level down into smaller and more manageable units. These are identifiable, measurable, and controllable units. The estimation of cost is also possible at this point. Generally, the elements at the lowest level of a WBS are work packages, however, these work packages are sometimes further divided into activities. The activity level of a WBS usually contains smaller units than work packages. The activity level is the measurable, costable, and controllable lowest level of a WBS. The estimate of the time duration of and resources required for an activity at the activity level is at the core of project management. However, the activity level is further divisible into the operational level, for planning day-to-day operations.

2.3.2 WBS: An Example

Let us consider a sample project that involves the construction of three residential and four industrial buildings. The construction of three residential and four industrial buildings is at the level-I. The first division of the project is at level-II, which involves the division of the project into seven separate buildings as shown in Figure 2.1. These are residential buildings I, II, and III. The industrial buildings at this level are I, II, III, and IV. Hence, level-II contains seven units which may be considered seven sub-projects. Level-III is the sub-division of the units of level-II. For example, residential building-II has the following sub-divisions: site survey; civil work; electrical work; heat, ventilation and air conditioning (HVAC); and sanitary fittings, as shown in Figure 2.2.

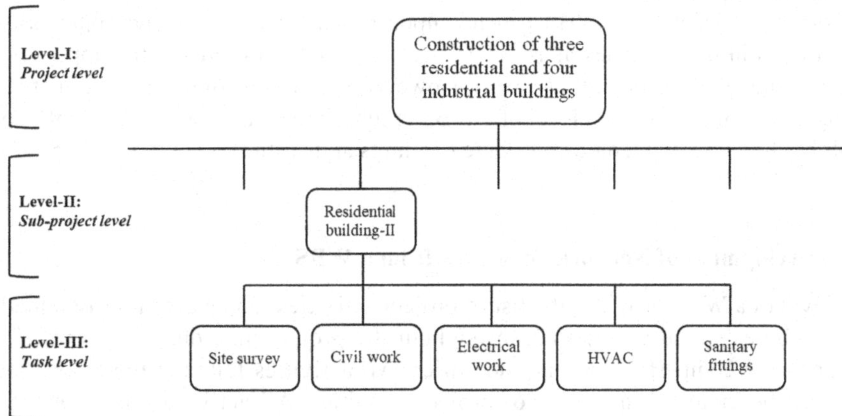

Figure 2.2 Three levels of a work breakdown structure

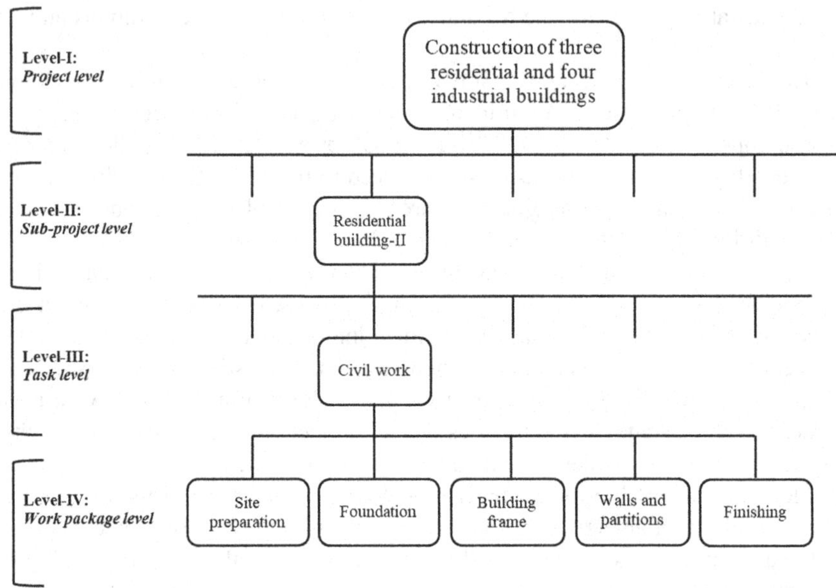

Figure 2.3 Four levels of a work breakdown structure

The next sub-division of the project is at level-IV which involves a further sub-division of the units of level-III. Civil work has further sub-divisions: site preparation, foundations, building frames, walls and partitions, and finishing, as shown in Figure 2.3. A further sub-division of the project is performed at level-V which involves a sub-division of the units of level-IV. For example, the unit *foundation* has sub-divisions: *site* marking, excavation, reinforcement, concreting, and curing, as shown in Figure 2.4.

The upper levels of a WBS may be decided by the planners or managers who may have less experience, but the planning of lower levels require good experience in handling such projects. Hence, planning at the lower levels of a WBS requires more in-depth knowledge. In the WBS shown in Figure 2.4, the sub-division of the foundation into smaller parts requires more prior experience than required in the upper levels.

The number of levels of a WBS depends upon project size, scope, and complexity. The number of units in the lower levels is higher than the number of units in the upper levels. In Figure 2.1, a single unit is present at level-I. However, the number of units is seven at level-II. Supposing that each unit on each level of a WBS is sub-divided into five smaller units, and that this WBS has five levels, it would contribute 625 units at its fifth level ($1 \times 5 \times 5 \times 5 \times 5$).

2.4 The Development of Network Diagrams from a WBS

The last level of a WBS provides the list of project activities. Some activities obtained from a WBS can be executed at any time throughout the project time duration, whereas some depend upon the completion of other activities. All activities found at the lowest level of a WBS must be included in the list of project activities. All activities are arranged in an execution sequence in the form of a bar chart or a network diagram. The execution sequence provides information about how activities are related to each other, and the project time

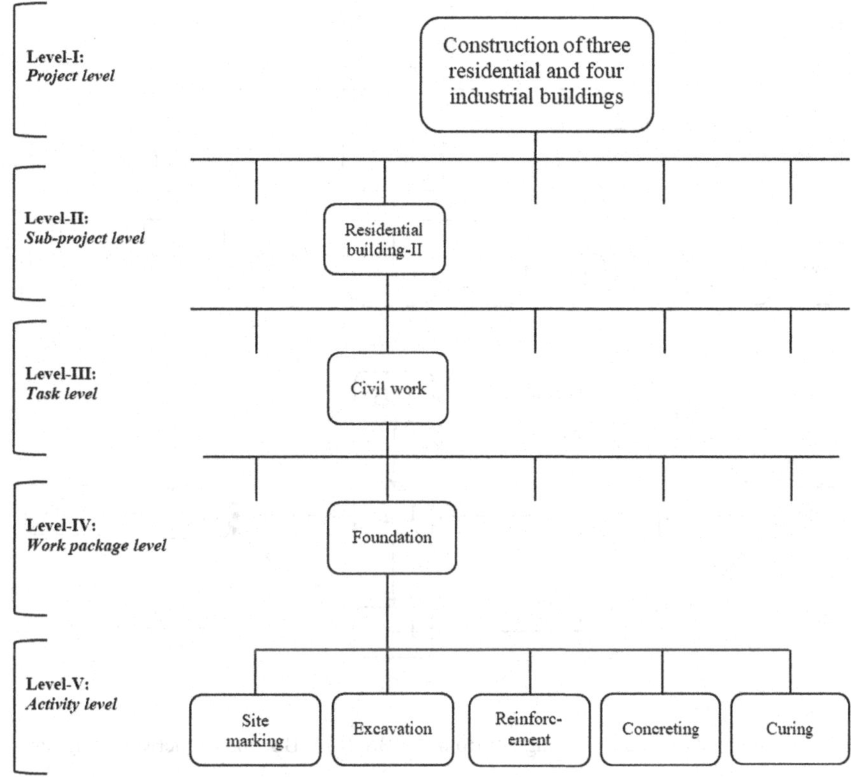

Figure 2.4 Five levels of a work breakdown structure

durations and start and finish times of each activity. The execution sequence, project time duration, estimated cost, and resource schedules change if a single activity is missed. Thus, missing an activity is a big issue.

Figure 2.5(a) represents the WBS of a sample project used to demonstrate the development of a project network diagram shown in Figure 2.5(b). It is an example of activity-on-node network representation in which activities are represented in rectangles/nodes and relationships among them are represented with the help of arrows. A network is a logical and chronological graphic representation of project activities. This is the representation of activities in a flow to represent the execution sequence. Network development will be covered in detail in subsequent chapters.

Units which lie on the same branch of a WBS are not connected to each other in a manner that would develop inter-dependency relationships in a network. For example, inter-dependency relationships between units L_1 and $L_{1.1}$, $L_{1.1}$ and $L_{1.1.1}$, $L_{1.1}$ and $L_{1.1.2}$, and $L_{1.1}$ and $L_{1.1.3}$ are not possible. The units on the last level of each tree of a WBS are considered as activities of a project. For example, the last level units $L_{1.1.1}$, $L_{1.1.2}$, $L_{1.1.3}$, $L_{1.2.1}$, $L_{1.2.2}$, $L_{1.2.3}$, $L_{1.3.1}$, $L_{1.3.2}$, and $L_{1.3.2}$ are used to form inter-dependency relationships in the project network diagram. When upper-level units of a branch are used, its lower-level units are not considered. For example, if unit $L_{1.1}$ is used then lower-level units $L_{1.1.1}$, $L_{1.1.2}$, and $L_{1.1.3}$ are not considered as activities in the network diagram because unit $L_{1.1}$ already contains units $L_{1.1.1}$, $L_{1.1.2}$, and $L_{1.1.3}$. In this case, project activities are $L_{1.1,}$ $L_{1.2.1}$, $L_{1.2.2}$, $L_{1.2.3}$, $L_{1.3.1}$, $L_{1.3.2}$, and $L_{1.3.2}$ only. Generally, the last level units of a branch are used for project network development. However, sometimes units from any one level of

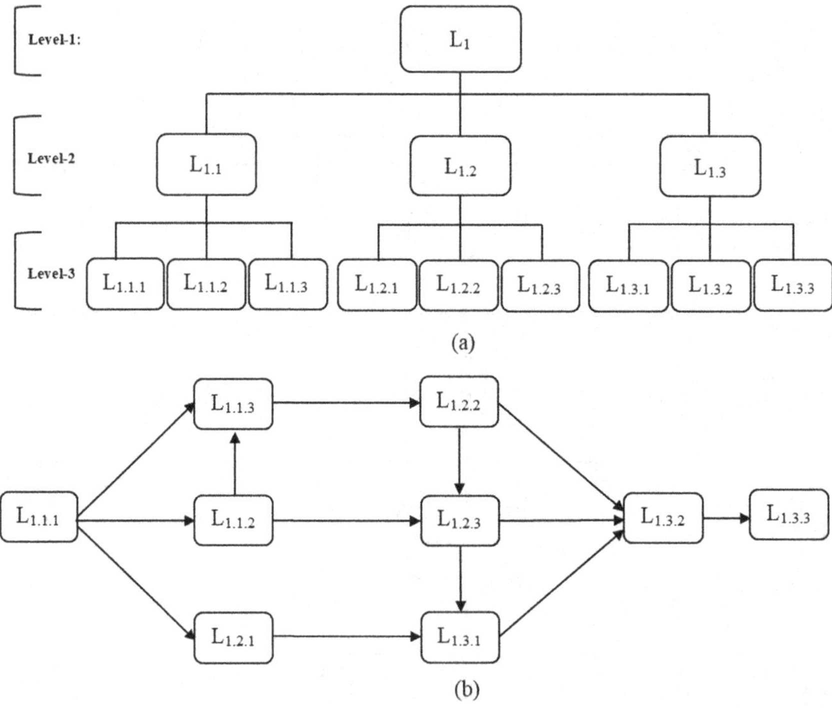

Figure 2.5 Development of a network diagram from a WBS, (a) WBS, and (b) network diagram

each branch of a WBS may be considered in the development of a project network diagram. For example, units $L_{1.1.1}$, $L_{1.1.2}$, $L_{1.1.3}$, $L_{1.2}$, $L_{1.3.1}$, $L_{1.3.2}$, and $L_{1.3.2}$ may also be used to form interdependency relationships in the project network diagram.

2.4.1 Network Diagrams from a WBS: An Example

In Figure 2.6, level-I is the project level; the first division is at level-II which contains three units: planning and design, project execution, and final check. Level-III is the sub-division of the units of level-II; planning and design has the following sub-divisions: feasibility study, data collection, and design, as shown in Figure 2.6(a). Project execution has the sub-divisions pilot study, actual execution, and controlling, however the final check has no sub-divisions. The last level of the WBS provides the list of activities that are used in project network development. The activities in the present case are the feasibility study, data collection, design, pilot study, actual execution, controlling, and final check. Different management levels need project information of different levels. For example, the upper management level is responsible for decision-making and does not require execution-level details. The decomposition at level-II and the corresponding network diagram shown in Figure 2.6(b) may be sufficient for upper-level management. However, project team members involved at the operational level require operational-level details. The seven activities at level-III are arranged in an execution sequence in the form of a network diagram, shown in Figure 2.6(c), providing operational-level details to the project team members.

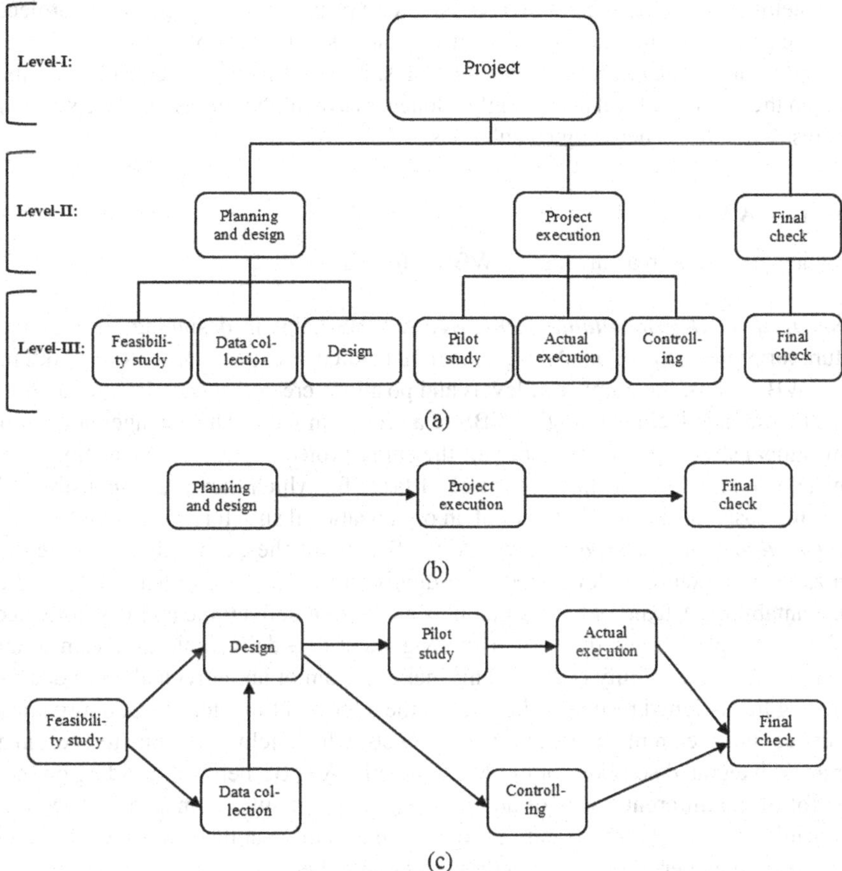

Figure 2.6 Development of a network diagram from a WBS, (a) WBS, (b) network diagram for management level, and (c) network diagram for operational level

2.5 WBS: First Step Toward Project Management

The development of a WBS is the first step toward project management. A WBS starts with the description of a project and the project objectives. The project is divided into small manageable parts at different levels. The parts are sub-divided into smaller and smaller sub-parts until all sub-parts take the form of small deliverables of a project. All small deliverables are seen as activities of a project.

Sub-division to the next level down is not required if the last level of a WBS allows for part deliverables to be converted into project activities. Further, it is also possible to accurately estimate the time duration, cost, and resources required for each activity. However, when the level reached by a WBS provides complex activities and it is not possible to estimate the time durations and costs involved, this level requires further sub-division into simpler sub-divisions on lower levels to allow for more accurate estimations of the time durations, costs, and resources requirements involved. The level of a WBS that provides good estimates of time durations, costs, and required resources is taken as the manageable level for the executing team, and its sub-divisions or units are taken as activities for the development of the project plan. This becomes the last level of a WBS.

A WBS helps in the development of a bar chart or a project network, as used in project management. The greater the number of levels or units in a single level of a WBS, the greater the number of activities in the execution schedule. Thus, it is necessary to decide on the number of levels and on the units in a level accordingly. Unnecessarily high numbers of levels and of units in a level result in more scheduling calculations.

2.6 Benefits of A WBS

The following are a few areas in which a WBS is helpful:

Development of an Organizational Structure: A WBS helps in designing an organizational structure for project execution. An organizational structure expands from top to bottom, similar to a WBS. Thus, the number of levels and positions created at each level of an organizational structure are decided with the WBS of a project in view. The manager at the top of an organizational structure is responsible for the entire project. An executive at any level of an organizational structure has his or her tree of tasks for which they are responsible. Thus, a WBS is the basis for the development of an organizational structure for a project.

Assigning Authority and Responsibility: A WBS lists all the deliverables required and an organizational structure is developed accordingly to assign work to team members and to set accountability. A team or an individual is assigned a deliverable and is wholly accountable for its completion and delivery. The assignment of a deliverable to a team or an individual reduces responsibility overlap. This makes a team or an individual more dedicated to completing the assigned responsibility within the allocated time duration. An organizational structure contains team members at different levels, which helps everyone to coordinate and generate an integrated solution for every deliverable. A WBS helps in deciding on the spans of control of the different levels of an organizational structure. A manager at the top level is ultimately responsible for the entire project. An executive at the lower level has authority over the team members working directly under him or her.

Manpower Requirements: A WBS helps in developing an organizational structure; the positions created in an organizational structure are then filled by the recruitment of people. This helps estimate the manpower requirements at different levels of an organizational structure.

Improve Information Flow: A WBS is a base for the development of an organizational structure that defines the responsibilities of executives at different levels. The executives taking responsibilities at different levels of an organizational structure require information regarding the deliverables for which they are responsible. A WBS helps with deciding on the relevant information and supplying it to the executives concerned, because full information is not required for every individual. Thus, the distribution of information is focused on the needs of team members.

Project Scheduling: The lowest level of a WBS yields activities to help develop an execution schedule. At the lowest level, activities become simple with regards to the estimation of time durations, resources required, and costs involved. The activities found at the lowest level are used directly in the project network development.

2.7 Conclusion

The decomposition of work into small manageable parts is called work breakdown. It is the splitting up of work into small divisions, sub-divisions, and further sub-divisions. A WBS is the hierarchical representation of all divisions and sub-divisions of a task or project. The level of a WBS that provides a good estimation of time durations, costs, and required resources is taken as the manageable level for the executing team, and its sub-divisions or units are taken as activities for the development of a project execution schedule. This is the last level of a WBS. A WBS does not provide the execution sequence of project activities. The next chapter introduces the simplest tool – called a bar chart – for representing the execution sequence of a project. the remaining chapters of the book, however, focus on techniques for developing and refining the plan and schedule of a project.

Exercises

Question 2.1: Develop a WBS for a project with which you are familiar.
Question 2.2: Develop a WBS for a small residential house construction project.
Question 2.3: Why is WBS a deliverable-oriented hierarchical decomposition of the work involved in a project, aimed at achieving project objectives?
Question 2.4: Draw a WBS for the construction of a two-room house.

3 Bar Charts

3.1 Learning Objectives

After completion of this chapter, readers will be able to:

- Draw bar charts to represent the execution schedules of projects,
- Do single and multiple calendar date scheduling, and
- Understand the shortcomings of bar charts and their remedial measures.

3.2 Introduction

A bar chart is a simple and widely used graphical technique for representing the execution schedule of a project. A schedule, in simple words, is the timetable for the execution of different parts of a project. It is generally said that a task will start at this point of time and finish at this point of time – putting the start and finish times of different parts of a project into a time duration is called scheduling. In a schedule, different parts of a project planned for execution are shown on a time duration with their start and end times. All parts of a project planned for execution always have well-defined start and finish times on calendar dates. A schedule is a timetable for the performance of all small/large parts of a project and a bar chart is a widely used graphical technique for depicting this schedule.

Suppose a bar chart is used to represent routine daily activities from morning to evening, like a morning walk, brushing one's teeth, bathing, having breakfast, and so on. Every individual has his or her own timetable for routine activities, and every individual has his or her own sequence for carrying out routine daily activities. In general, we brush our teeth before breakfast, however there may be an individual who likes to have their breakfast before brushing their teeth. In the same way, the planned execution sequence for carrying out various parts of a project also varies from planner to planner, and the time durations assigned to the execution of various parts of a project also vary.

3.3 Bar Charts

H. L. Gantt developed the bar chart around 1900. Thus, bar charts are also referred to as Gantt charts. The use of bar charts has continued to increase in the field of project management since their invention. To develop a bar chart representing an execution schedule, *in the first step*, a project is divided into small, generally manageable parts. The number of manageable parts in the same project may vary from planner to planner. In general, a planner with less experience may divide a project into a greater number of parts; on the other hand, an experienced planner may divide a project into fewer parts. It depends upon how much prior experience a planner

DOI: 10.1201/9781003428992-3

has in handling such projects. Manageable parts are parts for which a planner can estimate a time duration for the execution of a part of the project using the given resources, or, when the time duration is fixed, estimate what resources are required. An perfect technique for dividing a project into manageable parts is not available. A part may be large as the construction of the superstructure of a building, or as small as the plastering of its walls.

In the first step, a project is divided into manageable parts; these parts are called activities for tasks. If the time durations of all the activities are fixed, resources are estimated for each activity. If resources are fixed, time durations are estimated for each activity. There are various methods for executing different types of task, thus the time duration for an activity depends upon the chosen method. *In the second step*, methods for executing all the activities are chosen, and a time duration is assigned to each activity accordingly. The method employed for the execution of the activity determines its time duration.

The different activities in a project are related to each other. How these activities are related to each other is critically examined. Some activities can be performed simultaneously, and some activities follow others or are done in a sequence. How its various activities are related to each other determines the execution sequence of a project. *In the third step*, the execution sequence is determined. The identification of project activities, and the examination of inter-dependencies among them, requires an exhaustive analysis of the project. This is the most useful stage of project planning in the development of a bar chart.

In the last/fourth step, the execution schedule is represented graphically in the form of a bar chart. Figure 3.1 shows a simple bar chart. It has two axes; the x-axis is generally used as the time axis, and activities are shown on the y-axis in the form of horizontal bars. On the x-axis, the time scale may be measured in days, weeks, months, or calendar dates. The anticipated start and end times of activity are depicted on a bar chart as the start and end of the horizontal bar which represents the activity. The length of a horizontal bar represents the time duration of an activity. The number of horizontal bars in a bar chart is equal to the number of activities in the project.

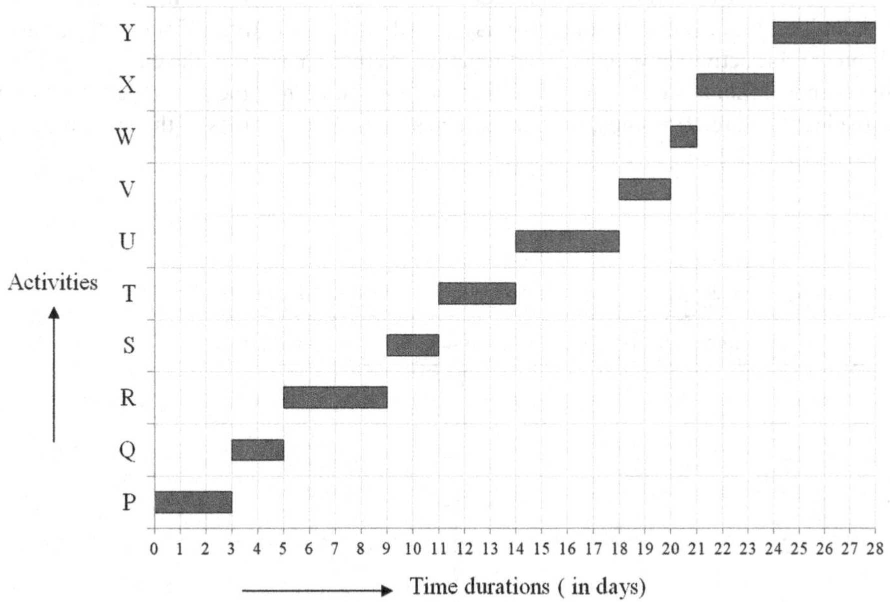

Figure 3.1 A bar chart of the project in which activities are in a series

3.4 Construction of Bar Charts

Consider a project that has ten activities – P, Q, R, S, T, U, V, W, X, and Y – as listed in Table 3.1. All activities must finish for the project to be completed. The time durations for the completion of the 10 activities are 3, 2, 4, 2, 3, 4, 2, 1, 3, and 4 days, respectively. If the activities are executed in a sequence or series according to the following finish-to-start relationships, the project will take 28 days to complete. The bar chart for the project, with 10 activities that are executed in a sequence by the following finish-to-start relationship, is shown in Figure 3.1. The ten activities need not follow a sequence to be executed. The activities of the same project were critically examined for their interdependencies, and the following relationships among the various activities were identified.

- Activities P and Q start together as the project starts; it is assumed that activities P and Q are independent of each other.
- Activity R starts when activity Q is completed; the start of activity R does not depend upon the completion of activity P.
- Activity S starts only when activities P and Q are both completed.
- Activities T and U start simultaneously when activity S is completed.
- Activity V starts when activities T and U are both completed.
- Activities W and X start concurrently when activity V ends.
- Activity Y is the last activity of the project, starting only after the completion of activity W.

The first relation between activities P and Q is that they start simultaneously; the activities are shown by two horizontal bars starting together at the start of the project, as shown in Figure 3.2. Activity P ends at the end of day 3, and activity Q ends at the end of day 2. The second relation is activity R starting when activity Q is completed. In the figure, activity R starts when activity Q is completed at the end of day 2. Activity S starts when activities P and Q are completed at the end of day 3. Activity S ends at the end of day 5; after this, activities T and U start simultaneously. Activities T and U end at the end of day 9, and after this, activity V starts. When activity V ends on day 11, activities W and X start together. Activity Y starts at the end of day 12 when activity W ends. From Figure 3.2 it is clear that the time duration of the project, after incorporating all the identified relationships, is 16 days. The start and finish times of the various activities are given in Table 3.2.

Table 3.1 Activities of the project, their time durations, and start and finish times

Activities	*Time durations (in days)*	*Start times*	*Finish times*
P	3	0	3
Q	2	3	5
R	4	5	9
S	2	9	11
T	3	11	14
U	4	14	18
V	2	18	20
W	1	20	21
X	3	21	24
Y	4	24	28

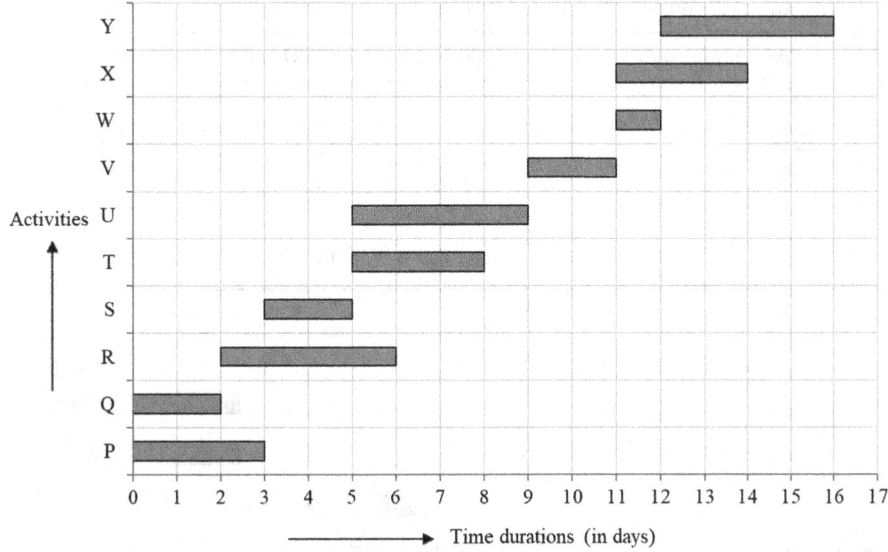

Figure 3.2 A bar chart of the project in which activities are not in a series

Table 3.2 Activities of the project, their time durations, and start and finish times

Activities	*Time durations (in days)*	*Start times*	*Finish times*
P	3	0	3
Q	2	0	2
R	4	2	6
S	2	3	5
T	3	5	8
U	4	5	9
V	2	9	11
W	1	11	12
X	3	11	14
Y	4	12	16

Another sample project has 7 activities. Their time durations are given in Table 3.3. The activities are related to each other as follows.

- Activity B and activity C are carried out simultaneously, and both start after the completion of activity A.
- Activity B must precede activity D.
- Activity E begins after the completion of activities A, B, and C.
- Activity F starts after activities D and E are completed.
- Activity G is the last activity, starting only after the completion of activity E.

The project has seven activities: A, B, C, D, E, F, and G. The project starts with the start of activity A. Activity A ends at the end of day 2, and activities B and C start simultaneously, as shown in Figure 3.3. Activity B must precede activity D; that is, activity D must start immediately after

Table 3.3 Activities of the project, their time durations, and start and finish times

Activities	*Time durations (in days)*	*Start times*	*Finish times*
A	2	0	2
B	3	2	5
C	2	2	4
D	4	5	9
E	5	5	10
F	3	10	13
G	4	10	14

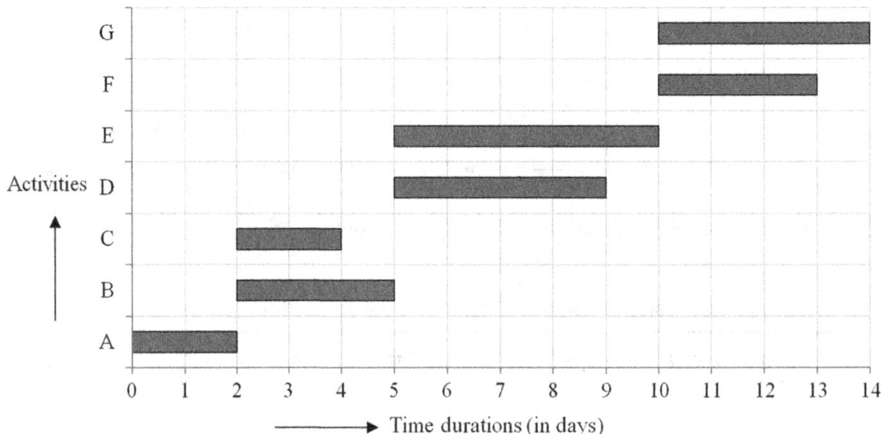

Figure 3.3 A bar chart of the project

the completion of activity B. Activity B ends at the end of day 5, and activity D starts at the start of day 6. Activity E starts at the end of activities A, B, and C at the end of day 5. Activities D and E end at the end of day 10; with the end of activities D and E, activity F starts. At the end of activity E, activity G starts (with the end of day 10). The duration of the project in this case is 14 days.

3.5 Calendar Date Scheduling

In bar charts, time is measured on the x-axis in days, weeks, months, or calendar dates. In case of calendar date scheduling, calendar dates are plotted along the x-axis, as shown in Figure 3.4. This figure shows a bar chart for the project discussed in Figure 3.3, in which all activities are scheduled for seven working days in a week. In single calendar date scheduling, all activities follow the same number of working days in a week. Figure 3.4 shows an example of single calendar date scheduling where a week has seven working days. However, there may be six working days in a week, or holidays in between. Dotted-line bars are generally used to represent breaks or holiday periods, and solid-line bars are used to represent days in which work is actually done. Figure 3.5 shows another bar chart for the same project in which all activities follow single calendar dates given in Table 3.4, in which a week consists of six working days, Sunday being taken as a non-working day. Sunday is shown by dotted-line bars between the solid-line bars.

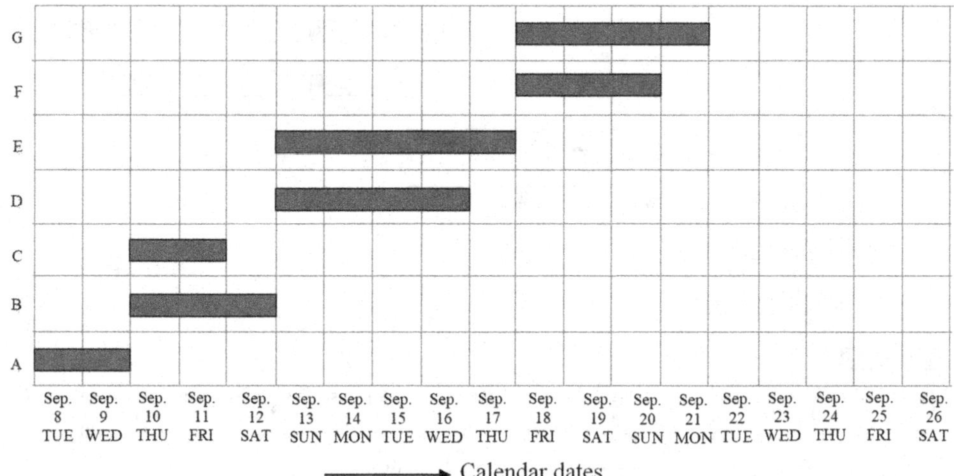

Figure 3.4 A bar chart of the project in which all activities follow a single calendar (7 working days in a week)

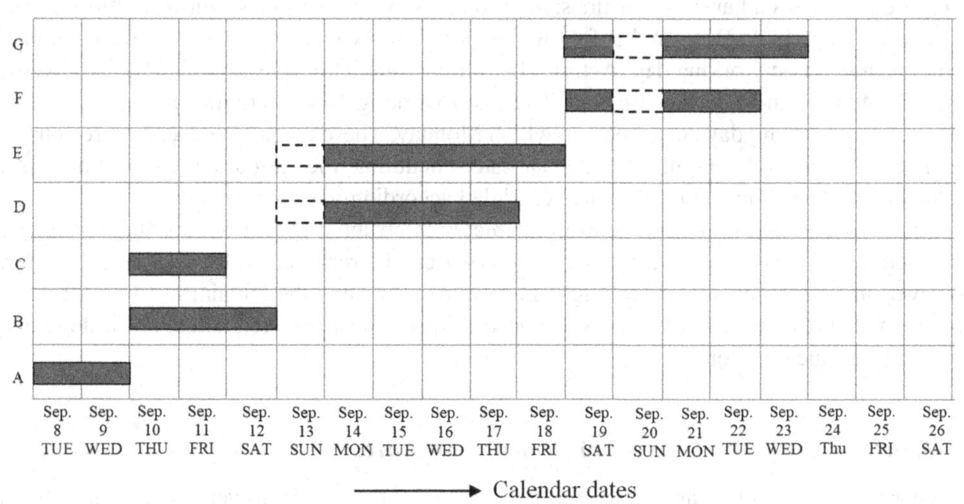

Figure 3.5 A bar chart of the project in which all activities follow a single calendar (6 working days in a week)

Table 3.4 Calendar dates in a month

SUN	*MON*	*TUE*	*WED*	*THU*	*FRI*	*SAT*
		1	2	3	4	5
6	7	8	9	10	11	12
13	14	15	16	17	18	19
20	21	22	23	24	25	26
27	28	29	30			

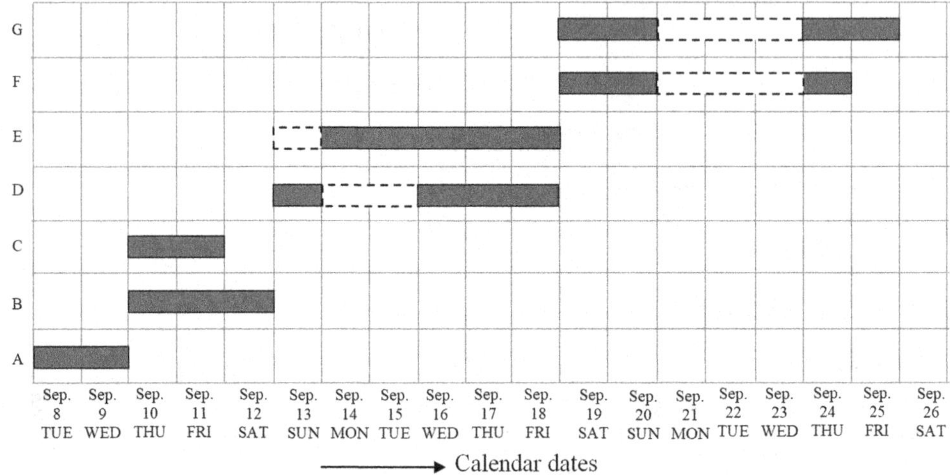

Figure 3.6 A bar chart of the project in which activities follow multiple calendars

Figure 3.6 shows a bar chart for the same project in which activities follow multiple calendar dates. Activities A, B, C, and E follow the model of six working days in a week, Sunday being taken as a non-working day. Activity D follows a model of five working days in a week, in which Monday and Tuesday are considered non-working days. Activities F and G follow a model of four working days in a week, in which Monday, Tuesday, and Wednesday are considered non-working days. In multiple calendar date scheduling, the various activities of a project follow different calendars, and these are scheduled accordingly.

A bar in a bar chart may sometimes not depict uninterrupted work from the start to the end of an activity. For example, the activity *concreting* can be represented by a 15 day–long bar. However, in most cases, the concreting is done in one day, and the remaining 14 days are for curing. In such cases, dotted-line bars are used to distinguish between actual work done and non-working time duration.

3.6 Shortcomings in Bar Charts and Remedial Measures

Bar charts have the following shortcomings; these shortcomings, however, can be partly fixed by suitable remedial measures, as discussed below.

3.6.1 Lack of Details

In bar charts, a large volume of work is generally placed under a single activity; hence, large-scale activities are made. A planner places a large volume of work under a single activity to decrease the number of activities in a project. Where activities include small volumes of work, the number of bars increases and the bar chart becomes difficult to understand. This makes bar charts difficult to use in projects which involve a large volume of work. In general, an activity, whether it is big or small, is represented by a single bar in a bar chart, without any details of how much work is involved in it. Consider, for example, the activities involved in the construction of a small house, the relationships among them, and the corresponding bar chart, as shown

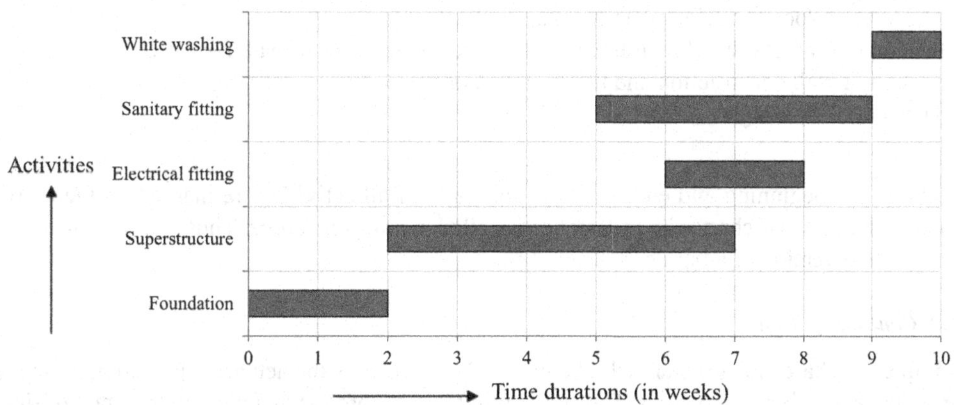

Figure 3.7 A bar chart of a house construction project

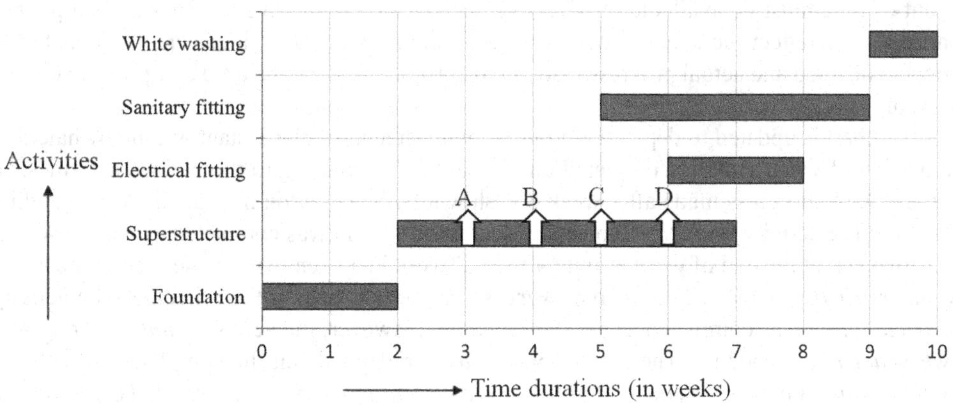

Figure 3.8 A bar chart of a house construction project with milestones

in Figure 3.7. The activity *superstructure* is represented by a single bar. However, this activity requires the following sub-parts for its completion.

- Making door and window frames.
- The constructing walls.
- Roofing.
- Flooring.

The activity *superstructure* is represented by a single bar that does not provide any detailed information about the work included in the main activity, as explained above. A remedial measure for the lack of detail about the work included in each activity is to give details about the sub-parts involved. The start and/or end of all the sub-parts of the activity can be marked on the bar corresponding to the activity, as shown in Figure 3.8. The beginning or ending marks of the sub-parts on a bar corresponding to an activity are called *milestones*, as shown in Figure 3.8. The activity *superstructure* has the following milestones.

A. The start of door and window frame manufacture.
B. the end of door and window frame manufacture and the start of wall construction.
C. the end of wall construction and the start of roofing.
D. The start of flooring.

Similarly, the beginning and end of all the sub-parts of all activities are marked on the corresponding bars. A bar chart with milestones is called a *milestone chart*. Thus, a *milestone chart* is a modified version of a bar chart.

3.6.2 Controlling Tool

A simple bar chart shows a planned schedule. It does not show the actual project progress made during the execution of a project. The actual progress of a project at a particular instant of time is compared with the planned progress at the same instance of time to allow for updates to the plan, after getting some knowledge or information by executing a part of the project. This helps one take suitable corrective measures to keep a plan on track. The progress made at a particular instant of time must be available for effective decisions on corrective measures. A simple bar chart does not reflect the actual progress made in the project. Thus, a bar chart is updated to show the planned and actual progress made in the project, so it can be used as a project controlling tool.

A bar chart is updated to depict the progress made at a particular instant of time by hatching the portion of work actually completed on the bars corresponding to the various activities. In Figure 3.9, the progress made after week 8 is shown by hatching on the bars of corresponding activities. The activity *foundation* took a total of 2 weeks and was completed as per the schedule. However, at the end of week 8, only 4 weeks' work had been done in the case of the activity *superstructure* – that is, the activity *superstructure* was behind the schedule. The activity *electrical fitting* was completed as per the schedule. However, the activity *sanitary fitting* was 1 week *ahead* of schedule. The activity *whitewashing* depends on the completion of activity *sanitary fitting*, thus the rescheduling of the activity *whitewashing* is essential. To control the project, planned and actual achieved progress are compared. This categorizes project activities into three categories: on-schedule activities, behind-schedule activities, and activities that are ahead of schedule. This classification helps one to take suitable corrective measures when required. The bar chart in which the actual progress of project activities is marked is used as a controlling tool.

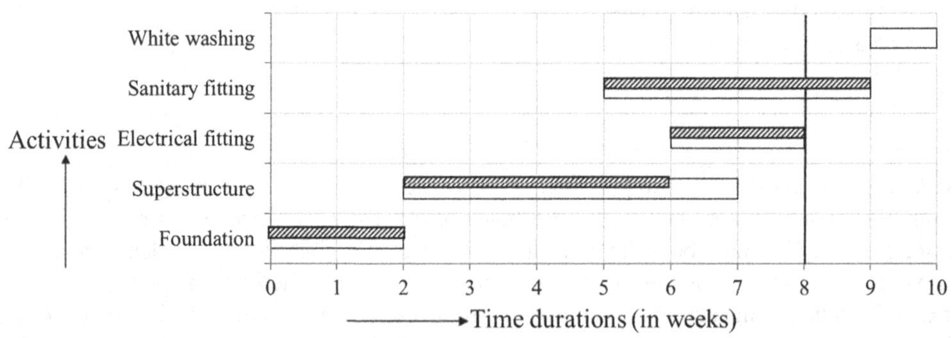

Figure 3.9 A bar chart in which work progress is shown on bars

3.6.3 Inter-dependencies between Activities

Some activities in a project are carried out concurrently, while some are carried out only after the completion of some activities. Concurrent activities are represented using bars that run parallel to each other. Activities whose start and end times depend on other activities are shown in a sequence. However, some activities in a project start with a certain degree of concurrency. These activities are also shown using parallel bars, although parallel bars do not clearly reflect the interdependencies among them. When activities are scheduled parallel to each other or with a certain degree of overlap, it becomes difficult to decide whether activities are interdependent or completely independent of each other.

Figure 3.10 shows an execution plan for three activities: the *plastering, flooring,* and *whitewashing* of a building of 20 rooms. The activity *whitewashing* depends upon the activity *flooring* and the activity *plastering*. Further, the activity *flooring* depends upon the activity *plastering*. The three activities may be done in a series but in that case, the time required for their completion would increase. As shown in Figure 3.10, the three activities are scheduled with a certain degree of concurrency, to reduce the time duration. Such a bar chart is difficult to interpret – either the activity *flooring* starts 2 weeks after the activity *plastering*, or it has 2 weeks of work left after the completion of *plastering*. The activity *whitewashing* has 2 weeks of work left after the completion of *flooring*. If, due to some circumstances, the completion time of *plastering* were to be delayed by 1 or 2 weeks, how would the activities *flooring* and *whitewashing* be affected? This is not reflected in the bar chart. How would the activities of *flooring* and *whitewashing* be affected if the progress rate assumed for the activity *plastering* were not achieved? How would *whitewashing* be affected if the progress rate assumed for *flooring* were not achieved or were to be delayed? This ambiguity exists because inter-dependencies among the activities are not indicated in the bar chart.

This shortcoming can be partly overcome by dividing each large-size activity into sub-activities or sub-parts. Sub-parts are shown by using milestones for various activities to depict interdependency. For example, in Figure 3.11, all the activities are divided into 4 sub-parts, each sub-part reflecting work corresponding to 5 rooms. Since the activity *plastering* is of the same time duration as the activity *flooring*, both activities have 4 parts of equal time durations. However, the activity *whitewashing* is faster than *flooring* and *plastering*, thus, in the case of *whitewashing*, the sub-parts will be of smaller size. The improved bar chart would depict the completion of each sub-part using milestones 1, 2, 3, and 4. In Figure 3.11, it is reflected that

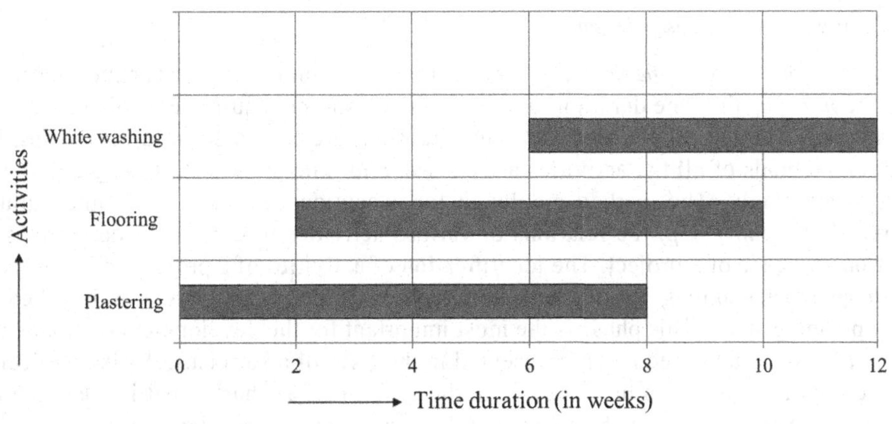

Figure 3.10 A bar chart in which activities have a certain degree of concurrency

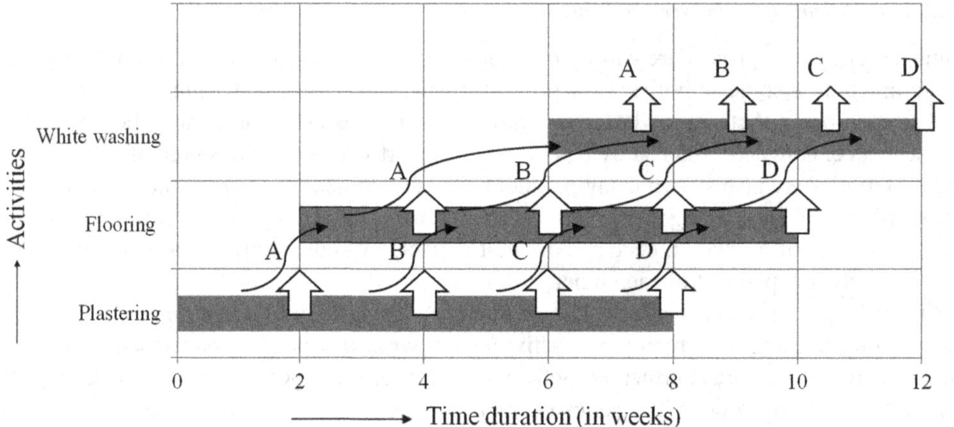

Figure 3.11 An improved bar chart using milestones in which activities have a certain degree of concurrency

when the *plastering* of the first five rooms is completed, *flooring* will start. When the *plastering* of 10 rooms is completed, the *flooring* of the first 5 rooms will need to be completed. When the *plastering* of the first 15 rooms and the *flooring* of the first 10 rooms are completed, then the activity *white washing* will start.

3.6.4 Useful for Routine Tasks

Bar charts are suitable for projects of routine nature. Bar charts are not suitable for projects not of routine nature, nor for research and development projects. In the case of projects of a non-routine nature or which regard research and development, the estimated time durations will lack a fair degree of accuracy. Bar charts are not suitable for projects in which the time duration estimations of various activities are uncertain. Thus, bar charts are suitable for routine projects, in which time duration estimates can be made with a fair degree of accuracy.

3.6.5 Planning and Scheduling Together

To develop a bar chart, *in the first step*, a project is divided into small parts called activities. *In the second step*, the time durations and individual resource requirements of activities are estimated, or, when resources are fixed, time durations are estimated for all activities. The execution methods of all the activities are chosen, and time duration are assigned to each activity accordingly. The method employed for execution determines the time duration required. *In the third step,* the relations of various activities to each other determines the execution sequence of a project. The identification of activities of a project and identifying inter-dependencies among various activities requires in-depth analysis; this is called the project planning stage. This phase is the most important for the development of a bar chart. *In the fourth step*, the schedule is represented in the form of a bar chart. A planner decides on an activity and puts it on the schedule in the form of a bar chart – that is, planning and scheduling are done together. In a bar chart, planning and scheduling are done together, which leads to less effective planning.

3.7 Milestone Charts

A milestone chart is a modified or upgraded version of a bar chart in which times at which key events should occur are highlighted on the respective bars of a bar chart. These key events may be the submission of first, intermediate, or final bills, the placement of purchase orders for materials, the approval of the owner on a particular decision, etc. These events can also be the start or finish times of various sub-parts of different activities. Figure 3.11 is an example of a milestone chart. In general, milestones are critical events that have serious implications for project completion. If a milestone does not impact a project's schedule, it may be eliminated from its position. Milestones are sometimes also used to reflect inter-dependencies among various activities of a project, as discussed earlier.

When an activity includes a large volume of work, it is represented by a lengthy bar in the bar chart. In lengthy bars, there is a low degree of detail. In such cases, large-scale activities are divided into sub-parts by milestones, each sub-part being easily identifiable during the execution of a project. The inter-dependencies between sub-parts are easily reflected using milestones in a bar chart. A milestone chart displays some key points in the form of milestones, through which control can be easily achieved.

3.8 Linked Bar Charts

Linked bar charts can be used to depict the successors and predecessors of each activity in a project. Linked bar charts use arrows to link various activities of a project, so as to represent the successors and predecessors of each activity. Links depict which activity/activities must be completed before a subsequent activity starts. Linked bar charts clearly show the inter-dependencies among the various activities of a project. This is a way of representing a project network in most scheduling programs. Figure 3.12 shows a linked bar chart. It shows that when the activity *foundation* is finished, the activity *superstructure* should start. On the third day after the start of the activity *superstructure*, the activity *sanitary fitting* should start. On the fourth day after the start of *superstructure,* the activity *electrical fitting* should start. The activity *whitewashing* should start when all the activities preceding it are finished.

A project contains hundreds of activities. If the finish time of one activity is delayed during the execution of a project, how might this delay impact other activities? A simple chart does not

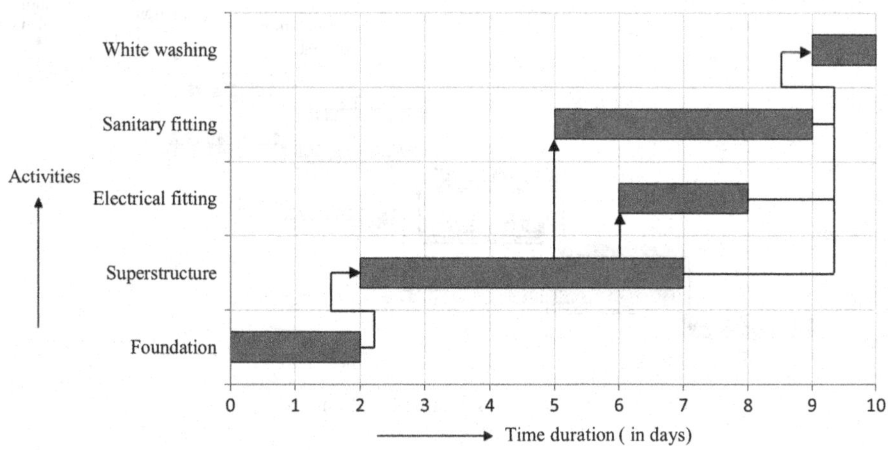

Figure 3.12 A linked bar chart

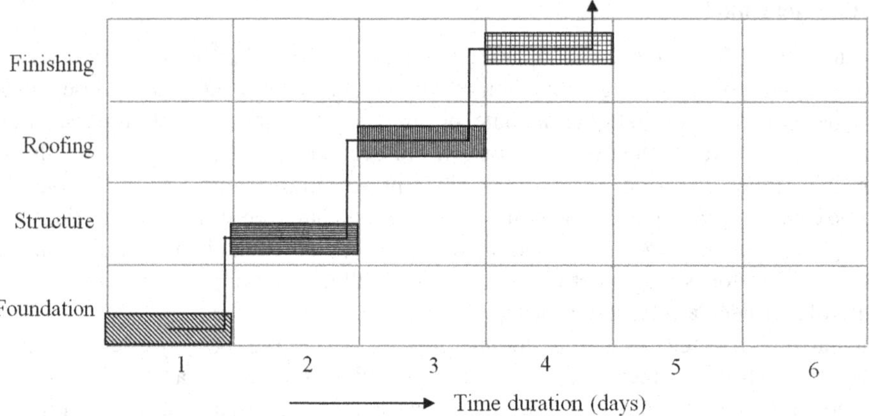

Figure 3.13 A linked bar chart representing the execution sequence for the construction of a parking shed

provide any information about the number of activities that may get affected. Linked bar charts are helpful in such cases.

3.9 Bar Charts for Repetitive Projects

Repetitive projects are projects in which various activities are repeated. Projects containing repetitive activities include the construction of roads, the laying of pipelines, tunneling, the construction of multi-story buildings, etc. Consider a project that involves the construction of three similar parking sheds. The project has three similar parking sheds, thus it is a repetitive project. The activities involved in constructing the parking sheds are the construction of the *foundation* and the *structure*, *roofing*, and *finishing*. The construction of one parking shed involves four activities which are repeated three times to construct three similar parking sheds, making a total of twelve activities. In summary, the project has four repetitive activities repeated three times, making a total of twelve activities. Figure 3.13 shows the execution sequence for the construction of the parking shed in the form of a bar chart in which all four activities are in a sequence. Figure 3.14

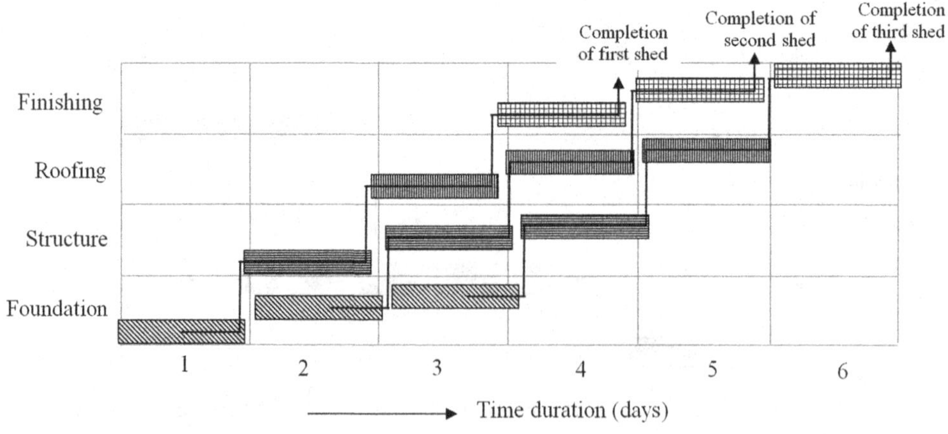

Figure 3.14 A linked bar chart representing the execution sequence for the construction of three similar parking sheds

shows the execution sequence for the construction of three parking sheds in the form of a bar chart representing all twelve activities.

3.10 Differences between Bar Charts and WBS

Repetitive projects are projects in which various activities are repeated. Projects containing repetitive activities include the construction of roads, the laying of pipelines, tunneling, the construction of multi-story buildings, etc.

- Bar charts use graphical representation to depict the execution schedule of a project, whereas a WBS is the division of a project into small manageable parts. A WBS uses hierarchical representation for manageable parts which are created before the development of a bar chart or project network;
- Bar charts use graphical representation to represent inter-dependencies among manageable parts. Further, the presence of milestones in a bar chart provides more details about key events. On the other hand, a WBS does not depict the execution sequence through time, but facilitates the estimation of time duration, cost, and the resources required for different deliverables;
- Bar charts depict the start and finish times of all manageable parts and are used throughout the lifecycle of a project, whereas a WBS is used at the start of a project to describe the deliverables needed to complete the project; and
- Bar charts are used to depict the plan and schedule of a project, whereas a WBS is used to represent the initial planning stage of a project in a hierarchical form.

3.11 Conclusion

Bar charts are simple graphical charts used to represent the plan and schedule of a project. Bar charts do not require special technical knowledge to be understood. A bar chart lists activities to be performed on the y-axis, and the time duration of a project on the x-axis. The activities in a bar chart are represented by bars. The lengths of the bars in a bar chart are equal to their time durations. The start of a bar depicts the start time and the end of a bar depicts the finish time of the activity. Inter-dependencies among various activities are also represented in bar charts. Bar charts are good for the planning and scheduling of simple and small projects, but are not good for complex and big projects.

Example 3.1: Activities, their time durations, and the inter-dependencies among them are given in Table 3.5. Draw the bar chart and find the time duration of the project.

Solution: The bar chart for the project is shown in Figure 3.15; the time duration of the project is 12 days.

Example 3.2: Activities and their time durations are given in Table 3.6. The inter-dependencies among various activities are as follows:

- The project starts with the start of the activity *feasibility*; the activity *research* follows *feasibility*;
- The activity *design* starts when 50 percent of the activity *research* is complete; once *design* is complete, the activity *pilot study* starts;
- The activity *implementation* follows *pilot study*; the activity *control* starts when 25 percent of *implementation* is complete; and

Table 3.5 Activities, their time durations, and inter-dependencies between them

Activities	*A*	*B*	*C*	*D*	*E*	*F*	*G*	*H*
Time durations	2	2	3	3	4	3	1	3
Subsequent activities	B, C	D	E, F	G	H	–	–	–

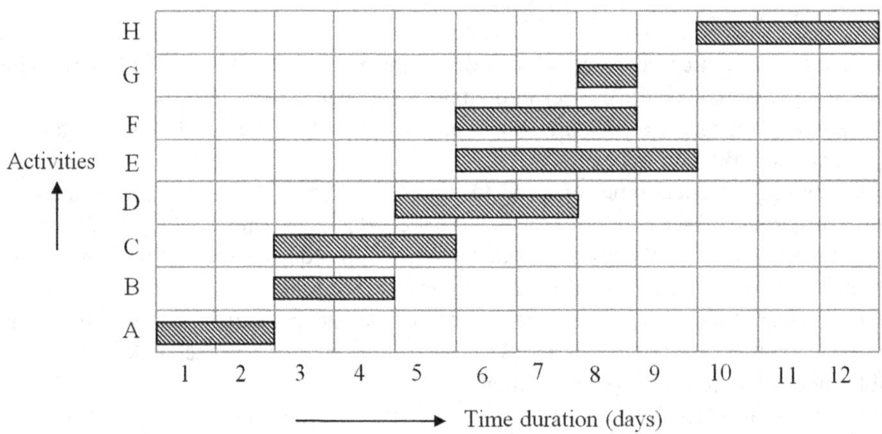

Figure 3.15 Bar chart representing the execution sequence for the project

Table 3.6 Activities and their time durations

Activities	*Feasibility*	*Research*	*Design*	*Pilot study*	*Implementation*	*Control*	*Final check*	*Handing over*
Time durations (weeks)	2	2	2	1	4	3	1	1

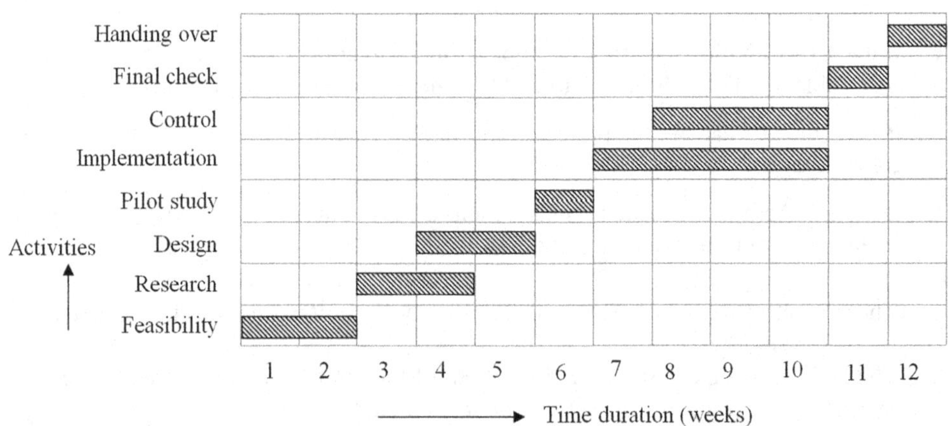

Figure 3.16 Bar chart representing the execution sequence for the project

- The activity *final check* starts when *control* is complete; the project ends with the end of the activity *handing over* which follows the activity *final check.*

Draw the bar chart and find the time duration of the project.

Solution: The bar chart for the project is shown in Figure 3.16; the time duration of the project is 12 days.

Exercises

Question 3.1: What is a bar chart? What are the main advantages that make it a popular tool among project planners?

Question 3.2: Draw a bar chart of your day-to-day activities from morning to evening.

Question 3.3: Draw a bar chart for the construction of a two-room house; break the project into 5 to 10 activities.

Question 3.4: Activities, their time durations, and the inter-relationships among them are given in Table E3.1. Draw the bar chart and find the time duration of the project.

Answer: The bar chart for the project is shown in Figure E3.1; the time duration of the project is 11 days.

Table E3.1 Activities, their time durations, and inter-relationships among them

Activities	*A*	*B*	*C*	*D*	*E*	*F*	*G*	*H*
Time durations	3	1	2	3	1	2	2	3
Subsequent activities	B	C, D	E, F	G	G	H	–	–

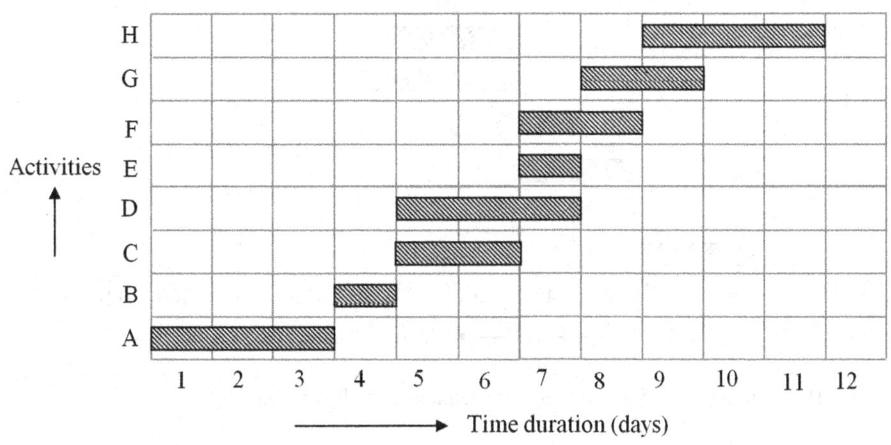

Figure E3.1 Bar chart representing the execution sequence for the project

Question 3.5: Activities, their time durations, and the inter-relationships among them are given in Table E3.2. Draw the bar chart and find the time duration of the project.

Answer: The bar chart for the project is shown in Figure E3.2; the time duration of the project is 120 days.

Question 3.6: A repetitive project involves the manufacturing of three similar units. The activities involved in manufacturing a unit are A, B, C, and D. The time durations of the four activities are 2, 1, 2, and 2 respectively. The four activities of a unit are in series and repeat three times, making a total of twelve activities. Draw the bar chart for the repetitive project.

Answer: Figure E3.3 shows the execution sequence for the manufacturing of three similar units in the form of a bar chart in which all twelve activities are depicted.

Table E3.2 Activities, their time durations, and inter-relationships among them

Activities	*A*	*B*	*C*	*D*	*E*	*F*	*G*	*H*	*I*	*J*	*K*	*L*
Time durations	20	10	20	20	30	20	10	20	10	20	20	30
Immediately preceding activities	–	A	A	B	C	C	D	G, F	E	E	H, I	J

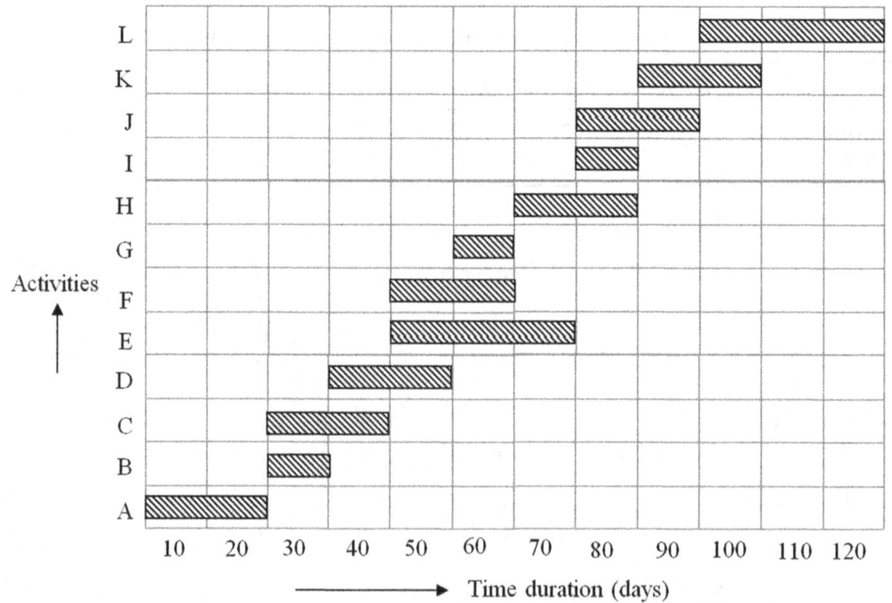

Figure E3.2 Bar chart representing the execution sequence for the project

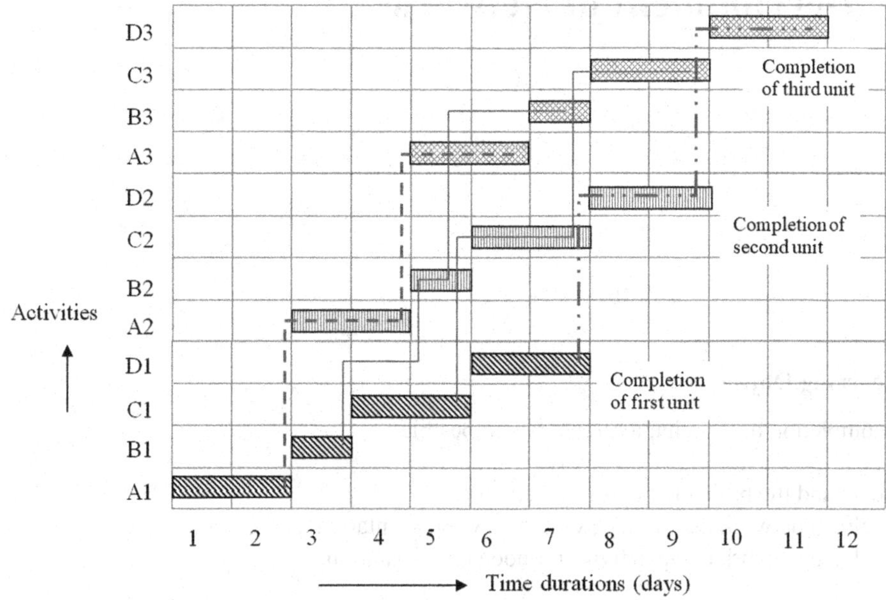

Figure E3.3 Bar chart representing the execution sequence of the project

4 Development of Network

4.1 Learning Objectives

After completion of this chapter, readers will be able to:

- Understand the basic elements of a network,
- Develop a network using activity-on-arrow representation, and
- Develop a network using activity-on-node representation.

4.2 Introduction

Network-based planning and scheduling techniques came into existence in 1931 with the development of the *Harmonygraph* by Karol Adamiecki. However, this work has not been cited much in histories of the development of network-based project management techniques. Network-based project management techniques have been developed in three or more areas simultaneously; these three widely explored areas have been covered in this book. The first of them is the application of network-based techniques in industry for routine planning and scheduling, which led to the invention of the *critical path method.* The second area is the application of network-based techniques in the military, which led to the invention of the *programme evaluation and review technique*. The third research area contributed to a widely used technique called the *precedence diagramming method.* There were some other techniques but they were never as widely accepted as the three mentioned above. The three techniques mentioned above are the most widely used network-based techniques for the planning and scheduling of projects.

A network is a logical and chronological graphic representation of project activities. It represents activities in a flow so as to represent the execution sequence. The two basic elements of a network are arrows and nodes. There are two types of network: *activity-on-arrow* (AOA) and *activity-on-node* (AON). In AOA networks, the activities of a project are represented by arrows. The description of an activity is written on the corresponding arrow, therefore such networks are called activity-on-arrow networks. AOA networks are also called arrow diagrams. In AON networks, the activities of a project are represented on nodes, therefore such networks are called activity-on-node networks. AON networks are also called node diagrams. AOA networks are more cumbersome than their counterpart, but are easier to understand.

4.3 Activity-on-Arrow Networks

In AOA networks, an arrow represents an activity. An activity is that portion of a project which consumes time and resources. It has well-defined start and end points, called events. An activity requires a time duration, labor, equipment, etc. for its execution. The tail of an arrow represents

DOI: 10.1201/9781003428992-4

the starting point and the head of an arrow represents the endpoint of an activity. An arrow may be straight or curved as necessary. Each arrow in a network connects two nodes as shown in Figure 4.1(a). Nodes are the start or end points of arrows. Nodes are called events in AOA networks. An event is the instant of time at which an activity starts or ends. The starting point is called the *from*, *start*, or *i* event and the end point is called the *to*, *end*, or *j* event. An arrow is denoted by the symbol *i-j*, an activity by a_{ij}, and an activity is identified by two events that define its start (*i*) and end (*j*). The arrows corresponding to activities are not drawn to scale like bars in a bar chart. The arrow corresponding to an activity may be of whatever length is necessary for the sake of better representing a network. The activity description is written above the arrow and the time duration is written below the arrow, as shown in Figure 4.1(a), so that the two are not confused. An event representing the completion of more than one activity at once is called a *merge* event, as shown in Figure 4.1(b). An event representing the start of more than one activity at once is called a *burst* event. This chapter is limited to the basic rules and procedures of network development.

4.3.1 AOA Networks

To develop a network, all the activities involved in a project must be identified. This is usually called the *project planning phase*, as the identification of project activities and their inter-dependencies requires a thorough analysis of the project. Decisions are also made regarding the resources to be used and the time duration for each activity in a project are determined accordingly. Once the activities, their time durations, and the inter-dependencies among them are identified, the execution sequence is represented in the form of a network. A network represents a plan comprehensively to show how a project should proceed. The accuracy of a network depends upon the prior experience of a planner in handling such projects.

In AOA representation, the various activities of a network must be taken to follow a finish-to-start relationship. In a finish-to-start relationship, before an activity begins all the activities preceding it must be completed. Activities with no preceding activities start at the beginning of a project. Arrows in an AOA network represent only the logical relationships between its various activities. The length of an arrow in a network has no significance.

In general, a human being reads the flow of an arrow from left to right, therefore, a network is drawn to show a consistent flow of arrows from left to right. However, it is not mandatory for arrows to flow from left to right; an arrow may flow from right to left, but this reverse flow

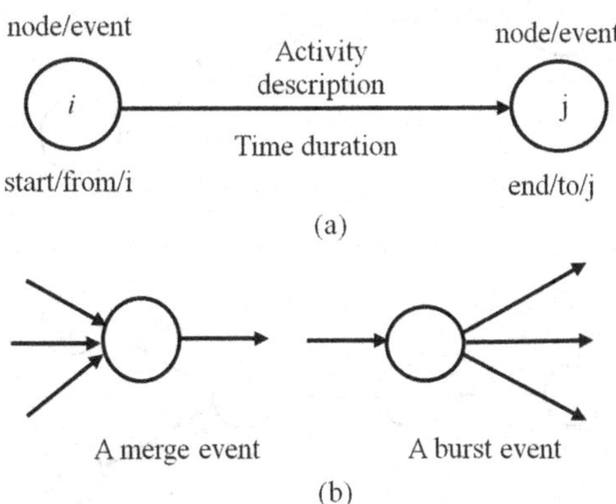

Figure 4.1 Activity-on-arrow representation

would have to be kept in mind when analyzing the network. The development of a network has been made clear using the following examples.

Activity Y is controlled by activity X. Activity Y cannot begin until activity X is complete. Activities X and Y occur in a series. The network diagram is shown in Figure 4.2.

Activity Z is controlled by activities X and Y. Activity Z cannot begin until both activities X and Y are complete. The network diagram is shown in Figure 4.3.

Activities Y and Z are controlled by activity X. Activities Y and Z cannot begin until activity X is complete. The network diagram is shown in Figure 4.4.

Activities C and D follow activity B (or activities C and D are controlled by activity B). Activities C and D cannot start unless activity B is complete. Activity A precedes activity B. The network diagram is shown in Figure 4.5.

Activities Y and Z are controlled by activities W and X. Activities Y and Z cannot start until both activities W and X are complete. However, activities Y and Z start independently of each other. The network diagram is shown in Figure 4.6.

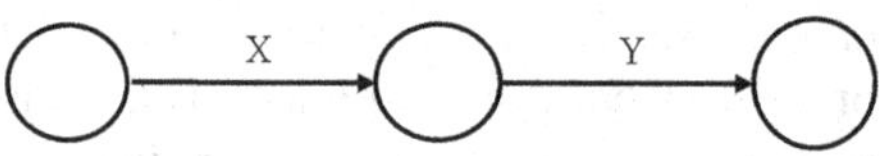

Figure 4.2 A network

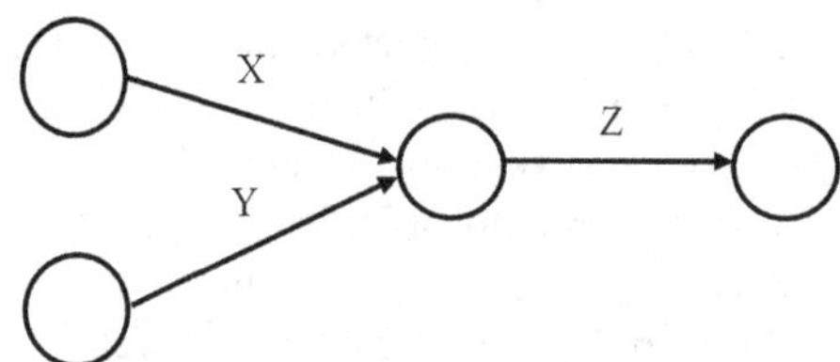

Figure 4.3 A network

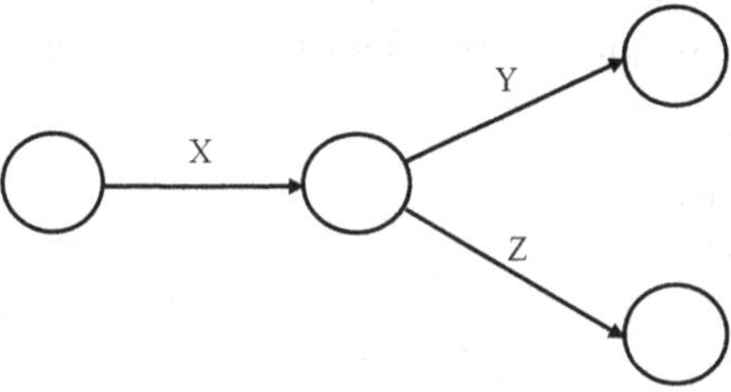

Figure 4.4 A network

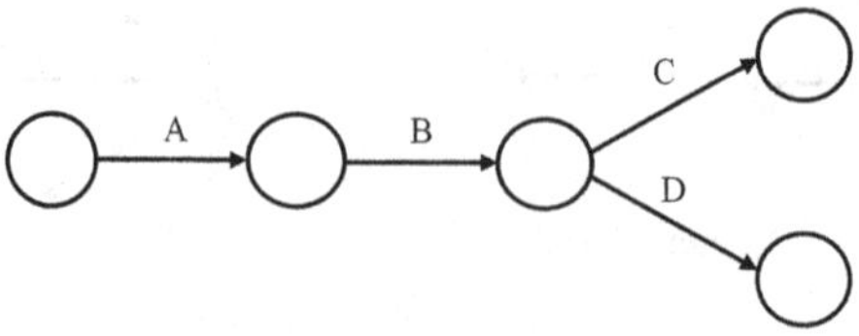

Figure 4.5 A network

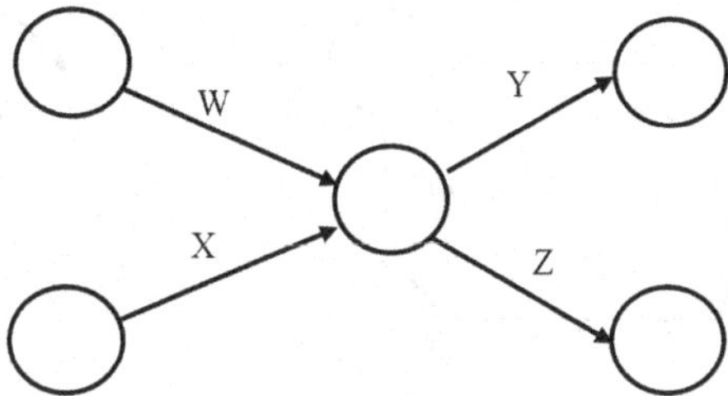

Figure 4.6 A network

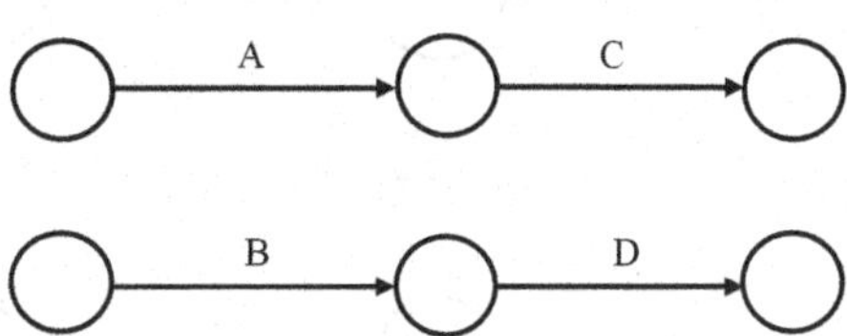

Figure 4.7 A network

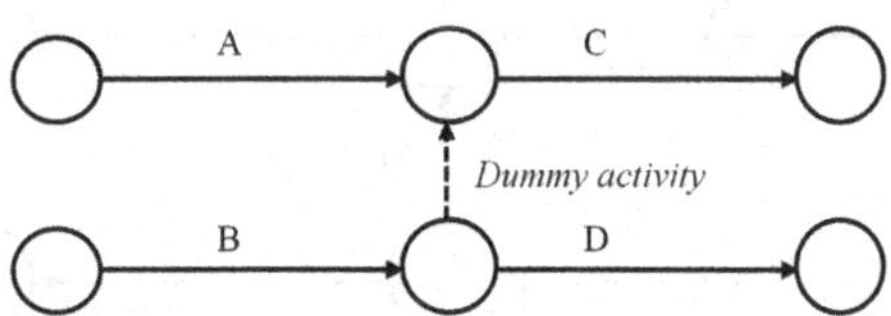

Figure 4.8 A network

Activity C follows activity A and activity D follows activity B; there is no connection between activity D and activity A, or between activity C and activity B. The network diagram is shown in Figure 4.7.

Activity C is controlled by activity A and activity B, however activity D is controlled by activity B only. Sometimes relationships between activities cannot be represented using arrows corresponding to regular activity. A *dummy activity*, represented by a dotted arrow, is used to represent relationships in such cases. A dummy activity has no time duration and involves no work or resources. In Figure 4.8, the dummy activity represents the dependency of activity C on activities A and B. The use of a dummy activity represents the dependency of activity C on activities A and B by facilitating the flow from activities A and B to activity C. Figure 4.8 shows how activity D depends upon activity B only by representing the flow from activity B to activity D. The network diagram is shown in Figure 4.8.

Activity Z is controlled by activities V and W, while activity Y is controlled by activities U and V. The network diagram is shown in Figure 4.9.

Activity D is controlled by activities A, B, and C. However, activity E is controlled by activity B and activity C. Activity F is controlled by activity C only. The network diagram is shown in Figure 4.10.

Activity A controls activities C and D, while activity B controls activities D and E. Activity D is controlled by both activity A and activity B. The network diagram is shown in Figure 4.11.

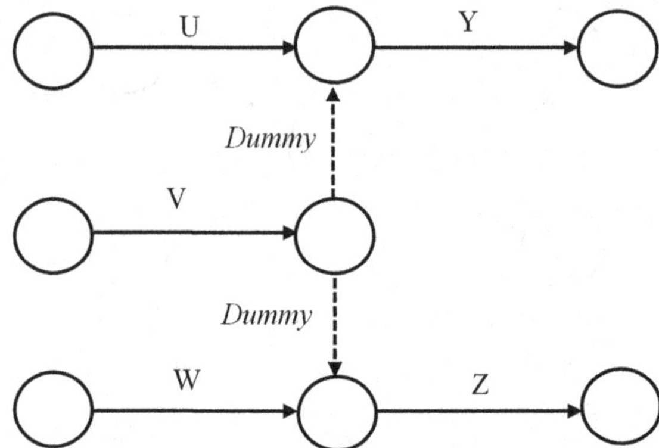

Figure 4.9 A network

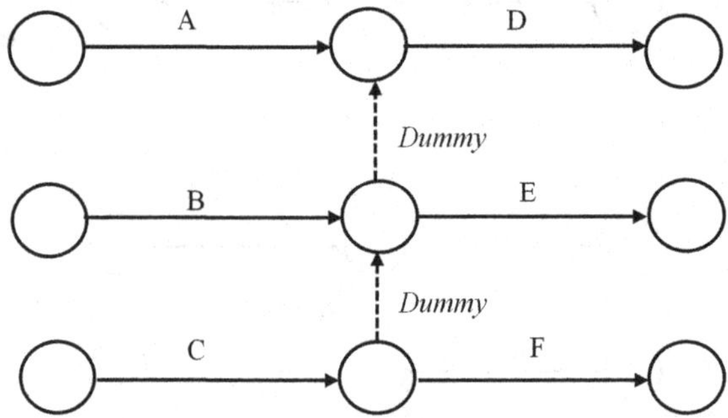

Figure 4.10 A network

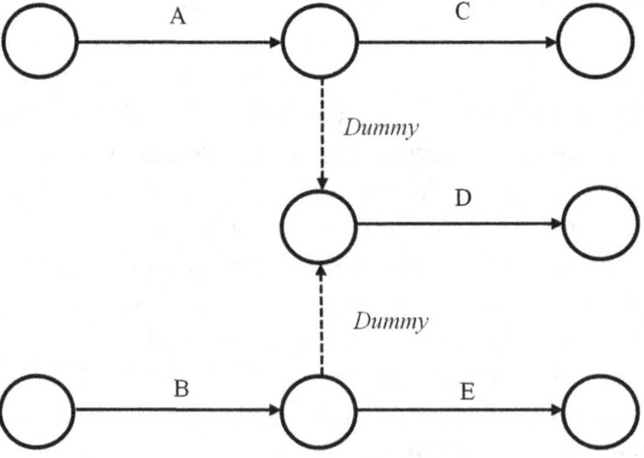

Figure 4.11 A network

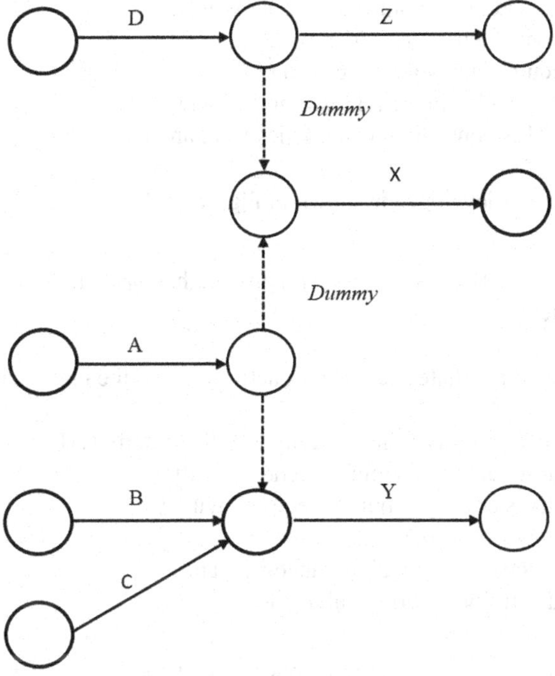

Figure 4.12 A network

Activity X is controlled by activities D and A. Activity Y is controlled by activities A, B, and C, while activity Z is controlled by activity D only. The network diagram is shown in Figure 4.12.

As shown in Figure 4.13, a network may have crossovers. In such cases, an initial draft of a network is prepared. After the preparation of the initial draft, the network is re-drawn to improve its representation by reducing the number of dummy activities and crossovers. In case it becomes impossible to eliminate crossovers, these are shown properly in Figure 4.13.

Example 4.1: Draw the network diagram for project activities with the following inter-relationships.

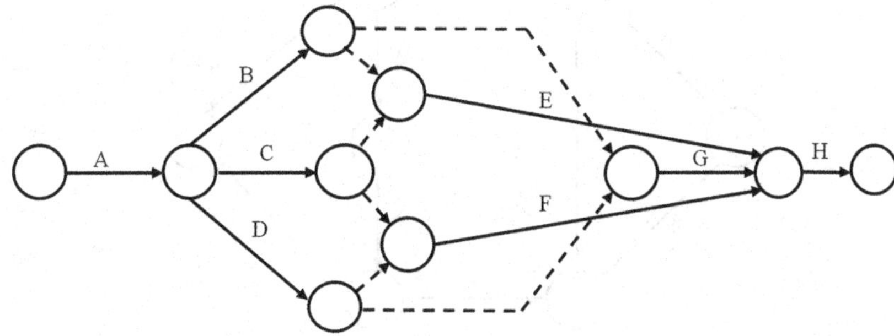

Figure 4.13 A network

- Activities B, C, and D are controlled by activity A,
- Activity E is controlled by activities B and C,
- Activity F is controlled by activities C and D,
- Activity G is controlled by activities B and D, and
- Last operation H is controlled by activities E, F, and G.

Solution: The network developed is shown in Figure 4.13.

Example 4.2: Draw a network diagram for a project that has 9 activities with the following inter-dependencies.

- Activity T is the immediate successor of activity U and the immediate predecessor of activity W,
- Activity T follows activity R and activity Y follows activity T,
- Activity X follows activity W but precedes activity Z,
- Activity V follows activity S but precedes activity Z,
- Activity U follows activity S,
- Activity Y and activity Z end simultaneously, and
- Activity R and activity S start simultaneously.

Solution: The network developed is shown in Figure 4.14.

Example 4.3: Draw a network diagram for a project with 9 activities, using the inter-relationships given in Table 4.1.

Solution: The network developed is shown in Figure 4.15.

Example 4.4: Draw a network diagram for a project that has 9 activities, using the inter-relationships given in Table 4.2.

Solution: The network developed is shown in Figure 4.16.

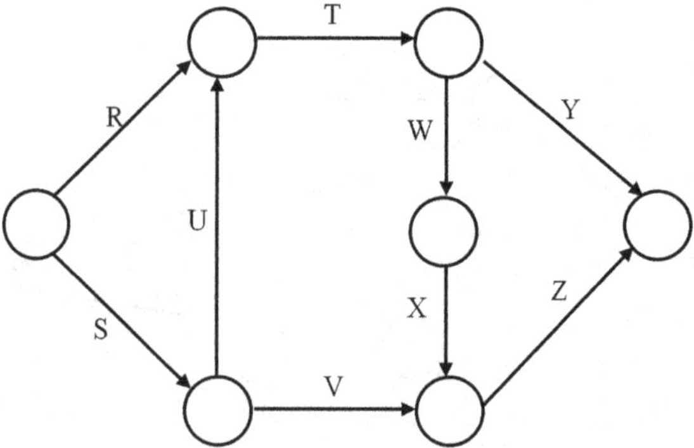

Figure 4.14 A network

Table 4.1 Events and their inter-dependencies

Events	*Immediate Predecessor events*
1	-
2	1
3	1, 2
4	3
5	4
6	2, 5
7	4, 6

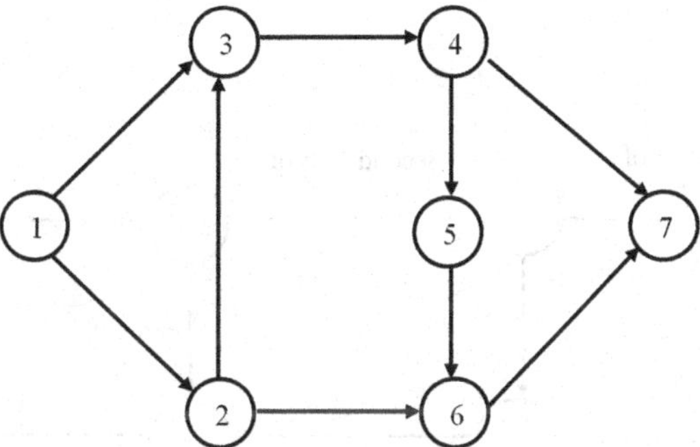

Figure 4.15 A network

Table 4.2 Activities and their inter-dependencies

Activities	*Immediately preceding activities*	*Immediately succeeding activities*
A	–	C
B	–	D, E
C	A, D	F, H
D	B	C
E	B	I
F	C	G
G	F	I
H	C	J
I	E, G	J
J	H, I	–

4.3.2 Network Issues

Consider a project consisting of 5 activities, A, B, C, D, and E. Activity C depends upon the completion of activity A and the first half of activity B. The completion of the second half of activity B is independent of activity C. Activity D depends upon the completion of the second half of activity B. Activity E depends upon the completion of activity C and the second half of activity B. To represent this situation, activity B has been divided into two parts as shown in Figure 4.17.

In Figure 4.18, activity B cannot begin until activity A and activity D are complete. Activity C depends upon the completion of activity B. Activity E depends upon the completion of activity C.

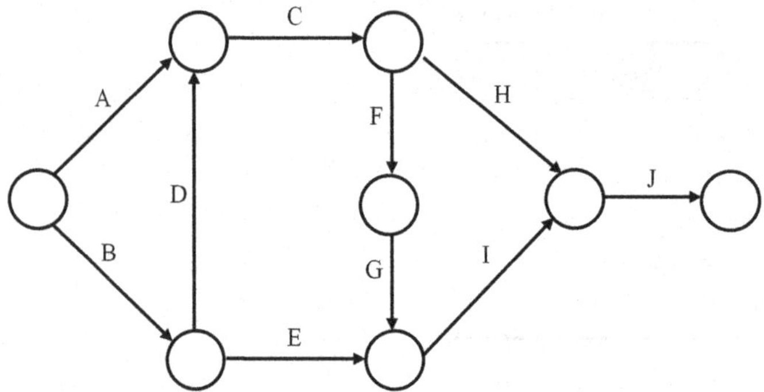

Figure 4.16 A network

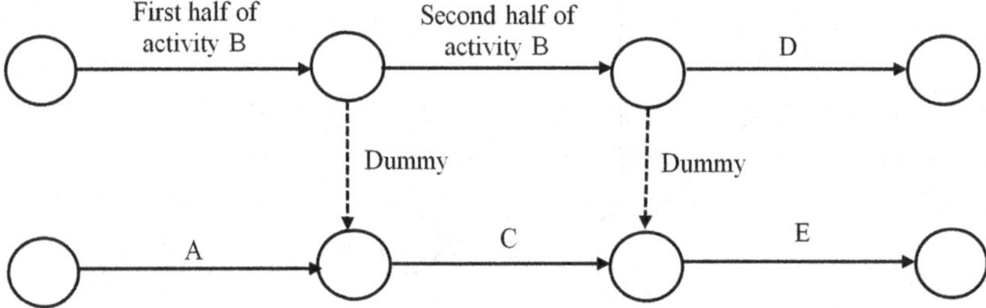

Figure 4.17 A network

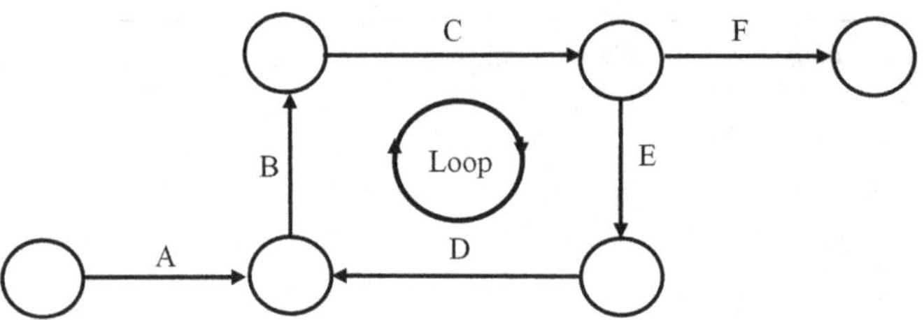

Figure 4.18 A network with a loop

Similarly, activity D cannot start until activity E is completed. However, activity B cannot begin until activity A and activity D are completed. In this way, the network flow returns to the earlier event. This is because of *closed looping*, the formation of a loop by activities B, C, E, and D. Activity F would never be able to get started because of the formation of the loop. The formation of loops occurs in a network because of incorrect logic. Negligence of loops results in an incorrect representation of the inter-dependencies between various activities. Loops formed in a network are corrected by redefining the dependencies among the activities of a project to relate them correctly.

4.4 Event Numbering for Identification of Activities

It is better to number the events or nodes of a network. An activity is represented by an arrow that joins the two events and is identified in a network by the event numbers on its tail and head.

In general, event numbering is not essential when a network is analyzed manually. However, when a network is analyzed using software, not all computer programs interpret the direction of an arrow in a network. Some developed computer programs interpret arrows' directions through the node numbers on their tails and heads. Generally, programs understand flow direction through movement from a lower to a higher node number. Therefore, event numbering is done in such a way that the event number on the tail of an arrow is lower than the event number on the head. Many computer programs can handle non-consecutive event numbers and also random event numbering. An activity (a_{ij}) is identified by its two nodes, that is, its start (i) and end events (j). The number on the tail event must be lower than the number on the head ($i<j$) to represent the flow from tail to head. To ensure $i<j$ for all activities of a network, the *D. R. Fulkerson rule* is used. According to the D. R. Fulkerson rule, the event number on the head (j) of an arrow must be greater than the event number on the tail (i). This is ensured by following the steps given below.

- In general, a project has a single starting event. This starting event has arrows coming out of it and none entering into it. This starting event is numbered as 1, as shown in Figure 4.19(a). When there are multiple starting events, these are numbered as 1, 2, 3, etc..

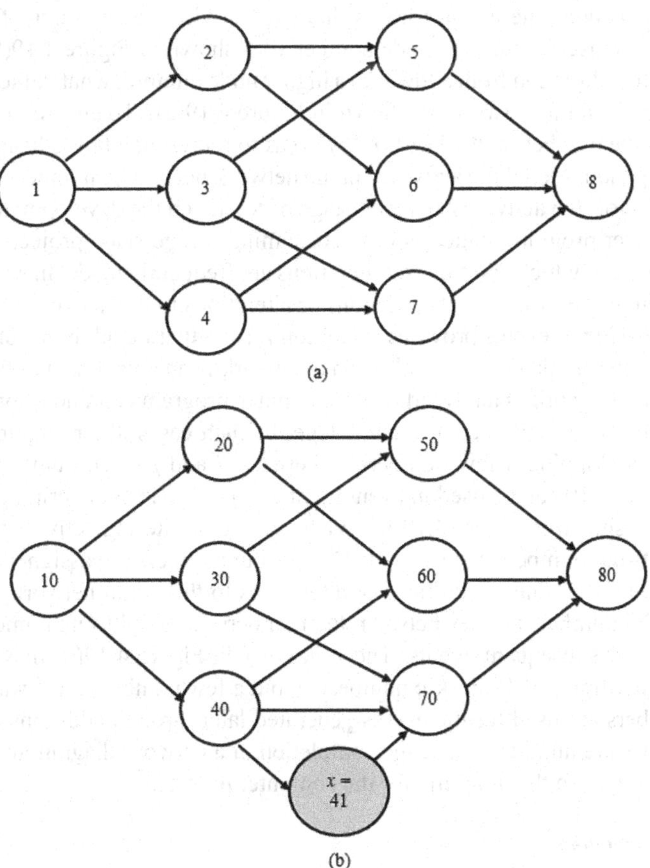

Figure 4.19 A network

- Once the starting event is numbered as 1, in the next step one can neglect all the arrows coming out of the starting event numbered as 1. This releases one or more new events which have arrows coming out and none entering into them;
- Number these newly released events as 2, 3, 4, etc..
- Neglect again all the arrows emerging from the newly released numbered events; this again releases a few more new events which have arrows coming out and none entering into them.
- Number these newly released events as 5, 6, 7, etc..
- Continue this event numbering till the last event of the project is numbered.

In general, a project network has only one initial event. From this initial event, arrows come out or burst; none enter it. Similarly, a project network has only one end event. In the end event, arrows come in or merge; none come out of it. To fulfill this requirement it is common practice to bring all loose start events to a single starting event, using dummies where necessary. Similarly, it is a common practice to bring all loose end events to a single terminal event using dummies where necessary.

4.4.1 Skip Numbering

Assume that the network shown in Figure 4.19(a) is an input to a computer program for analysis. Once the analysis is done the planner learns that two activities are missing. The two missing activities have been inserted through node number x, as shown in Figure 4.19(b). If a program understands the flow direction from a lower to a higher node number, what value of x will ensure $i < j$ for all activities in the network, as shown in Figure 4.19(a)? To ensure the number on the tail is lower than the number on the head ($i < j$), so as to represent a tail to head flow direction, the events corresponding to all the activities in the network need to be numbered again. In other words, the addition of two activities at a later stage necessitated the development of a new input file for the computer program, which is time-consuming. Large-scale projects contain a large number of activities, to which additions or alterations are frequently made. In such cases, repeatedly generating input files to facilitate additions or alterations to the network is not feasible.

The *skip numbering* of events provides a solution for facilitating additions or alterations to a network. Skip numbering is not essential when a network is analyzed manually. Skip numbering is used when a network is analyzed using computer programs and additions or alterations to the network are to be facilitated at a later stage. In such cases, the re-numbering of events can be avoided by skipping a few numbers in between i and j when numbering events. For example, multiples of 10 can be used in event numbering – that is, numbering events as 10, 20, 30, 40, 50, etc., as shown in Figure 4.19(b). When adding or altering activities at a later stage, the skipped or left-out numbers can be used. The left-out numbers are assigned to newly added events which have arisen due to additions or alterations to the initial network. A newly added event is assigned a number that lies between the numbers assigned to its immediately preceding and immediately subsequent events. The value of x in Figure 4.19(b) may be 41. It is not necessary to use multiples of 10 in skip numbering, but a few numbers are always skipped, and the skipped numbers are used for the events generated later through additions or alterations to the network. Nodes are numbered after the completion of a network diagram to avoid having to make repeated changes to the input file for the computer program.

4.4.2 Event Representation

When a network is analyzed using a computer program, it identifies each activity in a network by its event numbers. Each activity generally has a unique pair of start and end events. Event

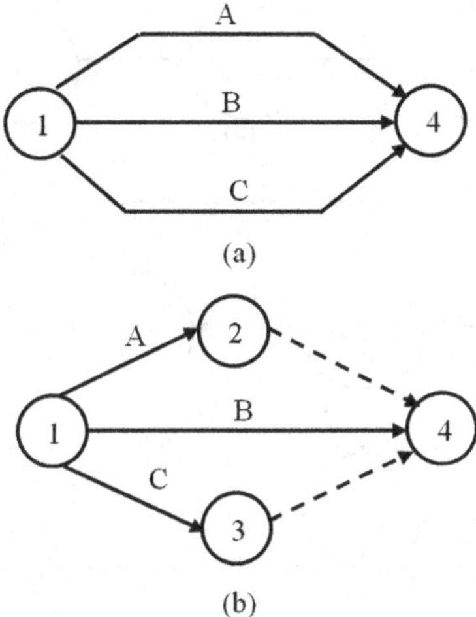

Figure 4.20 A network

numbers also help in the interpretation of flow directions. However, sometimes two events are connected by two or more arrows, representing two or more activities. For example, in Figure 4.20(a), events number 1 and 4 are connected by three arrows which represent three activities, A, B, and C. Activities A, B, and C (which have identical event numbers) are differentiated by their nomenclature in manual calculations. However, in some computer programs the differentiation of such activities is not possible – activities A, B, and C may be interpreted as duplicate activities, as computer programs generally use event numbers to identify activities. Such representations are thus corrected by adding dummies, as shown in Figure 4.20(b). Dummies provide unique pairs of event numbers for all activities, for easy identification in computer programs.

The event numbers in a network must not be repeated. This can create problems with regard to the identification of the activities in a network in many computer programs. For example, in Figure 4.21(a), the event numbers of a network are not unique, however the activity description is unique. There are no issues with the manual calculation or flow interpretation of the network shown in Figure 4.21(a); all its activities are in a sequence. Computer programs, however, may understand this flow differently because of the duplication of event numbers. The interpretation of the network by some computer programs may be like the network shown in Figure 4.21(b). In the figure, the network flow returns to node 20. The duplication of the event number forms a closed loop. Due to the formation of a closed loop, activity D can never get started. The formation of a closed loop in the network occurred because of the duplication of the event numbers. The negligence of such loops results in incorrect representations of the inter-relationships between the various activities on the part of computer programs.

4.5 Activity-on-Node Networks

In activity-on-node networks (AON), nodes represent activities. Each node represents one activity. All nodes are connected with arrows to represent the logical relationships between various activities. Let us redraw all of the logics which have been discussed with regard to the AOA diagrams from Figures 4.2 to 4.15. In AON networks, all logics are represented without dummy activities. AON representation of networks is much easier than AOA representation. The main

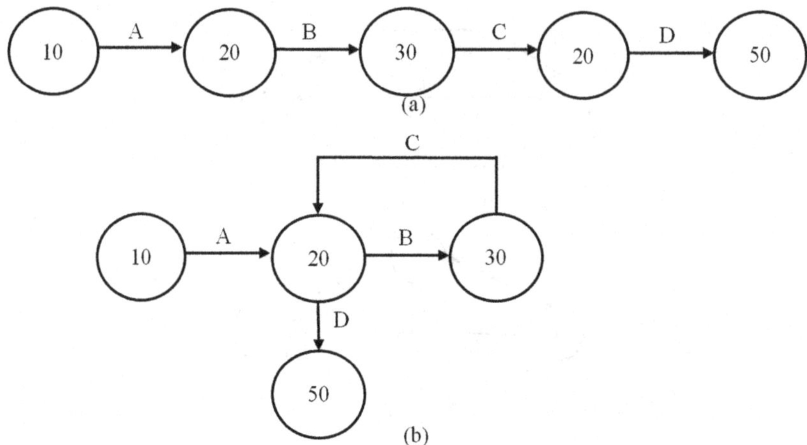

Figure 4.21 A network

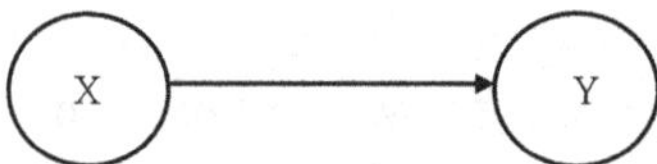

Figure 4.22 A network

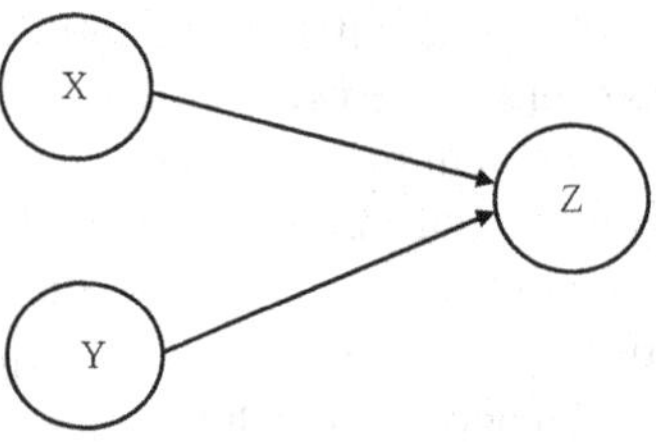

Figure 4.23 A network

advantage of AON is that it eliminates the need for dummies when representing inter-dependencies. This makes AON networks more efficient and easier to learn. In AOA networks the most difficult aspect is learning to make proper use of dummies. In AON networks all arrows are dummies; this is made clear in the following examples.

Activity Y is controlled by activity X. Activity Y cannot begin until activity X is complete. The network diagram is shown in Figure 4.22.

Activity Z is controlled by activities X and Y. Activity Z cannot begin until both activities X and Y are complete. The network diagram is shown in Figure 4.23.

Activities Y and Z are controlled by activity X. Activities Y and Z cannot start unless activity X is complete. The network diagram is shown in Figure 4.24.

Activities C and D follow activity B (or activities C and D are controlled by activity B); activities C and D cannot start unless activity B is complete. Activity A precedes activity B. The network diagram is shown in Figure 4.25.

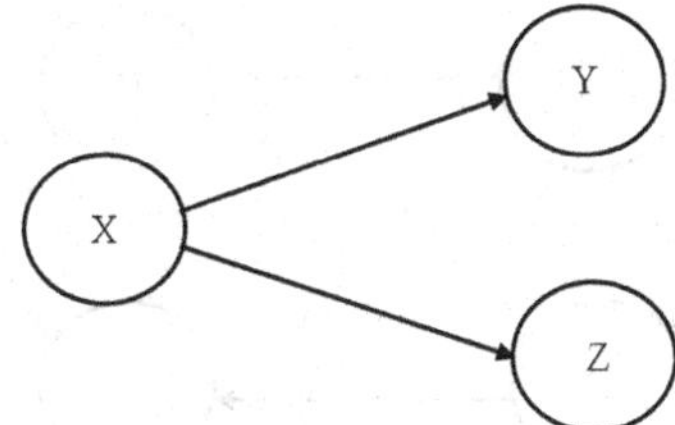

Figure 4.24 A network

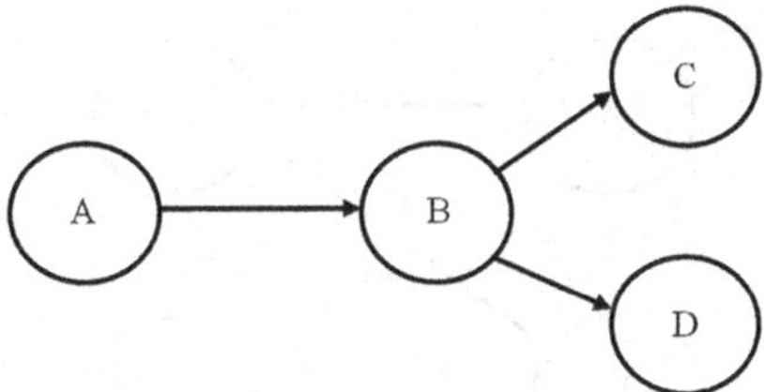

Figure 4.25 A network

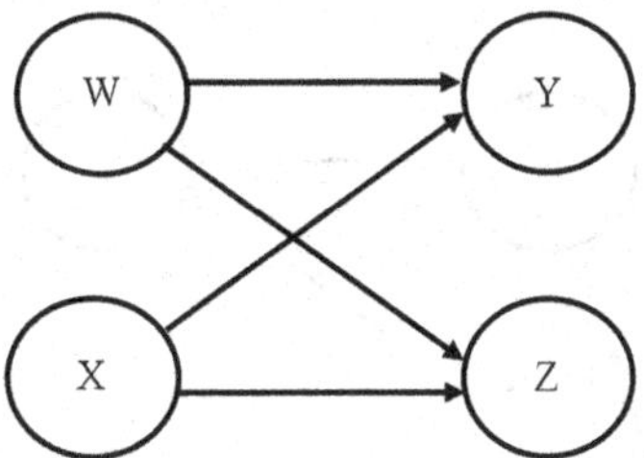

Figure 4.26 A network

Activities Y and Z are controlled by activities W and X. Activities Y and Z cannot start until activities W and X are both complete. However, activities Y and Z start independently of each other. The network diagram is shown in Figure 4.26.

Activity C follows activity A, and activity D follows activity B; there is no connection between activity D and activity A or between activity C and activity B. The network diagram is shown in Figure 4.27.

Activity C is controlled by activity A and activity B; however, activity D is controlled by activity B only. The network diagram is shown in Figure 4.28.

Activity Z is controlled by activities V and W, while activity Y is controlled by activities U and V. The network diagram is shown in Figure 4.29.

Activity D is controlled by activities A, B, and, C. However, activity E is controlled by activity B and activity C. Activity F is controlled by activity C only. The network diagram is shown in Figure 4.30.

Activity A controls activities C and D, while activity B controls activities D and E. Activity D is controlled by both activity A and activity B. The network diagram is shown in Figure 4.31.

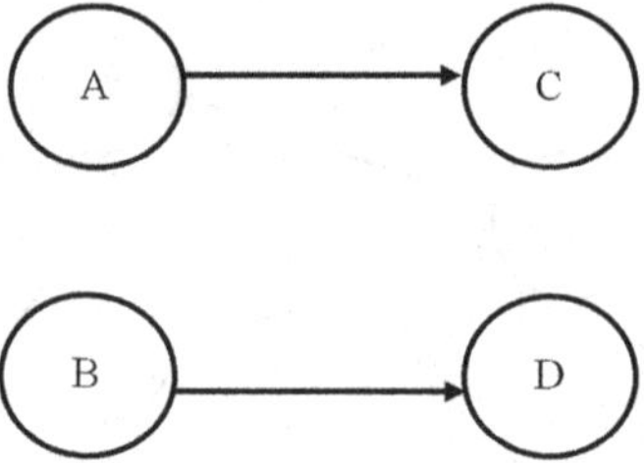

Figure 4.27 A network

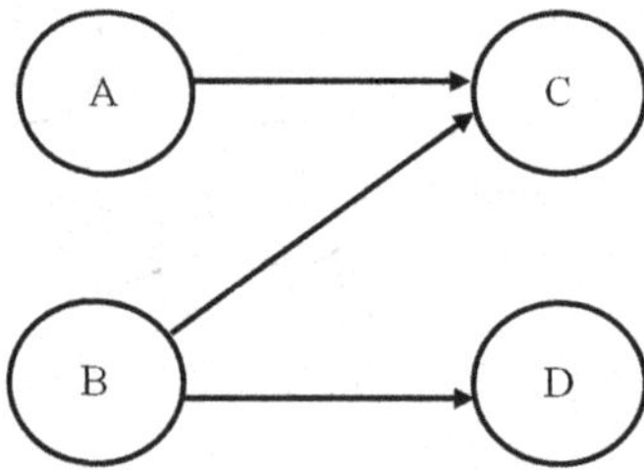

Figure 4.28 A network

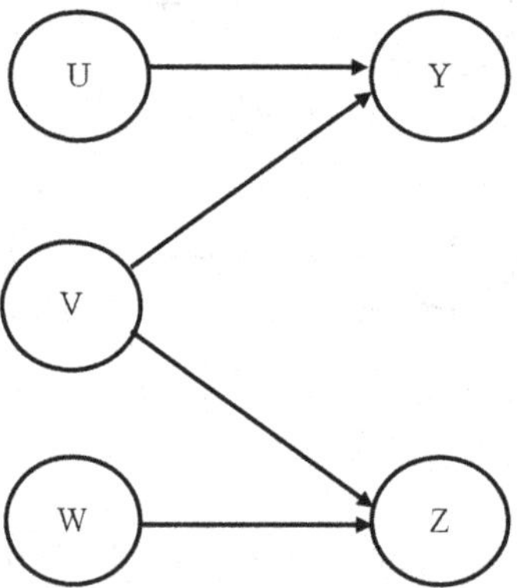

Figure 4.29 A network

Activity X is controlled by activities D and A; activity Y is controlled by activities A, B, and C while activity Z is controlled by activity D only. The network diagram is shown in Figure 4.32.

4.6 Comparison of AOA and AON Networks

In AOA or arrow-based network representation, an arrow represents an activity. The tail of an arrow represents the start point and the head of an arrow represents the endpoint of an activity. In AON or node-based network representation, nodes represent activities. Each node represents one activity and all nodes are connected with arrows to represent the logical relationships

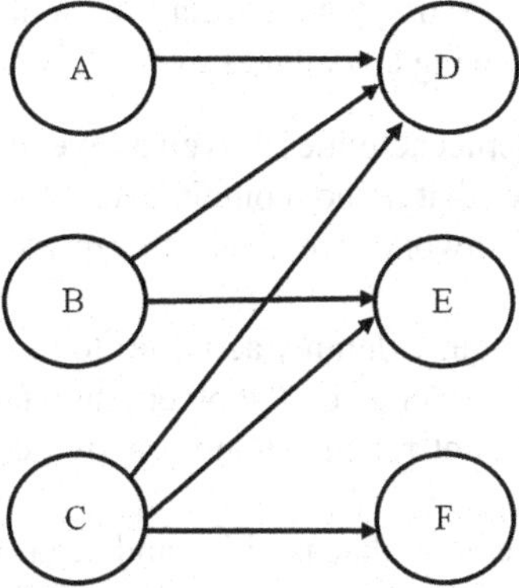

Figure 4.30 A network

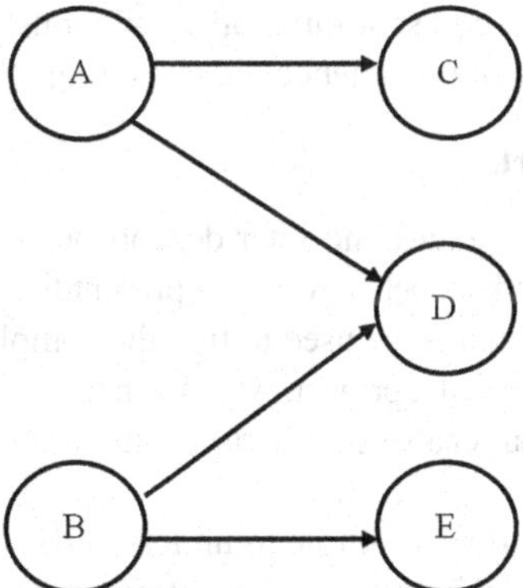

Figure 4.31 A network

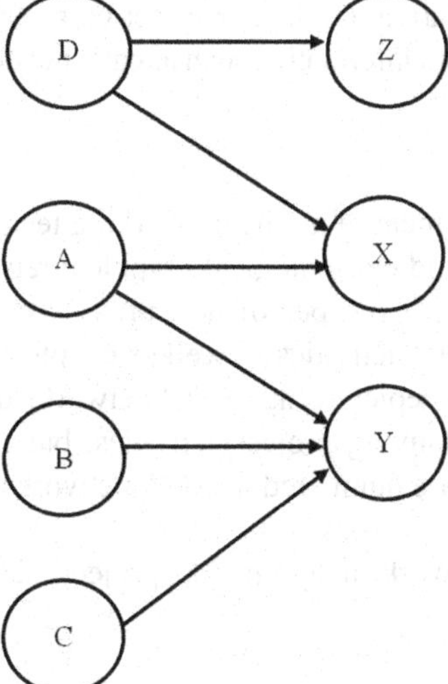

Figure 4.32 A network

between various activities. After discussions regarding AOA and AON networks, we can state that AON networks have the following advantages over AOA networks.

- Activity-on-arrow networks depict activities as well as events; activity-on-node networks, on the other hand, depict activities that do not contain events;
- Relationships in node-based networks are easier to represent than relationships in arrow-based networks;
- Node-based networks do not require dummy activities to represent the correct logic in a network diagram; activity-based networks, on the other hand, use dummy activities;
- Activity-on-node networks do not require dummy activities to identify activities with the same pair of nodes;
- Activity-on-node networks accommodate positive and negative lags between activities without the addition of more activities to represent the lags. This has been explained in detail in discussions of precedence diagramming; and
- Activity-on-arrow networks represent only finish-to-start relationships; activity-on-node networks, on the other hand, can represent three other relationships in addition to the finish-to-start, explained in discussions of precedence diagramming.

4.7 Networks versus Bar Charts

Network representations show the logics and inter-dependencies between activities more clearly than bar charts. This makes networks a better way to represent the plans and schedules of large and complicated projects. Networks can also be used to find the completion date of a project and other times critical to the various activities of a project, which is not possible using bar charts. Despite the strong benefits of networks over bar charts, bar charts are still used for the following reasons.

- Bar charts are easy to draw and very simple to understand;
- Bar charts are drawn to scale, with a bar representing an activity and the length of that bar representing the time duration of that activity, while AOA and AON networks are generally not drawn to scale;
- Bar charts are more useful for communicating plans or execution schedules to field staff and non-technical personnel who are unfamiliar with network techniques; and
- Bar charts may also include more information such as cash-flow diagrams and required resources.

4.8 Conclusion

A network is the primary requirement of all the networking techniques used in project management. A network is the logical and chronological graphical representation of project activities and events. The development of two types of networks – AOA and AON – are covered in this chapter. AOA networks depict activities as well as events, while AON networks generally depict activities only, and do not depict events. AON networks do not require dummy activities to represent inter-dependencies among project activities, but AOA networks do use dummy activities. How and why events are numbered in AOA networks is also discussed.

Example 4.5: Draw a network diagram for project activities with the following inter-dependencies.

- Activities B, C, and D are controlled by activity A,
- Activity E is controlled by activities B and C,

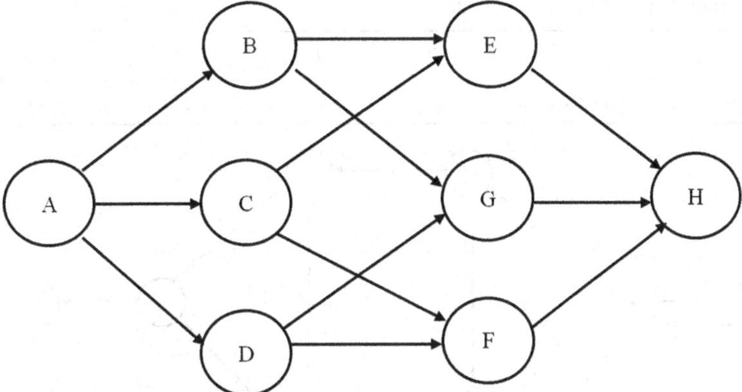

Figure 4.33 A network

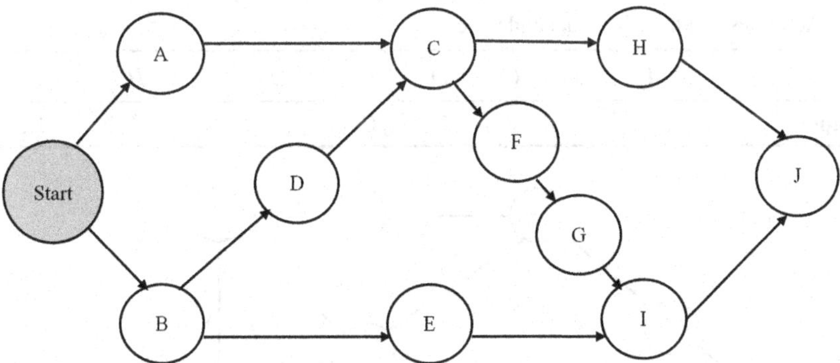

Figure 4.34 A network

- Activity F is controlled by activities C and D,
- Activity G is controlled by activities B and D, and
- Last operation H is controlled by activities E, F, and G.

Solution: The network developed is shown in Figure 4.33.

Example 4.6: Draw an activity-on-node diagram for a project with 9 activities using the inter-dependencies given in Table 4.2.

Solution: The network developed is shown in Figure 4.34.

Exercises

Question 4.1: The inter-dependencies between the activities of a project are given in Table E4.1; draw the network diagram.

Table E4.1 Activities and their inter-dependencies

Activities	*A*	*B*	*C*	*D*	*E*	*F*	*G*	*H*	*I*	*J*
Immediately preceding activities	–	A	A	C	C	B, D	E	F, G	E	H, I

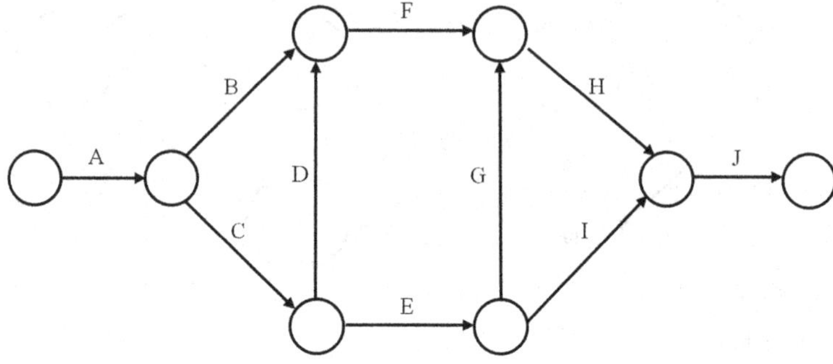

Figure E4.1 Developed network

Table E4.2 Activities and their inter-dependencies

Activities	*A*	*B*	*C*	*D*	*E*	*F*	*G*	*H*	*I*
Depends upon	–	A	B	B	C, B	–	E	D, E	H, G, F

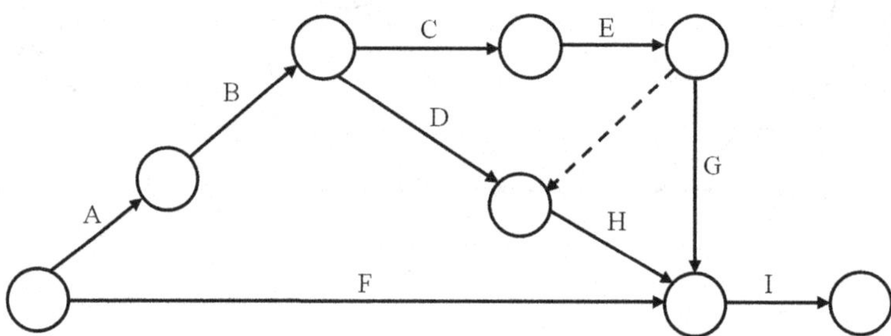

Figure E4.2 Developed network

Table E4.3 Activities and their inter-dependencies

Activities	*A*	*B*	*C*	*D*	*E*	*F*	*G*	*H*	*I*	*J*	*K*	*L*
Depends upon	–	A	A	B, C	C	B	F, E	E	D, G, H	G	H	J, I, K

Solution: The network developed is shown in Figure E4.1:

Question 4.2: The inter-dependencies between some activities are given in Table E4.2; draw the network diagram.

Solution: The network developed is shown in Figure E4.2:

Question 4.3: The inter-dependencies between some activities are given in Table E4.3; draw the network diagram.

Solution: The network developed is shown in Figure E4.3:

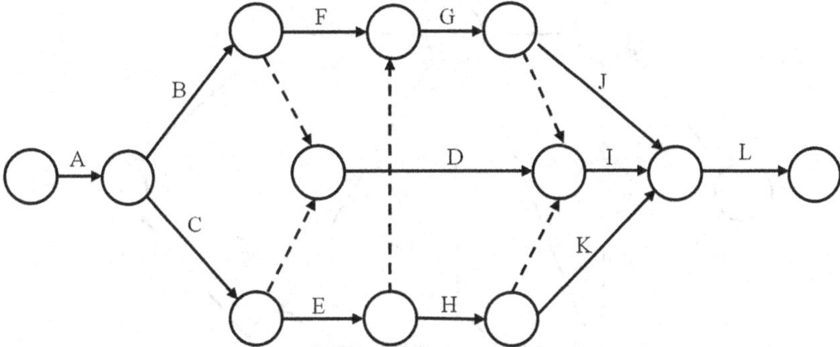

Figure E4.3 Developed network

Table E4.4 Activities and their inter-dependencies

Activities	*A*	*B*	*C*	*D*	*E*	*F*	*G*	*H*	*I*	*J*	*K*	*L*	*M*
Depends upon	–	A	A	A	C	B, C	B, C, D	F, G	E, G	G	E, G, J, H	I, H	L, K

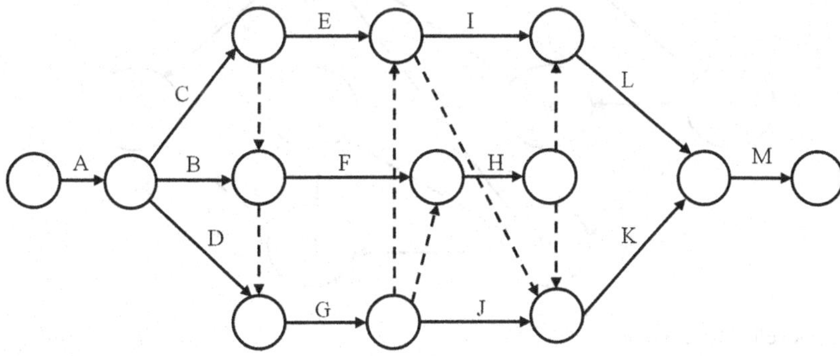

Figure E4.4 Developed network

Table E4.5 Activities and their inter-dependencies

Events	*1*	*2*	*3*	*4*	*5*	*6*	*7*	*8*	*9*
Immediate predecessor events	–	1	2	1, 3	2, 4	4, 5	5	5, 6	7, 8

Question 4.4: The inter-dependencies between some activities are given in Table E4.4; draw the network diagram.

Solution: The network developed is shown in Figure E4.4:

Question 4.5: The inter-dependencies between some activities are given in Table E4.5; draw the network diagram.

Solution: The network developed is shown in Figure E4.5:

Question 4.6: A project has the following dependencies between its various activities; draw the network diagram using AON representation:

- Activities A, B, and C are controlled by activity X,
- Activity D is controlled by activities A, B, and C,

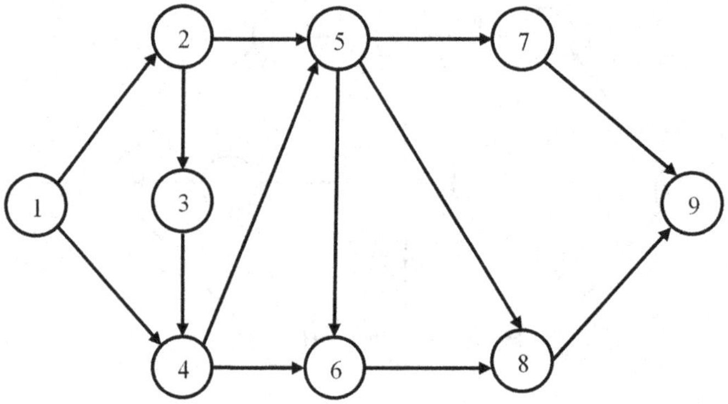

Figure E4.5 Developed network

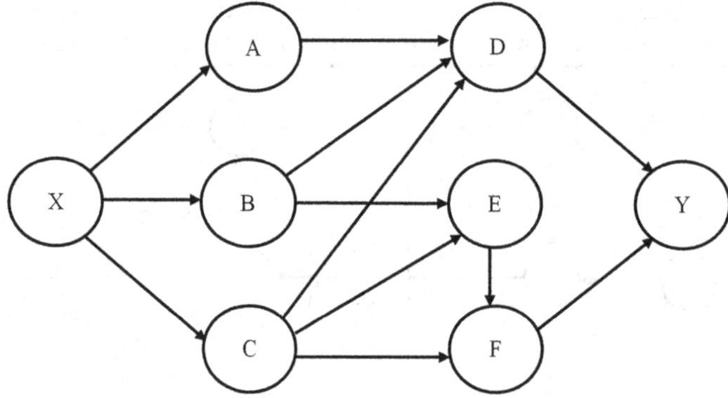

Figure E4.6 Developed network

- Activity E is controlled by activity B and activity C,
- Activity F is controlled by activities C and E, and
- Activity Y is controlled by activities D and F.

Solution: The network developed is shown in Figure E4.6:

Question 4.7: What are the main differences between bar charts and networks?

Question 4.8: What are the main differences between arrow- and node-based network representations?

Question 4.9: What are the main advantages of node-based network representation over arrow-based network representation?

5 Critical Path Method – I

Activity Times

5.1 Learning Objectives

After completion of this chapter, readers will be able to:

- Calculate the time duration of a project,
- Calculate the earliest start and finish times of the different activities in a project, and
- Calculate the latest start and finish times of the different activities in a project.

5.2 Introduction

The *critical path method* (CPM) was developed in the late 1950s by James E. Kelley of Remington Rand and Morgan R. Walker of DuPont. CPM uses a network that depicts activities in a flow in order to represent an execution sequence. In the previous chapter, the development of a network was discussed. The next step after the development of a network is the estimation of a time duration for each activity therein. The first draft of a network is prepared before the estimation of any activity's time duration. The draft helps one concentrate on the inter-dependencies or logics between the various activities in the network. After the completion of the first draft of a network, a time duration is estimated for each activity. An initial review of the network is performed after the addition of the different activities' time durations. This initial review may result in several modifications based upon the planners' prior experience in handling such projects. Once the initial review is completed, a simple manual calculation of the forward pass is carried out to find the start times of the various activities and the time duration of the project. It can reveal errors involved in the draft network or the need for further refinement before the preparation of the final draft of the network.

5.3 Activity Time Durations

The estimation of activity time duration and network refinement are closely related to each other and are usually done simultaneously. In many cases, the activities in a network are re-defined to display the desired level of detail while estimating the time durations of different activities. Project activities are re-defined, merged into fewer, or divided into more activities, so as to represent the project at the desired level of detail. This process is used when planners are experienced in handling such projects. The critical path method is a technique used for projects of routine nature. Thus, only a single time duration estimate is made for each activity in the network. Decisions are also made regarding the resources to be used in various activities and the time duration of each activity in a project is estimated accordingly. Once activities, their time

DOI: 10.1201/9781003428992-5

durations, and the inter-dependencies between them are finalized, the final execution sequence is represented in the final version of the network. A network represents a plan comprehensively to show how a project should proceed. The accuracy of the network depends upon the prior experience of a planner in handling such projects.

The time duration estimate made for each activity in CPM is the mean or average time an activity takes to be executed. It is also called a normal time duration. An activity's time duration estimate is called its *activity time duration.* The unit of an activity time duration can be hours, working days, weeks, etc. The chosen unit is used consistently for each activity in the network. The efficiency of the manpower, equipment, or other resources used is assumed to be normal when estimating an activity time duration. Activity time durations do not include uncontrollable situations such as strikes or delays due to legal issues, floods, fires, etc. All such situations are unpredictable, and so no safety factors are used to consider such emergencies when estimating time durations. In estimating an activity time duration, an activity is considered independently of preceding or subsequent activities.

5.3.1 Accuracy of Activity Time Duration

The main factor that contributes to the accuracy of time duration estimation is the person doing the estimation. It is always recommended that the most knowledgeable personnel, those who have sufficient experience in executing the activities under consideration, must be involved in the estimation of time durations. The objective of the time duration estimation process should be to obtain the most realistic estimates possible. It is further recommended that, during the planning stage, meetings with all experienced personnel must be called to estimate the time durations of activities. It is also desirable to involve knowledgeable and experienced personnel in the preparation of the first draft of a network. They can also be helpful in the review of the execution sequence finalized in the first draft of a network.

The participation of key members of a project team in network review meetings also has major advantages. Discussions among the planners, execution-level supervisors, and members of the project team could lead to early solutions to many problems which may arise during the execution of a project. The discussions in these meetings can be helpful in identifying and resolving planning problems before a project starts, rather than dealing with them when they actually occur during the execution of a project. Corrections during the planning stage are always smoother than corrections during the execution stage of the project. Corrections during the execution stage can be very expensive or may sometimes become impossible to implement and require the approval of a higher authority.

Weather is one of the greatest sources of uncertainty in the estimation of activities' time durations. The effects of the weather should be considered while scheduling a project. There are two ways to consider the effects of the weather when estimating time durations. The first way is not to consider the effects of the weather in the process of estimating the time durations of individual activities. The effects of weather are instead considered at the level of the total time duration of a project. Consideration is given to seasonal temperature, precipitation, and other weather-related aspects of the season in which the work is to be done. The total project time duration is increased accordingly to consider the effects of the weather. However, this does not take the advantage of the work breakdown structure of the project.

The second way considers the effects of weather on the time durations of each activity, taking advantage of the project's work breakdown structure. Each activity is evaluated for its weather sensitivity. The experience of handling various activities in different weather conditions provides useful information for understanding possible weather delays. The main advantage of this approach is that it applies weather adjustments to the portions of a network that would actually be affected. This results in a more accurate schedule for each activity, with reference to calendar dates. In general, both normal and weather-adjusted time duration estimates are prepared for each activity.

5.4 Activity Times

Consider the execution of an activity. The activity is a trip from *station A* to *station B* and the estimated time duration (journey time) is 15 hours, as shown on the time scale in Figure 5.1. However, the actual time available is 20 hours. The available time runs from 12.00 midnight to 08.00 PM, making the total available time 20 hours. The activity can be started at the earliest at 12.00 midnight; this is the *earliest start time* of the activity.

> ***Earliest Start Time (EST):*** This is the earliest possible time at which an activity can start. In the case of AOA representation, an activity starts at its tail, therefore EST is the earliest event time on the tail node/event of an activity. It is the earliest time by which all predecessor activities can be completed to allow for the starting of the activity under consideration. It is calculated in the forward pass by moving from the first to the last event of the network.

EST_{ij} = Earliest event time of the tail node/event of an activity (a_{ij})

In Figure 5.1, the estimated time duration of the activity of journeying from *station A* to *station B* is 15 hrs. If the activity starts at 12.00 midnight, that is the earliest start time of the activity under consideration. When an activity is started as early as possible it will end as early as possible. If the activity is started at the earliest start time of 12.00 midnight, it will end as early as possible at 3.00 PM. Hence, 3.00 PM is the earliest finish time of the activity.

> ***Earliest Finish Time (EFT):*** This is the earliest possible time at which an activity can finish. If an activity starts at its earliest start time and takes its time duration (d_{ij}) then it will

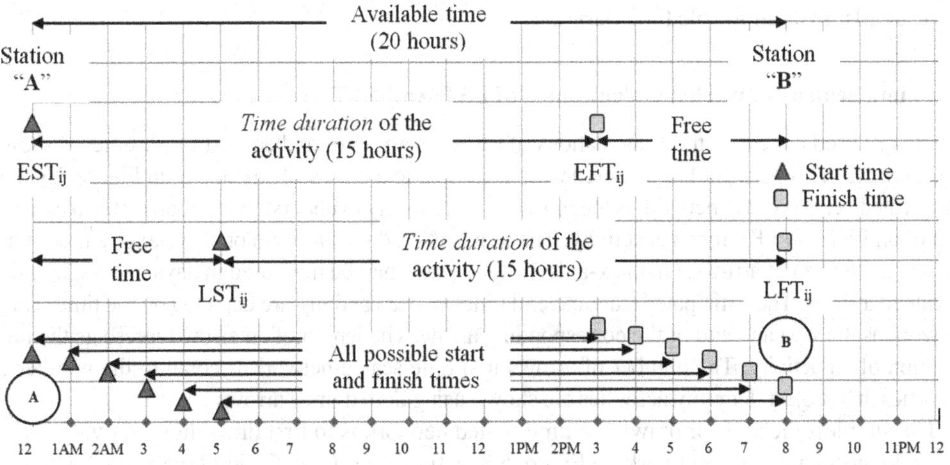

Figure 5.1 Time scaled network for the calculation of earliest start time, earliest finish time, latest start time, and latest finish time

finish at its *earliest finish time*. Mathematically this is calculated by adding its time duration to the earliest start time as below.

$$EFT_{ij} = EST_{ij} + d_{ij} \quad 5.1$$

In Figure 5.1, the available time is 20 hours and the estimated time duration of the activity is 15 hrs. If the activity starts at its earliest start time of 12.00 midnight, it will end at its earliest finish time of 3.00 PM. There is a free time interval of 5 hours from 3.00 PM to 8.00 PM following the earliest finish time of the activity. If the start time of the activity under consideration is delayed by 5 hours, moving from 12.00 midnight to 5.00 AM, the activity will start at 5.00 AM and end at 8.00 PM without changing the available or allowable time duration. In this case, the activity has started as late as possible. This is called the *latest start time* of an activity.

Latest Start Time (LST): This is the latest possible time at which an activity can start without any delay to the completion of a project. Mathematically, the latest start time is calculated by subtracting the time duration (d_{ij}) from the latest finish time as below.

$$LST_{ij} = LFT_{ij} - d_{ij} \quad 5.2$$

If the start time of the activity is delayed by 5 hours, the activity will start at 5.00 AM at its latest start time and the activity will end at 8.00 PM without any change to the available or allowable time duration. In this case, the activity starts as late as possible and ends as late as possible; this is called the *latest finish time* of an activity.

Latest Finish Time (LFT): This is the latest time at which an activity can finish without any delay to the project's completion. In the case of AOA representation, an activity ends at its head, therefore the LFT is the latest event time of an event/node on the head of an arrow. It is calculated in the backward pass by moving from the terminal to the start event of a network.

As shown in Figure 5.1, the start time of the activity under consideration may be delayed by a maximum of up to 5 hours from 12.00 midnight to 5.00 AM. This will not result in any change to the available or allowable time duration.

5.5 Time Scaled Networks: Calculations of EST And EFT

A network used to demonstrate calculations of earliest start times and earliest finish times is shown in Figure 5.2. This network has been redrawn according to a time scale as shown in Figure 5.3. The time scaled version of the network is the most convenient way to understand the concept and calculation of an EST and EFT for each activity. The x-axis is used for time and on the y-axis activities are shown in the form of arrows. On the x-axis the time scale may be measured in days, weeks, months, or calendar dates. The anticipated start and end times of each activity are depicted on the time scaled network as the start and end of the corresponding arrow. The length of an arrow represents the time duration of an activity. The number of arrows in a time scaled network is equal to the number of activities in a project. Dummy activities are shown using dotted lined arrows.

The simplest method for drawing a time scaled network is to first draw those activities lying on the longest path of the network. In Figure 5.3, activities A, C, E, G, and I make up the longest path, of length of 12 (2 + 3 + 2 + 2 + 3) days. All these activities are represented, as shown in the figure, by a straight line. The length of the time scaled version of the network is equal to the

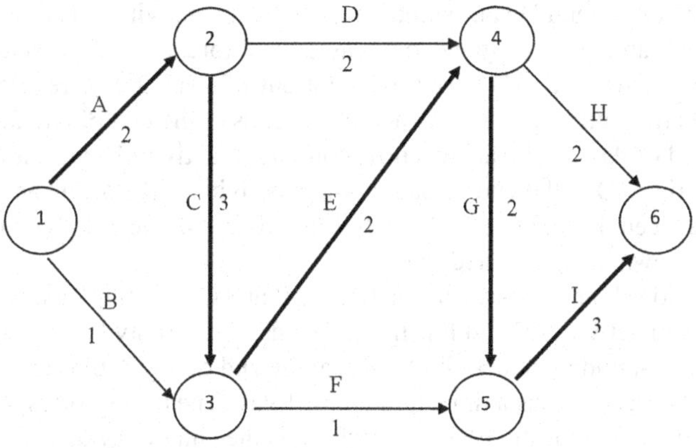

Figure 5.2 The sample network

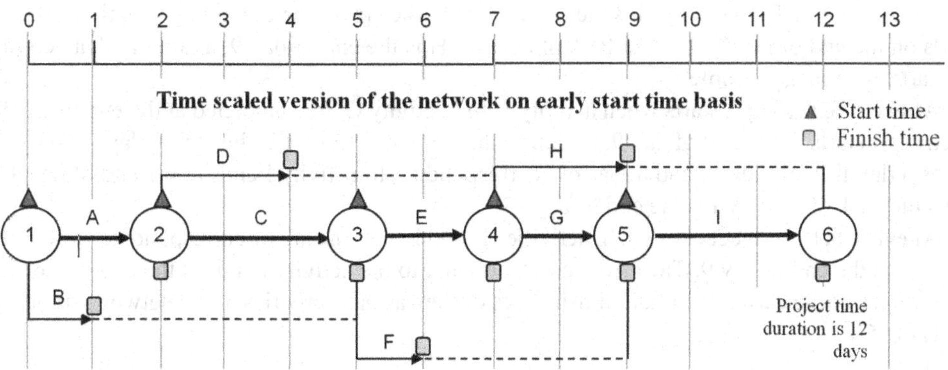

Figure 5.3 Time scaled version for the calculation of the earliest start time and earliest finish time of each activity

project time duration or the sum of the time durations of all the activities which lie on the longest path. After the longest path, one can draw the path just below the longest path in length. In this way, all the paths/activities of the network can be represented on a time scale. It must be noted that when activity D is drawn between nodes 2 and 4, the available time between nodes 2 and 4 is 5 days; however, activity D has a time duration of 2 days. Activity D runs from the end of day 2 to the end of day 4. The remaining 3 days on the path between node numbers 2 and 4 are shown using a dotted line. The starting event, numbered as 1, starts at time zero. The start time of the project or the start time of event 1 is taken as zero.

At **event 1**, the project starts with the simultaneous start of activities A and B. The starting event, numbered as 1, starts at time zero. Therefore, the EST of activities A and B is 0. Activities A and B have time durations of 2 and 1 day respectively, as shown in Figure 5.3. The arrow corresponding to activity A ends at the end of day 2, therefore the EFT of activity A is the end of day 2. The arrow corresponding to activity B ends at the end of day 1, and so the EFT of activity B is the end of day 1. It is assumed that activity B will start as soon as possible. The 4 days on the path between nodes 1 and 3 are shown using a dotted line.

At **event 2**, activities C and D start simultaneously when activity A is completed at the end of day 2. Activities C and D start at the end of day 2, therefore, the EST of activities C and D is the end of day 2. Activities C and D have time durations of 3 and 2 days respectively, as shown in Figure 5.3. The arrow corresponding to activity C ends at the end of day 5, thus the EFT of activity C is the end of day 5. The arrow corresponding to activity D ends at the end of day 4, thus the EFT of activity D is the end of day 4 – here activity D is starting as soon as possible. Activity D runs between the end of day 2 and the end of day 4. The 3 days on the path between nodes 2 and 4 are shown using a dotted line.

At **event 3**, activities E and F start when activities B and C are completed at the end of day 5. Therefore, the EST of activities E and F is the end of day 5. Activity E has a time duration of 2 days. The arrow corresponding to activity E ends at the end of day 7, thus the EFT of activity E is the end of day 7. Activity F has a time duration of 1 day. The arrow corresponding to activity F ends at the end of day 6, thus the EFT of activity F is the end of day 6.

At **event 4**, activities G and H start simultaneously when activity D and activity E are completed at the end of day 7. Therefore, the EST of activities G and H is the end of day 7. Activities G and H have a time duration of 2 days. The arrow corresponding to activity G ends at the end of day 9, thus the EFT of activity G is the end of day 9. The arrow corresponding to activity H also ends on the end of day 9, thus the EFT of activity H is the end of day 9, assuming that activity H starts as soon as possible.

At **event 5**, activity I starts when activity F and activity G are completed at the end of day 9. Activity I starts at the end of day 9, therefore the EST of activity I is the end of day 9. Activity I has a duration of 3 days, and the arrow corresponding to activity I ends at the end of day 12, thus the EFT of activity I is the end of day 12.

At **event 6,** the project ends when activities H and I end. The arrow corresponding to activity H ends at the end of day 9. The arrow corresponding to the activity I ends at the end of day 12. The values of the earliest start and finish times of the various activities in the network are listed in Table 5.1.

5.6 Time Scaled Networks: Calculations of LST and LFT

The network used in the last section can also be used to demonstrate the calculation of the *latest start times* and *latest finish times* of all the activities therein. This network has been redrawn according to a time scale, as shown in Figure 5.4, activities starting in this example as late as

Table 5.1 The earliest start time, earliest finish time, latest start time, and latest finish time of each activity

Activities	*Time durations*	*EST*	*EFT*	*LST*	*LFT*
A	2	0	2	0	2
B	1	0	1	4	5
C	3	2	5	2	5
D	2	2	4	5	7
E	2	5	7	5	7
F	1	5	6	8	9
G	2	7	9	7	9
H	2	7	9	10	12
I	3	9	12	9	12

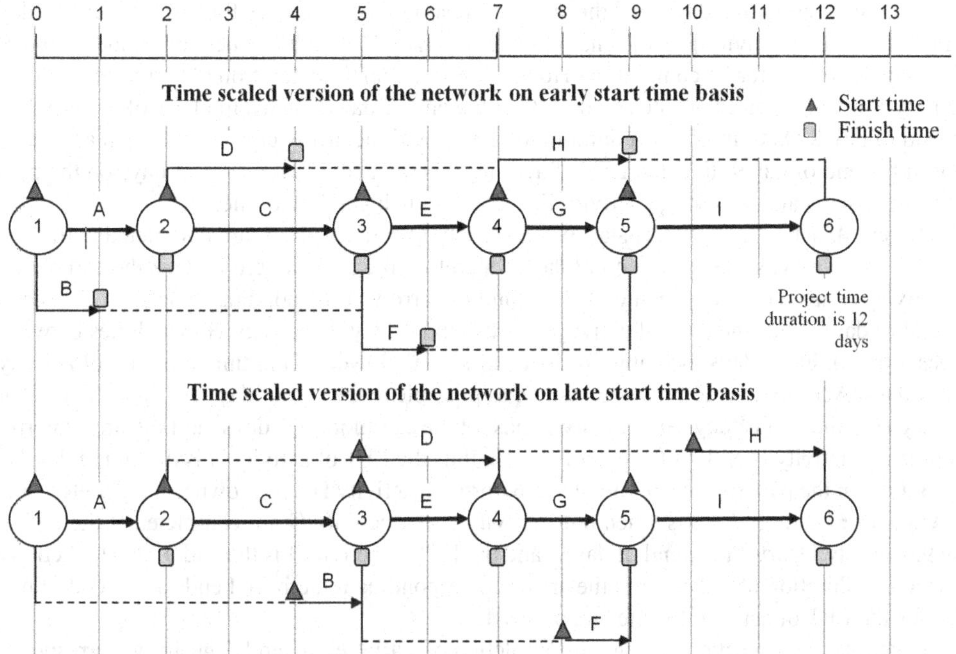

Figure 5.4 Time scaled version for the calculation of the earliest start time, earliest finish time, latest start time, and latest finish time of each activity

possible. It must be noted that activities B, D, F, and H are drawn between the corresponding nodes, keeping their start times as late as possible.

At **event 1**, the project starts when activities A and B start. The starting event, numbered as 1, starts at time zero. Therefore, the LST of activity A is 0. Activity A has a time duration of 2 days, and the arrow corresponding to activity A ends at the end of day 2, thus the LFT of activity A is the end of day 2. Activity B lies between nodes 1 and 3, and has 5 days available for its completion. However, the actual duration of activity B is 1 day. Activity B has to be started as late as possible, that being at the end of day 4. Thus, the LST of activity B is the end of day 4. Activity B has a time duration of 1 day, and the arrow corresponding to activity B ends at the end of day 5, thus the LFT of activity B is the end of day 5. It is assumed that activity B will start as late as possible. The 4 days on the path between nodes 1 and 3, through activity B, are shown using a dotted line.

At **event 2**, activities C and D start when activity A is completed at the end of day 2. Activity C starts at the end of day 2, therefore the LST of activity C is the end of day 2. Activity C has a time duration of 3 days. The arrow corresponding to activity C ends at the end of day 5, thus the LFT of activity C is the end of day 5. Activity D lies between nodes 2 and 4, and has 5 days available for its completion. However, the actual time duration of activity D is 2 days. Activity D is to be started as late as possible at the end of day 5. Thus, the LST of activity D is the end of day 5. Activity D has a time duration of 2 days, and the arrow corresponding to activity D ends at the end of day 7, thus the LFT of activity D is the end of day 7. The 3 days on the path between nodes 2 and 4, through activity D, are shown using a dotted line.

At **event 3**, activities E and F start when activities B and C are completed at the end of day 5. Activity E starts at the end of day 5, therefore the LST of activity E is the end of day 5. Activity

E has a time duration of 2 days, and the arrow corresponding to activity E ends at the end of day 7, thus the LFT of activity E is the end of day 7. Activity F, which lies between nodes 3 and 5, has 4 days available for its completion. However, the actual time duration of activity F is 1 day. Activity F is to be started as late as possible at the end of day 8. Thus, the LST of activity F is the end of day 8. Activity F has a duration of 1 day, and the arrow corresponding to activity F ends at the end of day 9, thus the LFT of activity F is the end of day 9. The 3 days on the path between nodes 3 and 5, through activity F, are shown using a dotted line.

At **event 4**, activities G and H start when activity D and activity E are completed at the end of day 7. Activity G starts at the end of day 7, therefore the LST of activity G is the end of day 7. Activity G has a time duration of 2 days, and the arrow corresponding to activity G ends at the end of day 9, thus the LFT of activity G is the end of day 9. Activity H, which lies between nodes 4 and 6, has 5 days available for its completion. However, the time duration of activity H is 2 days. Activity H is to be started as late as possible at the end of day 10. Thus, the LST of activity H is the end of day 10. Activity H has a time duration of 2 days, and the arrow corresponding to activity H ends at the end of day 12, thus the LFT of activity H is the end of day 12. The 3 days on the path between nodes 4 and 6, through activity H, are shown using a dotted line.

At **event 5**, activity I starts when both activity F and activity G are completed at the end of day 9. Activity I starts at the end of day 9, and the LST of activity I is the end of day 9. Activity I has a time duration of 3 days, and the arrow corresponding to activity I ends at the end of day 12, thus the LFT of activity I is the end of day 12.

At **event 6,** the project ends with the completion of activities H and I. The arrow corresponding to activity H ends at the end of day 12. The arrow corresponding to the activity I also ends at the end of day 12. The values of the latest start times and latest finish times of different activities are listed in Table 5.1.

Example 5.1: For the network shown in Figure 5.5, calculate the earliest start time, earliest finish time, latest start time, and latest finish time for each activity by drawing the time scaled version of the network.

Solution: The earliest start time, earliest finish time, latest start time, and latest finish time for each activity of the network have been calculated by drawing the time scaled version of the network as shown in Figure 5.6 and the values have been listed in Table 5.2.

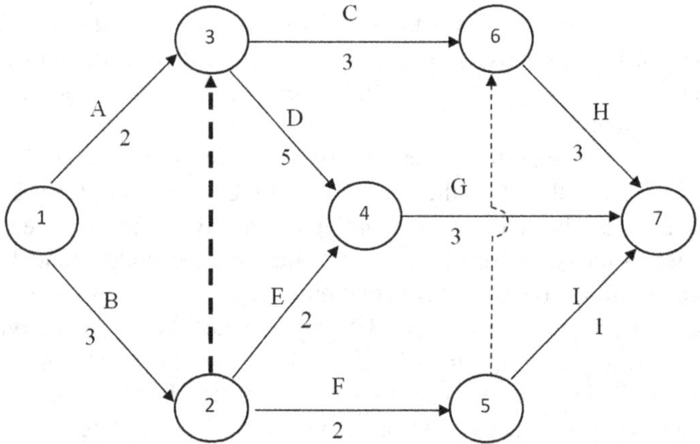

Figure 5.5 A project network

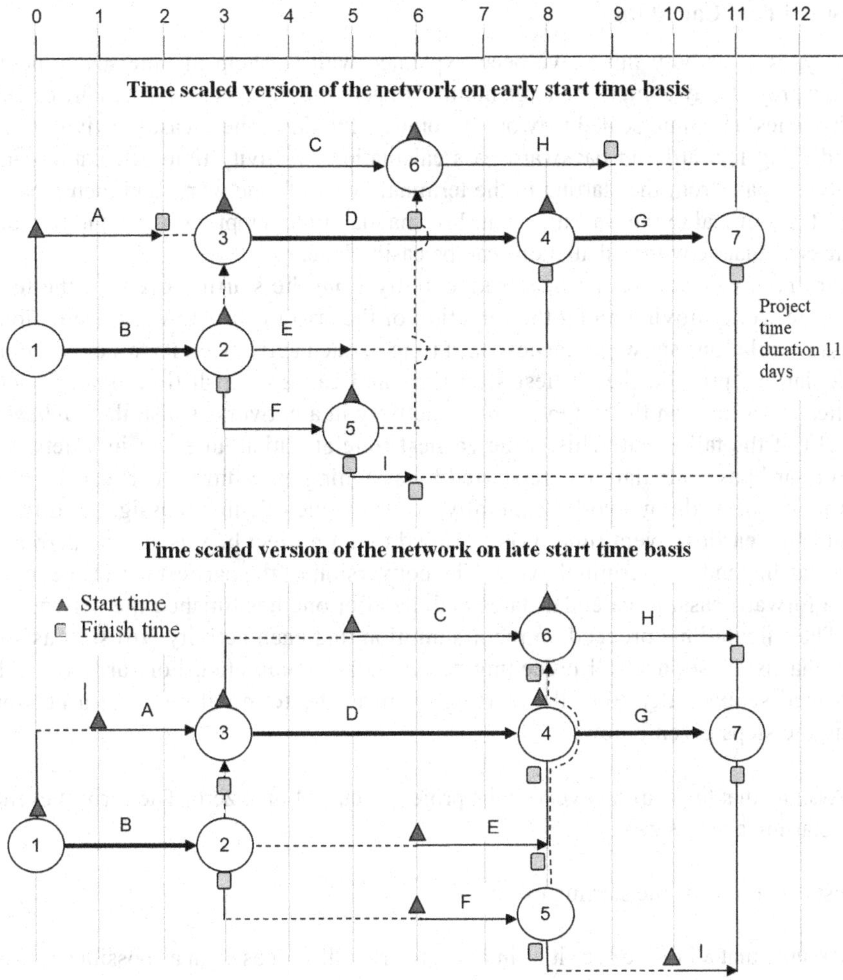

Figure 5.6 Time scaled version for the calculation of the earliest start time, earliest finish time, latest start time, and latest finish time of each activity

Table 5.2 The earliest start time, earliest finish time, latest start time, and latest finish time of each activity

Activities	*Time durations*	*EST*	*EFT*	*LST*	*LFT*
A	2	0	2	1	3
B	3	0	3	0	3
C	3	3	6	5	8
D	5	3	8	3	8
E	2	3	5	6	8
F	2	3	5	6	8
G	3	8	11	8	11
H	3	6	9	8	11
I	1	5	6	10	11

5.7 Forward Pass Calculations

The four types of activity time have been explained with the help of time scaled networks. However, a project may comprise a large number of activities, and in such cases the calculation of activity times on a time scaled network becomes difficult. In these cases, activity times are calculated using forward and backward pass calculations. Activity time calculations involve first a forward pass from the starting to the terminal event of a network, and then a backward pass from the terminal to the starting event. It is mandatory to adopt one time duration unit for all activities so that network calculations can be easily done.

Forward pass calculations proceed sequentially from the starting event to the terminal event of a network, moving along the direction of the arrows. In AOA representation, forward pass calculations show the earliest start time of each activity on its tail event. Forward pass calculations provide the earliest start time and earliest finish time of each activity. The earliest start time on the tail event of an activity in a network is also the *earliest event time* (EET) of the tail event. This is the earliest time at which an event in a network can occur. Forward pass calculations are started by assigning an arbitrary earliest start time to the starting event of the network. Generally, a zero value of time is assigned to the starting event as its earliest event time. It is assumed that a project begins at time zero and has only one starting and one terminal event. The conversion of the earliest start time, obtained through a forward pass, to calendar dates is done after one has finished the network calculations. The calculations proceed on the assumption that each activity will start as soon as possible, that is, as soon as all preceding activities are completed. Forward pass calculations proceed sequentially from the starting event to the terminal event of a network by following the steps given below:

Step 1: Assume that the starting event of the project occurs at time zero. The earliest event time of the starting event is zero.

Earliest event time of the starting event = 0

Step 2: Assume that all of the activities in the network will start as soon as possible. As soon as all activities preceding the activity under consideration are completed, the activity under consideration will start without any delay. That is, for an activity to start, the activities preceding it must be completed. The earliest start time (EST_{ij}) of an activity (a_{ij}) under consideration is given by:

EST_{ij} = Maximum EFT of the activities immediately preceding (all activities ending at node *i*) the activity (a_{ij}) under consideration

Step 3: The earliest finish time (EFT_{ij}) of activity a_{ij} is the algebraic sum of its earliest start time (EST_{ij}) and its time duration (d_{ij}). For activity a_{ij} it is given by equation 5.1:

$$EFT_{ij} = EST_{ij} + d_{ij}$$

The calculations proceed sequentially from the starting event to the terminal event of a network following the direction of the arrows. In AOA representation, forward pass calculations show the earliest start time of each activity on its tail.

5.7.1 Calculation of EST_{ij} and EFT_{ij} Using Forward Pass

Consider the network shown in Figure 5.7 for the demonstration of forward pass calculations as used to find the EST and EFT of all of the activities in the network. In the forward pass calculations, the starting event, numbered as 1, occurs at time zero. The start time of the project and the EET of event 1 is 0 (as discussed in step 1).

At **event 1**, the project starts with the simultaneous start of activities A and B. The starting event starts at time zero. Therefore, the EST of activities A and B is 0. The EFT of an activity is calculated by adding the time duration to the EST. Therefore, the EFTs of activities A and B are 2 (0 + 2) and 1 (0 + 1) respectively.

At **event 2**, activities C and D start simultaneously when activity A is completed at the end of day 2. Activities C and D start at the end of day 2, therefore the EST of activities C and D is the end of day 2. The EFTs of activities C and D are 5 (2 + 3) and 4 (2 + 2) respectively.

At **event 3**, activities E and F start simultaneously when activities B and C are completed. Activity B takes 1 day to complete and activity C takes 5 days to complete. Activities E and F start simultaneously when activities B and C are both completed at the end of day 5. This is the maximum EFT of the activities immediately preceding at node 3 (discussed in step 2), that is, the point at which activities B and C end. The shared EST of activities E and F is the end of day 5. The EFTs of activities E and F are the ends of days 7 (5 + 2) and 6 (5 + 1) respectively.

The crux of the forward pass calculation occurs at the merge event, where the maximum EFT of the preceding activities is considered as the EST of the subsequent activities. The calculation of EST and EFT are easily handled manually on a network by working along each path as far as possible, that is, by moving along the arrows from the start nodes to the end nodes of each path in a network.

At **event 4**, activities G and H start simultaneously when activities D and E are completed. Activity D takes 4 days to complete and activity E takes 7 days to complete. The EFTs of activities D and E are the ends of days 4 and 7 respectively. Activities D and E are completed at the end of day 7. This is the maximum EFT of the activities immediately preceding node 4. The shared EST of activities G and H is the end of day 7. The EFTs of activities G and H are the ends of days 9 (7 + 2) and 9 (7 + 2) respectively.

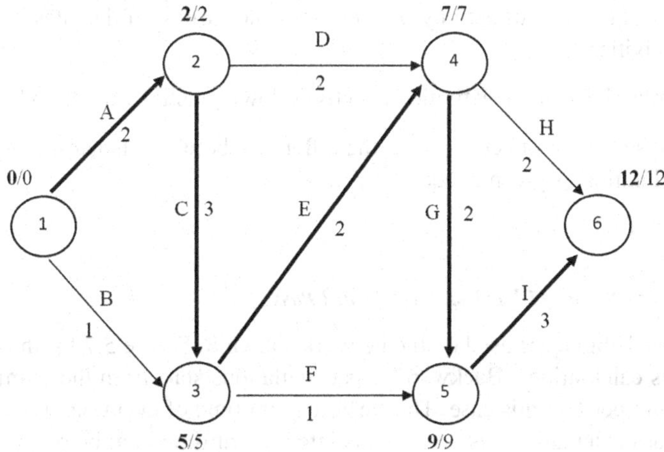

Figure 5.7 Forward and backward pass calculations for the determination of the earliest start time, earliest finish time, latest start time, and latest finish time of each activity

At **event 5**, activity I starts when activities F and G are completed. Activity F takes 6 days to complete and activity G takes 9 days to complete. Activities D and G are completed at the end of day 9. This is the maximum EFT of the activities immediately preceding node number 5. The EST for the activity I is the end of day 9. The EFT of activity I is the end of day 12 (9 + 3).

At **event 6,** the project ends when activities H and I end. Activities H and I are completed at the ends of days 9 and 12 respectively. The maximum EFT of the activities immediately preceding node number 6 is the end of day 12. Hence, the time duration of the project is 12 days.

5.8 Backward Pass Calculations

Backward pass calculations are carried out in sequentially from the terminal event back to the starting event of a project, moving against the direction of the arrows. In AOA representation, backward pass calculations show the latest finish times of each activity in a network on the head of the arrow representing that activity. Backward pass calculations provide the latest start times and latest finish times for each activity in a project and the *latest event time* (LET) of each event. In an AOA representation of a network, the latest finish time, as shown on the head of an arrow representing an activity, is also the latest event time of that arrow's head event.

Backward pass calculations start by deciding the *scheduled time duration* (d_s) for the completion of a project. A scheduled time duration is assigned to the terminal event as its latest event time. A project's terminal event must occur on or before its scheduled time duration. If the scheduled time duration for the completion of a project is not specified, the latest event time of the terminal event is taken to be equal to its earliest event time, as calculated through forward pass calculation. Backward pass calculations proceed sequentially from the terminal event to the starting event of a network by following the steps given below.

Step 1: The latest event time of a project's terminal event is taken to be equal to its scheduled completion time duration or its earliest event time as computed through forward pass calculation.

Latest event time of the terminal event (LET) = Scheduled completion time duration of the project (d_s) or earliest event time (EET) of the terminal event

Step 2: The latest finish time of activity a_{ij} is equal to the earliest of the latest start times of its subsequent activities.

LFT_{ij} = Minimum LST of the activities directly following activity a_{ij} on node j

Step 3: The latest start time of activity a_{ij} is the difference between its latest finish time and the activity time duration as given in equation 5.2.

$$LST_{ij} = LFT_{ij} - d_{ij}$$

5.8.1 Calculation of LST_{ij} and LFT_{ij} Using Backward Pass

The steps discussed above are used on the network shown in Figure 5.7 for the demonstration of backward pass calculations. Backward pass calculations start from the terminal event of a network, event number 6 in this case. The earliest event time of event number 6, as computed in the forward pass calculations, is taken as its latest event time, that being the end of day 12 (step 1).

At **event 6**, the project finishes with the simultaneous end of activities H and I at the end of day 12. Hence, the LFT of activities H and I is the end of day 12, as shown on the heads of the arrows representing activities H and I. The LST of activity H is the end of day 10 (12 – 2), and that of activity I is the end of day 9 (12 – 3).

At **event 5**, activities F and G are complete. The LFT is the minimum LST of the activities directly following activity i-j on node j (step 2). The LFT of activities F and G is the LST of the activity directly following node 5, that being the LST of activity I, the end of day 9. Thus, activities F and G share an LFT at end of day 9. The LST of activity F is the end of day 8 (9 – 1) and that of activity G is the end of day 7 (9 – 2).

At **event 4**, the LFT of activities D and E is equal to the earliest LST of the subsequent activities G or H. The LST of activity G is the end of day 7, and the LST of activity H is the end of day 10. The earliest, LST of the subsequent activities is the end of day 7. Thus, the LFT of activities D and E, as ending on node 4, is the end of day 7. The LST of activity D is the end of day 5 (7 – 2) and that of activity E is also the end of day 5 (7 – 2).

At **event 3**, the LFT of activities B and C is equal to the earliest LST of the subsequent activities E or F. The LST of activity E is the end of day 5, and the LST of activity F is the end of day 8. The smallest LSTs of the subsequent activities E or F is the end of day 5. Thus, the LFT of activities B and C, as ending on node 3, is the end of day 5. The LST of activity B is the end of day 4 (5 – 1) and that of activity C is the end of day 2 (5 – 3).

At **event 2**, the LFT of activity A is equal to the earliest LSTs of the subsequent activities C or D. The LST of activity C is the end of day 2, and the LST of activity D is the end of day 5. The earliest LSTs of the subsequent activities C or D is the end of day 2. Thus, the LFT of activity A, as ending on node 2, is the end of day 2. The LST of activity A is 0 (2 – 2).

At **event 1**, the LSTs of activities A and B are 0 and 4 respectively. The latest event time of node 1 is the earliest LST of the subsequent activities A or B. The earliest LSTs of the subsequent activities A and B are 0. The latest event time of the starting event is 0.

5.9 Conclusion

The time duration estimate required for the completion of a project activity mainly depends upon the prior experience of a planner in handling similar projects. Thus, it is always recommended that the most knowledgeable personnel, those who have sufficient experience in executing the kinds of activity under consideration, must be involved in estimating time durations. In CPM a deterministic model is used for estimating time durations. The deterministic model uses a single value time duration estimate for each activity. The time duration of an activity does not consider uncontrollable situations such as strikes or delays due to legal issues, floods, fires, etc. Calculations of EST, EFT, LST, and LFT for each activity of a project, using a time scaled version of a network, clearly describe their physical significance. The calculation of EST, EFT, LST, and LFT for each activity of a project, using forward and backward passes, is an easy mathematical alternative to the use of a time scaled network.

Example 5.2: For the network shown in Figure 5.5, calculate the earliest start time, earliest finish time, latest start time, and latest finish time of each activity using forward and backward pass calculations.

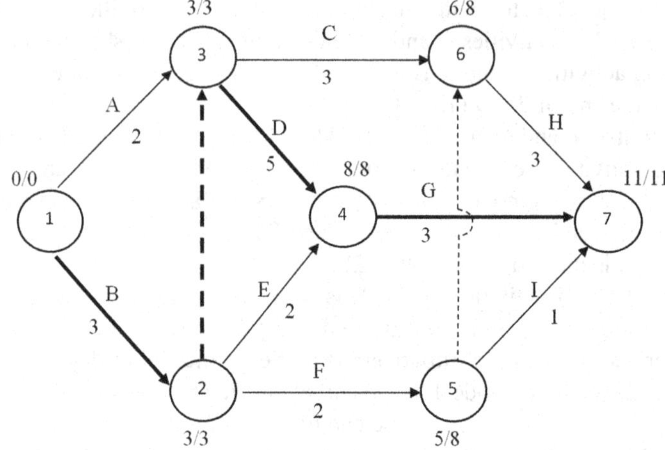

Figure 5.8 Forward and backward pass calculations for the determination of the earliest start time, earliest finish time, latest start time, and latest finish time of each activity

Solution: The earliest start time, earliest finish time, latest start time, and latest finish time of each activity of the network have been calculated using forward and backward pass calculations as shown in Figure 5.8, and values have been listed in Table 5.2.

Exercises

Question 5.1: The network of a sample project is shown in Figure E5.1; the estimated time duration of each activity is marked in the network. For each activity in the project, calculate the EST, EFT, LST, and LFT values.

Solution: The appropriate forward and backward pass calculations are shown in Figure E5.2; the EST, EFT, LST, and LFT values of the various activities in the project are in Table E5.1.

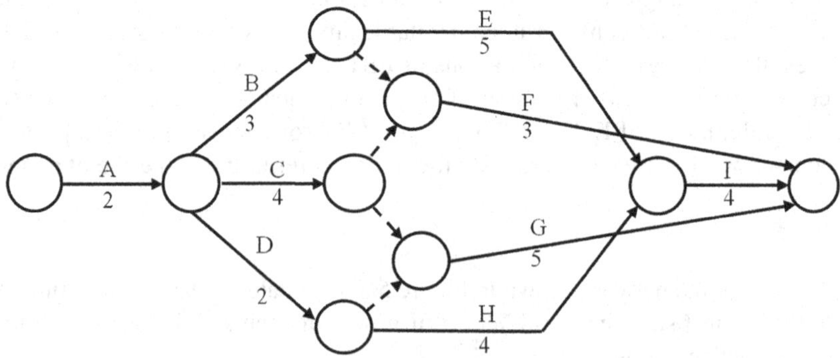

Figure E5.1 Network of a sample project

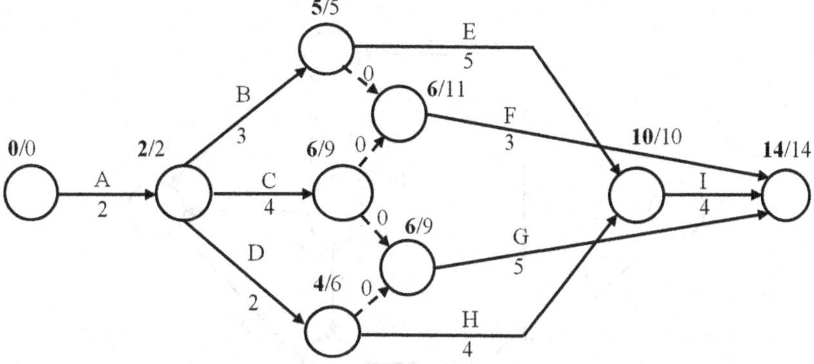

Figure E5.2 Forward and backward pass calculations

Table E5.1 EST, EFT, LST, and LFT values of various activities

Activities	*Time durations*	*EST*	*EFT*	*LST*	*LFT*
A	2	0	2	0	2
B	3	2	5	2	5
C	4	2	6	5	9
D	2	2	4	4	6
E	5	5	10	5	10
F	3	6	9	11	14
G	5	6	11	9	14
H	4	4	8	6	10
I	4	10	14	10	14

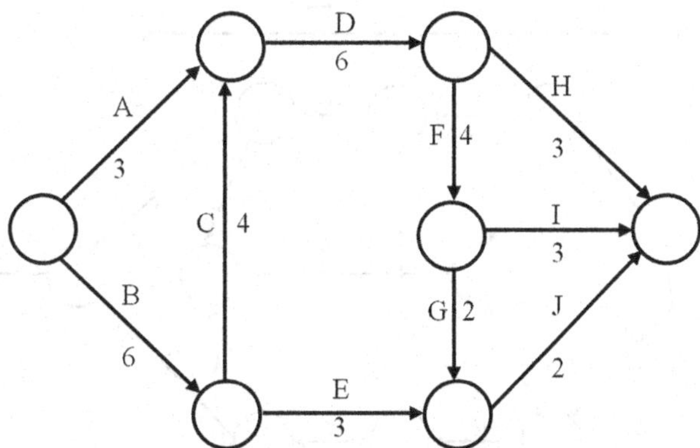

Figure E5.3 Network of the sample project

Question 5.2: The network of a sample project is shown in Figure E5.3; the estimated time duration of each activity is marked in the network. For each activity in the project, calculate the EST, EFT, LST, and LFT values.

Solution: The appropriate forward and backward pass calculations are shown in Figure E5.4; the EST, EFT, LST, and LFT values of the various activities in the project are in Table E5.2.

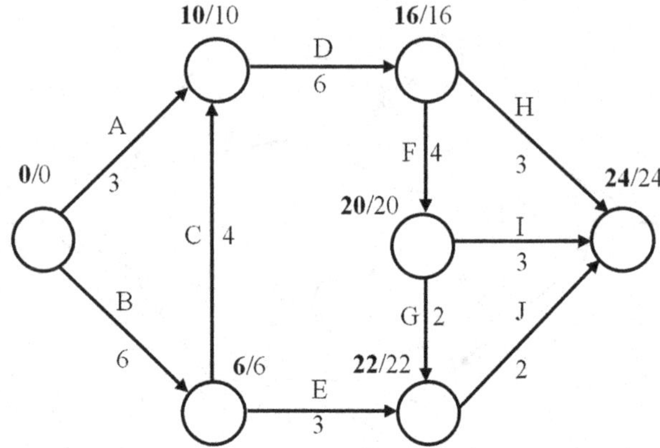

Figure E5.4 Forward and backward pass calculations

Table E5.2 EST, EFT, LST, and LFT values of various activities

Activities	*Time durations*	*EST*	*EFT*	*LST*	*LFT*
A	3	0	3	7	10
B	6	0	6	0	6
C	4	6	10	6	10
D	6	10	16	10	16
E	3	6	9	19	22
F	4	16	20	16	20
G	2	20	22	20	22
H	3	16	19	21	24
I	3	20	23	21	24
J	2	22	24	22	24

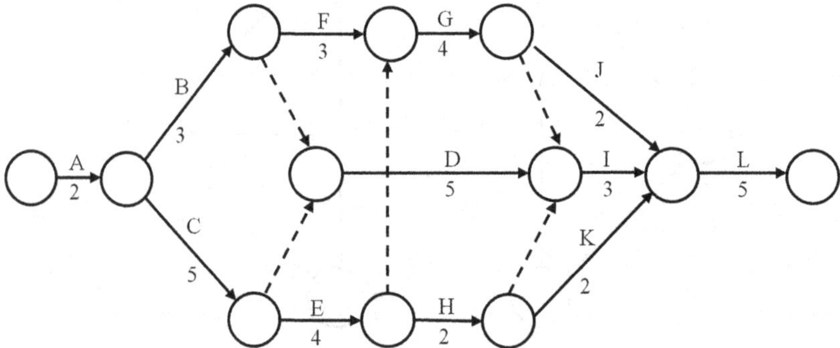

Figure E5.5 Network of the sample project

Question 5.3: The network of a sample project is shown in Figure E5.5; the estimated time duration of each activity is marked in the network. For each activity of the project, calculate the EST, EFT, LST, and LFT values.

Solution: The appropriate forward and backward pass calculations are shown in Figure E5.6; the EST, EFT, LST, and LFT values of the various activities in the project are in Table E5.3.

Question 5.4: The network of a sample project is shown in Figure E5.7; the estimated time duration of each activity is marked in the network. For each activity in the project, calculate the EST, EFT, LST, and LFT values.

Solution: The appropriate forward and backward pass calculations are shown in Figure E5.8; the EST, EFT, LST, and LFT values of the various activities of the project are in Table E5.4.

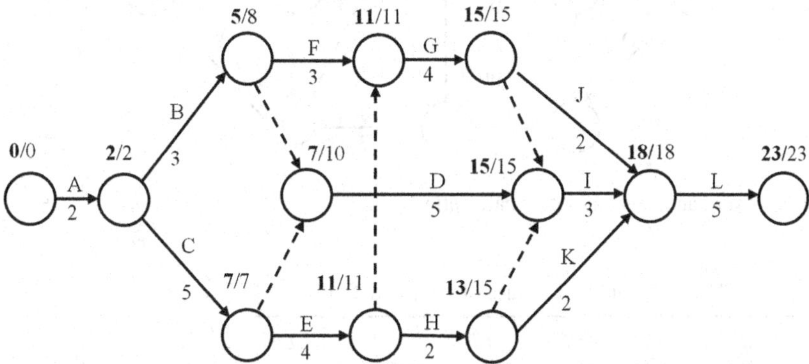

Figure E5.6 Forward and backward pass calculations

Table E5.3 EST, EFT, LST, and LFT values of various activities

Activities	*Time durations*	*EST*	*EFT*	*LST*	*LFT*
A	2	0	2	0	2
B	3	2	5	5	8
C	5	2	7	2	7
D	5	7	12	10	15
E	4	7	11	7	11
F	3	5	8	8	11
G	4	11	15	11	15
H	2	11	13	13	15
I	3	15	18	15	18
J	2	15	17	16	18
K	2	13	15	16	18
L	5	18	23	18	23

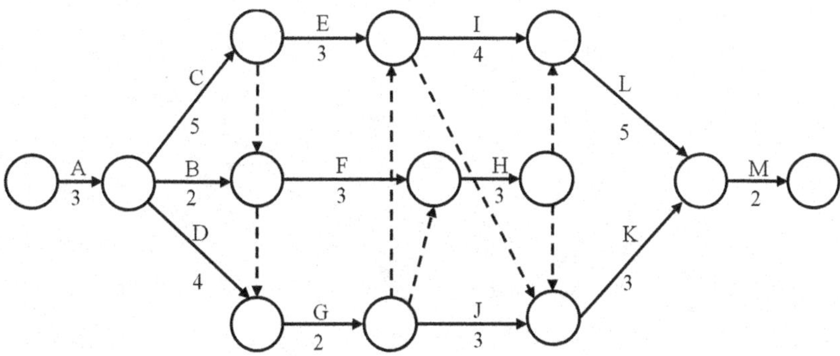

Figure E5.7 Network of the sample project

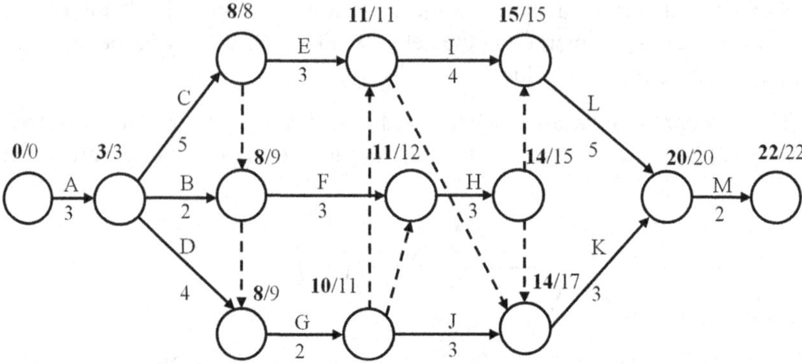

Figure E5.8 Forward and backward pass calculations

Table E5.4 EST, EFT, LST, and LFT values of various activities

Activities	*Time durations*	*EST*	*EFT*	*LST*	*LFT*
A	3	0	3	0	3
B	2	3	5	7	9
C	5	3	8	3	8
D	4	3	7	5	9
E	3	8	11	8	11
F	3	8	11	9	12
G	2	8	10	9	11
H	3	11	14	12	15
I	4	11	15	11	15
J	3	10	13	14	17
K	3	14	17	17	20
L	5	15	20	15	20
M	2	20	22	20	22

Question 5.5: Why are time-scaled network diagrams necessary? Discuss their practicality for project management.

6 Critical Path Method – II

Activity Floats

6.1 Learning Objectives

After completion of this chapter, readers will be able to:

- Determine the critical path and critical activities of a project,
- Understand different types of activity floats, and
- Understand the physical significance of different types of float.

6.2 Introduction

The allocated time duration or allowed duration of a project may be shorter than, equal to, or longer than its estimated time duration. When the allocated time duration is shorter than the estimated time duration, this indicates that more resources are required to complete the project within the allocated time duration. When the allocated time duration is equal to the estimated time duration, this indicates that the available resources are adequate for completion of the project within the allocated time duration. However, when the allocated time duration is longer than the estimated time duration, this indicates that fewer resources than allocated are required for completion of the project within the estimated time duration. When the allocated time duration is longer than the estimated time duration of an activity, this difference in time is considered surplus time. The surplus time available within an activity is called *float*.

The availability of float offers a planner flexibility in scheduling the start of an activity within its EST and LST. The availability of float also offers a planner flexibility in scheduling the finish time of an activity within its EFT and LFT. Float denotes the limit or range within which the start or finish time of an activity can fluctuate without delaying the completion of a project. Different types of float are discussed in the various sections of this chapter.

6.3 Critical Path

A network consists of many paths between its starting event and its end event. The different possible paths running from the starting to the end events of a network have different path lengths. The length of a path is the time duration required to complete the path in question. This is calculated by adding up the time durations of all the activities that lie along that path. Out of all of the possible paths in a network, the *critical path* is the longest path connecting the starting event and the end event of a network. A critical path is usually marked on a network by using thick-lined arrows. Out of all of the possible paths in a network, at least one path is critical. A network may have more than one critical path; in the most extreme situation, all of the paths in a network may become critical paths, with the lengths of all critical paths being equal.

DOI: 10.1201/9781003428992-6

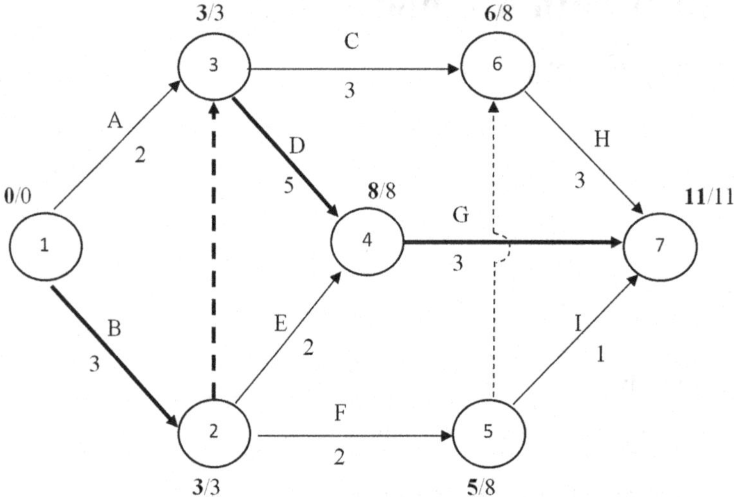

Figure 6.1 Sample network used to demonstrate the critical path

In the network shown in Figure 6.1 there are seven possible paths: A-C-H (path length = 8), A-D-G (path length = 10), B-C-H (path length = 9), B-D-G (path length = 11), B-E-G (path length = 8), B-F-I (path length = 6), and B-F-H (path length = 8). The longest path in the network is B-D-G. Path B-D-G is the network's critical path; the length of the path is 11 days. Forward pass calculations are carried out to find the earliest event time (written on every event as a bold number) as shown in Figure 6.1. The latest event time is calculated through backward pass calculations (written on every event as a normal number). In events which lie along critical paths, the earliest event time is equal to the latest event time. In the network under consideration, the critical path is path B-D-G, and the events that lie along the critical path are events 1, 2, 3, 4, and 7. In these events the earliest event time is equal to the latest event time.

6.4 Critical Activities

All of the activities along the critical path of a network are called *critical activities*. The reason for this criticality is that any delay to the time duration of a critical activity results in a delay to the time duration of the project under consideration. The identification of a critical path is useful for effective planning and project control. It helps a planner allocate the necessary resources to the critical activities to ensure their timely completion.

In the network shown in Figure 6.1, the critical path is B-D-G and the activities along the critical path are B, D, and G. Thus, activities B, D, and G are the network's critical activities. The sum of the time durations of the activities (3+5+3) along a critical path is equal to the critical path's length. The length of the critical path of the network shown in Figure 6.1 is 11 days. The length of the critical path is equal to the *project time duration*.

6.5 Total or Path Float

The *total float* (TF) is also called *path float* because it is the float associated with non-critical paths. TF is the total amount of time by which an activity lying on a non-critical path can exceed its EFT without affecting any time associated with any other activity or event lying on the critical path.

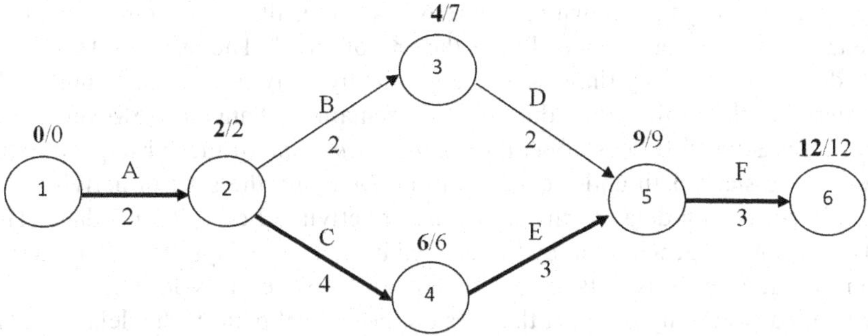

Figure 6.2 Sample network used to demonstrate the concept of *total float* or *path float*

A network shown in Figure 6.2 is used to describe the calculation of total float. The time-scaled version of the same network is shown in Figure 6.3. The total float or path float of path 2-3-5 is 3 days. Activities B and D lie on path 2-3-5, thus their path float has a value of 3 days, as shown in Figure 6.3. Activities B and D lie on the non-critical path 2-3-5. Path 2-3-5 is also called the *slack path*; slack path 2-3-5 is three days shorter than path 2-4-5. Path 2-4-5 lies on the critical path (1-2-4-5-6). In this case, the start of activity B can be

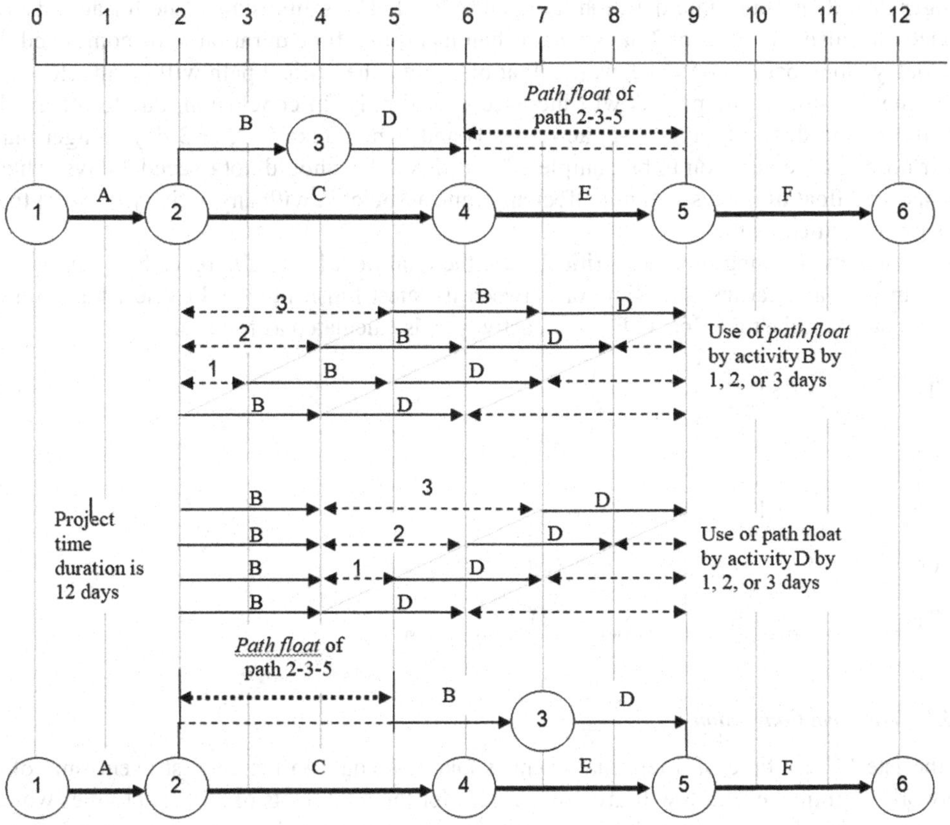

Figure 6.3 The concept of *total float* or *path float*

delayed by 1, 2, or 3 days as shown in Figure 6.3. This results in the completion of activity B at the end of day 5, 6, or 7, instead of at the end of day 4. The use of a TF of 3 days for activity B does not affect any time associated with any activity or event lying on the critical path, therefore the project time duration will remain unchanged. However, the use of a TF of 3 days for activity B does affect the *earliest start time* of the subsequent activity, D, which lies on the slack path under consideration. Delaying the start of activity B by 1, 2, or 3 days will result in a delay to the start time of activity D of 1, 2, or 3 days. The TF of activity D, on path 2-3-5, will thereby be reduced by 2, 1, or 0 days. The TF of activity D is reduced in case activity B uses its total or path float. However, this does not affect any time associated with any activity or event that lies on the critical path. If the delay on slack path 2-3-5 exceeds its path float of 3 days, then the critical path will be affected, and the time duration of the project will increase accordingly.

In Figure 6.3, the path float of activity D is also three days. Activity D lies on path 2-3-5. The start of activity D may also be delayed by 1, 2, or 3 days without changing the starting time of activity B. This would result in the completion of activity D at the end of day 7, 8, or 9, instead of at the end of day 6. The use of a TF of 3 days for activity D does not affect any time associated with any activity or event lying on the critical path, therefore the project time duration will remain unchanged. However, the TF of activity B is reduced in case activity D uses its path float.

Due to the availability of path float on path 2-3-5, activity B may take 1, 2, or 3 days longer than its normal time duration to be completed. The same is also true for activity D, which may also take 1, 2, or 3 days longer than its normal time duration to be completed. If the delay along path 2-3-5 exceeds the float of 3 days, the critical path will be affected and the time duration of the project will increase accordingly. In conclusion, due to the availability of path float on path 2-3-5, activities B and D may take 1, 2, or 3 days longer than their normal time duration to be completed, but this delay should not exceed 3 days. Thus, the use of a float of 3 days will not affect any time associated with any activity or event that lies on the critical path.

For activity a_{ij}, lying on a non-critical path, the total float is the difference between its latest start time and earliest start time or between its latest finish time and earliest finish time. Mathematically, the total float (TF_{ij}) of an activity a_{ij} is calculated as follows.

$$TF_{ij} = LST_{ij} - EST_{ij} \qquad 6.1$$

or

$$TF_{ij} = LFT_{ij} - EFT_{ij} \qquad 6.2$$

or

$$TF_{under\text{-}consideration} = LST_{under\text{-}consideration} - EST_{under\text{-}consideration} \qquad 6.3$$

6.5.1 Zero-Float Convention

If the latest event time of a terminal event is taken as equal to the earliest event time of a network's terminal event, it will give zero values for the total floats of all the activities which lie on the critical path. The total float values of all the activities along all paths other than the critical paths are positive. When the scheduled completion time duration (d_s) of a project is

Table 6.1 The total float values of critical activities B, D, and G in the zero-float convention are zero

Activities	*Durations*	*EST*	*EFT*	*LST*	*LFT*	*Total floats*
A	2	0	2	1	3	1
B	3	0	3	0	3	**0**
C	3	3	6	5	8	2
D	5	3	8	3	8	**0**
E	2	3	5	6	8	3
F	2	3	5	6	8	3
G	3	8	11	8	11	**0**
H	3	6	9	8	11	2
I	1	5	6	10	11	5

used as the latest event time of the terminal event of a project, the total float values along the critical path will be positive, zero, or negative, depending upon whether d_s is greater than, equal to, or less than the earliest event time of the terminal event as obtained through forward pass calculations.

When the latest event time of a terminal event in a backward pass calculation is taken as equal to the earliest event time of the terminal event of a network as determined in a forward pass calculation (LET = EET), it is called a *zero-float convention.* The zero-float convention is generally used to calculate four times for all the activities in a network. The LFT of an activity in the zero-float convention is the length of time by which the completion of an activity can be delayed without directly causing any increase in the project time duration.

The *zero-float convention* is used to demonstrate a backward pass calculation for the network shown in Figure 6.1, where terminal event number 7 occurs at the end of day 11. in this backward pass calculation, the latest event time of event number 7 is taken as equal to the earliest event time (LET = EET = 11) of event number 7 as determined through a forward pass calculation. LETs are determined, for all events, from network logic in backward pass calculations. For the network shown in Figure 6.1, the critical path is B-D-G and the activities which lie on the critical path are B, D, and G. The total float values of critical activities B, D, and G, in the zero-float convention, come out equal to zero, as listed in Table 6.1. This is the most suitable way to identify the *critical activities* and the *critical path* of a given network.

6.6 Free or Activity Float

Free float is the float that an activity is free to use without affecting any time associated with any other activity in a network. An activity is free to use free float, hence the adjective *free*. In Figure 6.4, on slack path 2-3-5 activities B and D have path floats of 3 days. If activity B uses this path float (not exceeding 3 days) it affects the times associated with activity D. However, if activity D uses this path float without exceeding 3 days, no time associated with any other activity in the network will be affected. The use of activity D's path float will not affect any time associated with its preceding activity B, or any time associated with its subsequent activity, F.

From Figure 6.4, it is clear that activity D is free to use the path float by an amount of up to 3 days without affecting any time associated with any other activity in the network. Only activity D is free to use this path float, therefore, this is called activity D's *free float*. This is also called *activity float* because it is associated only with an activity, and not with the path. Activity D, which lies on the slack path, has an activity float or free float of 3 days. However, this is not true

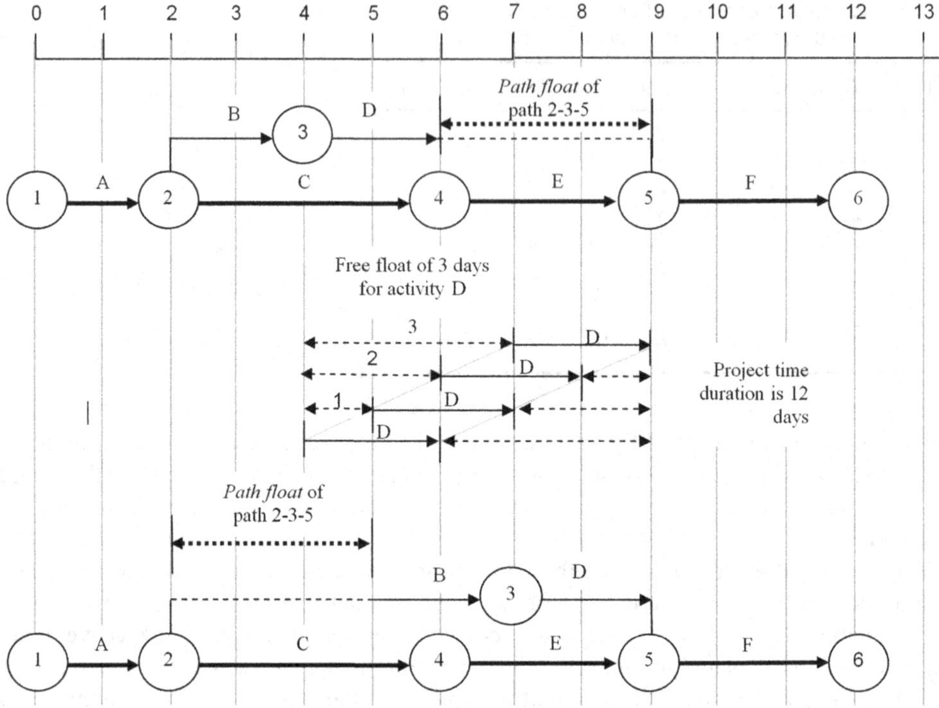

Figure 6.4 The concept of an activity's *free float*

for activity B. Activity B is not free to use its path float because if of activity B were to use it, it would affect the times associated with the subsequent activity, D. Therefore, activity B has no free float because its use of a path float of 3 days would delay the earliest start time of the subsequent activity, D. Thus, activity D has an activity float or path float of 3 (9–6) days. In this case, it is assumed that the activities preceding that under consideration are completed by their earliest finish times as shown in Figure 6.4 – that is, that the network is drawn on an EST basis. The relevant total float and free float values have been listed in Table 6.2.

Free float is equal to the amount of time by which the completion time of an activity can be delayed without affecting any time associated with any other activity in the network. Activity float is associated with individual activities, whereas path or total float is shared by all the

Table 6.2 Total, free, and interfering float values

Activities	*Durations*	*EST*	*EFT*	*LST*	*LFT*	*Total floats*	*Free floats*	*Interfering floats*
A	2	0	2	0	2	0	0	0
B	2	2	4	5	7	3	0	3
C	4	2	6	2	6	0	0	0
D	2	4	6	7	9	3	3	0
E	3	6	9	6	9	0	0	0
F	3	9	12	9	12	0	0	0

activities on a slack path. Mathematically, activity float is the difference between the earliest start time of the subsequent activity or activities and the earliest finish time of the activity under consideration. Thus, for activity a_{ij} free float (FF_{ij}) is given as follows.

$$FF_{ij} = EST_{jk} - EFT_{ij} \quad 6.4$$

or

$$FF_{\text{under-consideration}} = EST_{\text{succeeding}} - EFT_{\text{under-consideration}} \quad 6.5$$

Where jk denotes the activity following the activity (ij) under consideration.

6.7 Interfering Float

In Figure 6.4, on slack path 2-3-5 activities B and D both have path floats of 3 days. If the use of path float by activities B and D is critically observed, it is clear from the figure that, if activity B uses a path float of up to 3 days, it will affect the times associated with activity D. However, if activity D uses a path float of up to 3 days, the times associated with all the other activities in the network will not be affected. Activity D is free to use a path float of up to 3 days without interfering with the times associated with any other activity in the network. In other words, activity D does not interfere with the times associated with any other activity in the network, and so activity D has no *interfering float*. However, this is not true in the case of activity B. If activity B uses its path float, it will interfere with the times associated with the subsequent activity, D, by up to 3 days. Therefore, activity B has an interfering float of 3 days.

In this case, it is assumed that the activity preceding that under consideration is completed at its earliest finish time as shown in Figure 6.4, the network being drawn on an EST basis. Three types of float are listed in Table 6.2. The interpretation follows from the name. Activity B has both a path float and an interfering float of 3 days; if activity B were to use its path float, it would interfere by the same amount with the earliest start time of the subsequent activity, D. Activity D, on the other hand, has no interfering float but has both a path float and a free float of 3 days. Activity D can use its activity or path float by up to 3 days without interfering with any other activity times in the network. Mathematically, this is the difference between total float and free float or the difference between path float and activity float. For activity a_{ij}, the interfering float ($I_{int}F_{ij}$) is given as follows:

$$I_{int}F_{ij} = TF_{ij} - FF_{ij} \quad 6.6$$

or

$$I_{int}F_{\text{under-consideration}} = TF_{\text{under-consideration}} - FF_{\text{under-consideration}} \quad 6.7$$

6.8 Independent Float

Independent float does not occur frequently in any kind of network. It represents float that an activity always possesses; no other activity in a network can take independent float from a given activity. An activity with independent float can independently use this float. Float of an activity is the time, who takes the time of an activity either its preceding activity or its succeeding

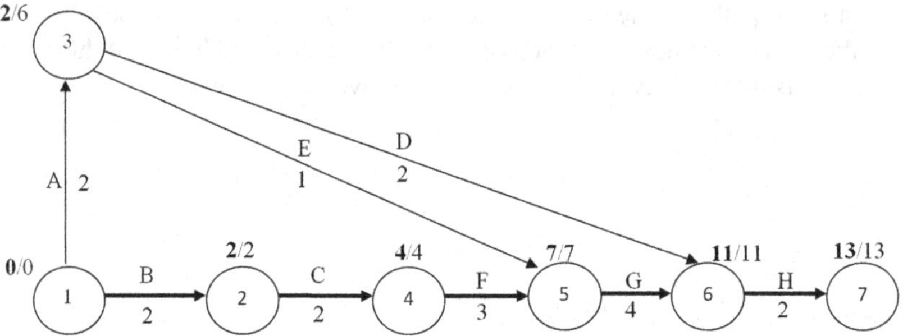

Figure 6.5 Network used to explain *independent float*

activity. A preceding activity can eat into the activity under consideration's time by ending at its latest finish time. Subsequent activities can take the time of the activity under consideration by starting at its earliest start time.

Independent float is determined by assuming the worst-case scenario, in which preceding activities end as late as possible and subsequent activities start as soon as possible. This is the worst case scenario for the activity under consideration. In this case, when the time interval available exceeds the time duration of the activity under consideration, this excess of time can be called independent float.

The network shown in Figure 6.5 is used to demonstrate the calculation of independent float. The time-scaled version of the same network is shown in Figure 6.6. Let us calculate the independent float of activity D. The worst-case scenario for activity D would occur if the preceding activity, A, were to end as late as possible (on its LFT), and the subsequent activity, H, were to start as soon as possible (on its EST). On the time-scaled version, two worst-case scenarios for the same network have been drawn. In the first case, the subsequent activity, H, starts at its earliest start time, that is, at the end of day 11. In the second case, the preceding activity, A, ends at its latest finish time, that is, at the end of day 6. In the worst-case scenario for activity D, the available time interval would be 5 days (11–6). This 5 day available time interval exceeds the time duration of the activity under consideration by 3 (5–2) days. This excess time duration of 3 days is the *independent float* of activity D, as shown in Figure 6.6. In contrast to this, activity E has no *independent float*, as shown in Table 6.3. For activity a_{ij}, independent float ($I_{ind}F_{ij}$) is written as:

$$I_{ind}F_{ij} = (EST_{jk} - LFT_{hi}) - d_{ij} \tag{6.8}$$

or

$$I_{ind}F_{\text{under-consideration}} = (EST_{\text{succeeding}} - LFT_{\text{preceding}}) - d_{\text{under-consideration}} \tag{6.9}$$

6.9 Simple Method for Calculating Floats

Figure 6.7 illustrates a simple method for the calculation of four types of floats. Forward and backward pass calculations yield two numbers for each node of a network. The first number, obtained from forward pass calculations, is the earliest event time and the second number, obtained from backward pass calculations, is the latest event time of a node. For a given activity,

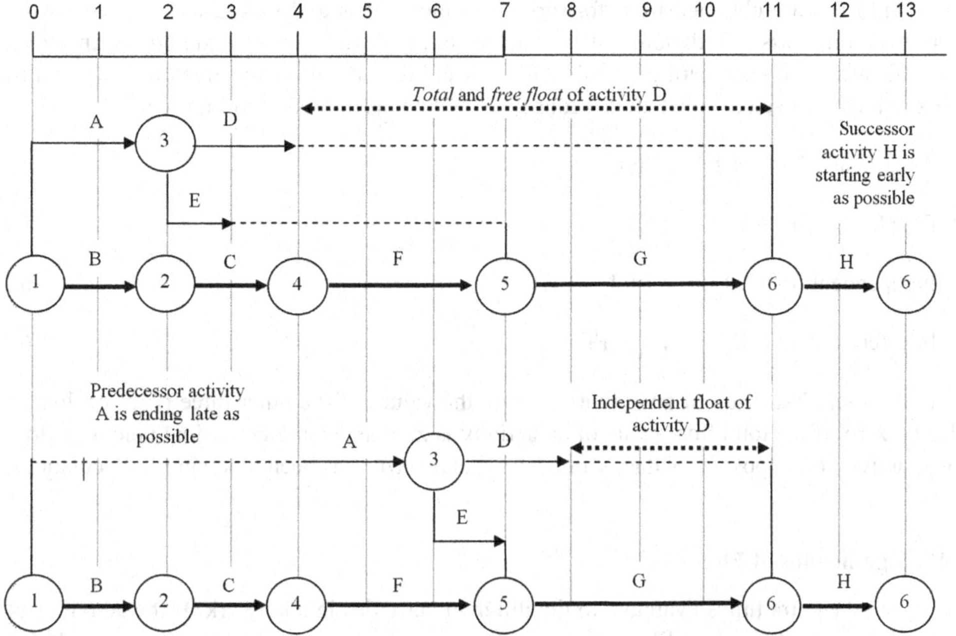

Figure 6.6 The concept of an activity's *independent float*

Table 6.3 Independent float values

Activities	*Durations*	*EST*	*EFT*	*LST*	*LFT*	*Total floats*	*Free floats*	*Interfering*	*Independent*
A	2	0	2	4	6	4	0	4	0
B	2	0	2	0	2	0	0	0	0
C	2	2	4	2	4	0	0	0	0
D	2	2	4	9	11	7	7	0	3
E	1	2	3	6	7	4	4	0	0
F	3	4	7	4	7	0	0	0	0
G	4	7	11	7	11	0	0	0	0
H	2	11	13	11	13	0	0	0	0

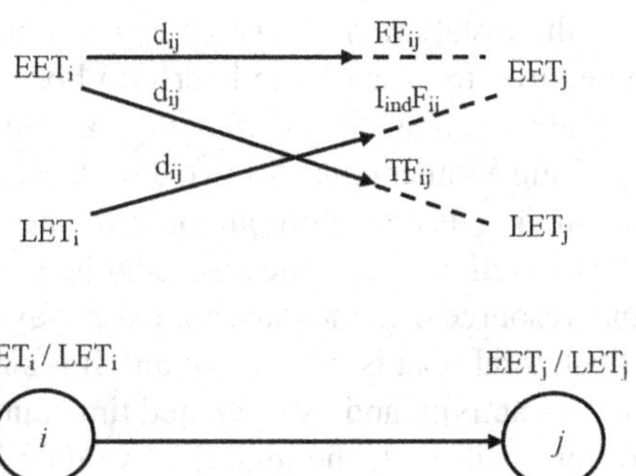

Figure 6.7 Calculations of the four floats

EET_i and EET_j are obtained from forward pass calculations and LET_i and LET_j are obtained from backward pass calculations, where *i* is the start node and *j* is the end node of an activity. Thus, forward and backward pass calculations yield four numbers for given activity in a network, which are used to calculate the four types of float as discussed in Figure 6.7.

$$\text{Total float} = TF_{ij} = LET_j - EET_i - d_{ij} \qquad 6.10$$

$$\text{Free float} = FF_{ij} = EET_j - EET_i - d_{ij} \qquad 6.11$$

$$\text{Independent float} = I_{ind}F_{ij} = EET_j - LET_i - d_{ij} \qquad 6.12$$

$$\text{Interfering float} = I_{int}F_{ij} = TF_{ij} - FF_{ij} \qquad 6.13$$

If the total float value of an activity is zero, the values of the other three types of float will also be zero. If the total float value of an activity is zero and some other float belonging to the same activity has a positive value, it indicates that an error has been made in the calculations.

6.10 Significance of Floats

Floats are the extra times available to the different activities in a network. If the activities of a network have no floats, it indicates that there is no surplus time available in that network. Such networks cannot accommodate any delay to their activities. The availability of float provides a planner with the opportunity to manage resources efficiently. Available float also reduces the risk of time overruns through the management of resources. Available float is considered the time contingency for the non-critical activities in a network and provides a planner with flexibility for scheduling the starting times of non-critical activities within their ESTs and LSTs.

The availability of float facilitates delays to non-critical activities, within float limits, to improve resource efficiency. For example, consider the network shown in Figure 6.4. Two activities (B and C) start together; activity C is critical and activity B is non-critical. Assume both require a crane for their execution. The contractor executing the activities, however, has only one crane available, thus the two activities cannot be executed together as per the resource availability constraint. To manage resources efficiently, firstly, the crane could be assigned initially to the critical activity, C, and the start of the non-critical activity, B, could consequently be delayed, provided that the extent of the delay is within the available float limit of activity B. However, in the present case, the required delay exceeds the available float limit by one day. This may result in a delay to the project time duration of one day. Further, hiring a crane for one day is not advisable, because the installation of a crane is not an easy task. Secondly, to avoid delay, another crane could be hired for one day and perform both activities simultaneously on day 6. However, if the availability of floats is not critically examined, a planner may start both activities simultaneously by hiring another crane for two days to execute activity B. This is more costly than the first suggestion. The last suggestion could be the re-scheduling of the project activities before the project's execution starts. The rescheduling of the project activities may be carried out in such a way that resource demand does not exceed availability.

Of the various types of float, total float is very important. It is simply the difference between the time duration allocated to an activity and its estimated time duration. The value of the total float may be negative, zero, or positive. If the total float value of an activity is negative, the activity is called a *supercritical activity*. Such activities need more resources than allocated and require much more attention during their execution. An attempt must be made to employ more resources so as to make the value of the total float zero or positive. If the total float value of an activity is zero, the activity is called a *critical activity*. Such activities have adequate resources and require continuous attention during their execution. However, the resources assigned are sufficient, neither too little nor too much. But no time freedom is available in such activities. If the total float value of an activity is positive, the activity is called a *non-critical activity*. Such activities have more resources than necessary and require less attention during their execution. Such activities allow for a certain degree of freedom, depending upon the value of the available float.

6.11 Network Development Procedure

The network development procedure used in CPM is discussed step-by-step below.

Step 1: The activities involved in a project are defined clearly and brief descriptions of the various activities involved are provided.

Step 2: A single value estimate is made for the time duration of each activity in the network.

Step 3: The network is developed directly by experienced planners, so as to represent the execution sequence of the various activities defined earlier. The inter-dependencies/logics between the various activities are sometimes listed in tabular form, and a network is subsequently developed from the table.

Step 4: The developed network is analyzed to determine activity times. The network time analysis has two passes: a forward pass for the calculation of earliest times and a backward pass for the calculations of latest times. The critical path of the network and the different floats of each individual activity are determined. Finally, a time-scaled version of the network is developed.

6.12 Conclusion

The estimated time durations of project activities, and the execution sequence finalized in previous chapters in the form of a network, are used to find the project's time duration and identify its critical path. The critical path is the longest path in a network between the start and the end event. The activities which lie on the critical path are called the project's critical activities. Any delay to the time durations of the critical activities causes the same length of delay to the time duration of the project. If the time allocated to an activity is longer than the estimated time duration of an activity, the extra amount of time is considered surplus time. The surplus time available to an activity is called float. The float available to an activity provides a planner with flexibility in scheduling the starting time of the activity between its EST and LST. Float denotes the limit or range within which the start or finish of an activity can fluctuate without delaying the completion of a project. Different types of float have been covered in the chapter in detail.

Example 6.1: Draw a network diagram for the relationships between the activities given in Table 6.4 and find the critical path length, the four types of time, and the floats of the various activities in the project.

Solution: The network for a given set of logics is shown in Figure 6.8. The critical paths are A-K-L-I-D-E and A-K-L-J-H-E. The critical path length or time duration of the project is 15 days. The four types of time and floats for the various activities in the project are given in Table 6.5.

Table 6.4 Time durations and relationships between various activities

Activities	*Durations (days)*	*Depends upon*
A	2	-
B	2	A
C	2	B, F
D	3	C, I
E	2	D, H
F	1	A
G	1	F
H	2	G, J
I	1	L
J	2	L
K	4	A
L	3	F, K

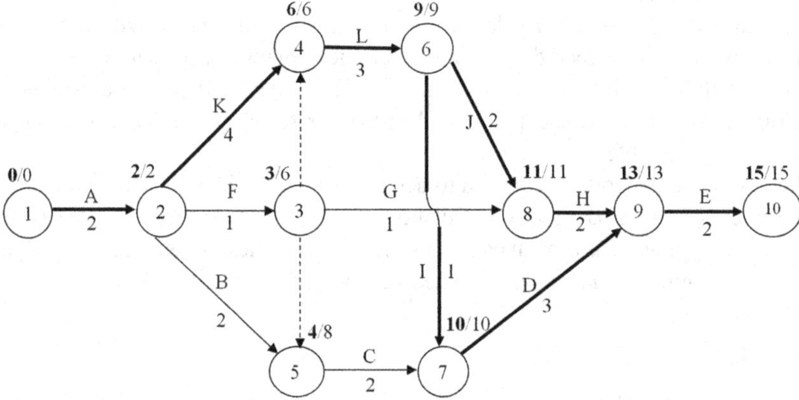

Figure 6.8 A network diagram

Table 6.5 The four types of time and floats of activities

Activities	*Time Duration*	*Depends upon*	*EST*	*EFT*	*LST*	*LFT*	*Total floats*	*Free floats*	*Interfering*	*Independent*
A	2	-	0	2	0	2	0	0	0	0
B	2	A	2	4	6	8	4	0	4	0
C	2	B, F	4	6	8	10	4	4	0	0
D	3	C, I	10	13	10	13	0	0	0	0
E	2	D, H	13	15	13	15	0	0	0	0
F	1	A	2	3	5	6	3	0	3	0
G	1	F	3	4	10	11	7	7	0	4
H	2	G, J	11	13	11	13	0	0	0	0
I	1	L	9	10	9	10	0	0	0	0
J	2	L	9	11	9	11	0	0	0	0
K	4	A	2	6	2	6	0	0	0	0
L	3	F, K	6	9	6	9	0	0	0	0

Table 6.6 Time durations and relationships between various activities

Activities	*Durations (days)*	*Depends upon*
A	2	-
B	2	A
C	2	B, M
D	3	C, L
E	2	D, H, J
F	1	I, E
G	1	M
H	2	G, L
I	1	L
J	2	L
K	4	A
L	3	K, M
M	1	A

Example 6.2: Draw a network diagram for the relationships between the activities given in Table 6.6 and find the critical path length, the four types of time, and the floats of the various activities in the project.

Solution: The network for a given set of logics is shown in Figure 6.9. The critical path is A-K-L-D-E-F. The critical path length or time duration of the project is 15 days. The four types of time and the floats of the various activities in the project are given in Table 6.7.

Example 6.3: Draw a network diagram for the relationships between the activities given in Table 6.8 and find the critical path length, the four types of time, and the floats of the various activities in the project.

Solution: The network for a given set of logics is shown in Figure 6.10. The critical path is M-D-Z-P-Q. The critical path length or time duration of the project is 13 days. The four types of time and the floats of the various activities in the project are given in Table 6.9.

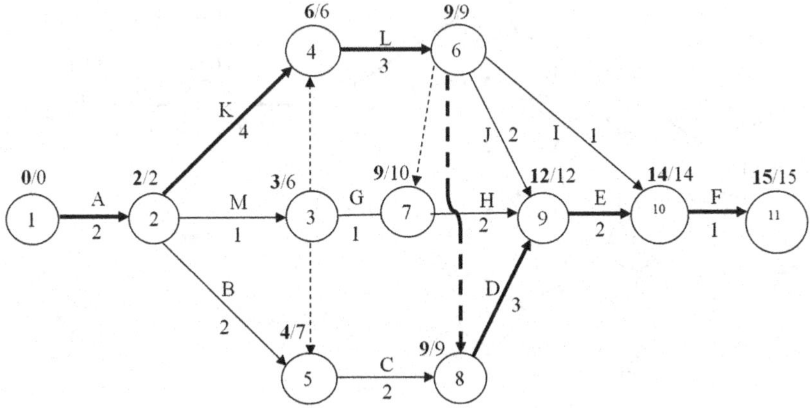

Figure 6.9 A network diagram

Table 6.7 The four types of time and floats of activities

Activities	*Time Duration*	*Depends upon*	*EST*	*EFT*	*LST*	*LFT*	*Total floats*	*Free floats*	*Interfering*	*Independent*
A	2	-	0	2	0	2	0	0	0	0
B	2	A	2	4	5	7	3	0	3	0
C	2	B, M	4	6	7	9	3	3	0	0
D	3	C, L	9	12	9	12	0	0	0	0
E	2	D, H, J	12	14	12	14	0	0	0	0
F	1	I, E	14	15	14	15	0	0	0	0
G	1	M	3	4	9	10	6	5	1	2
H	2	G, L	9	11	10	12	1	1	0	0
I	1	L	9	10	13	14	4	4	0	4
J	2	L	9	11	10	12	1	1	0	1
K	4	A	2	6	2	6	0	0	0	0
L	3	K, M	6	9	6	9	0	0	0	0
M	1	A	2	3	5	6	3	0	3	0

Table 6.8 Time durations and relationships between various activities

Activities	*Durations (days)*	*Inter-relationships*
A	1	M and N are starting operations,
B	2	D and A succeeds M,
C	3	B and C succeeds N,
D	4	X is controlled by D and A,
M	2	Y is controlled by A, B, and C,
N	1	Z is controlled by D only,
P	2	P succeeds Z and X, and
Q	2	Q succeeds P and Y
X	1	
Y	2	
Z	3	

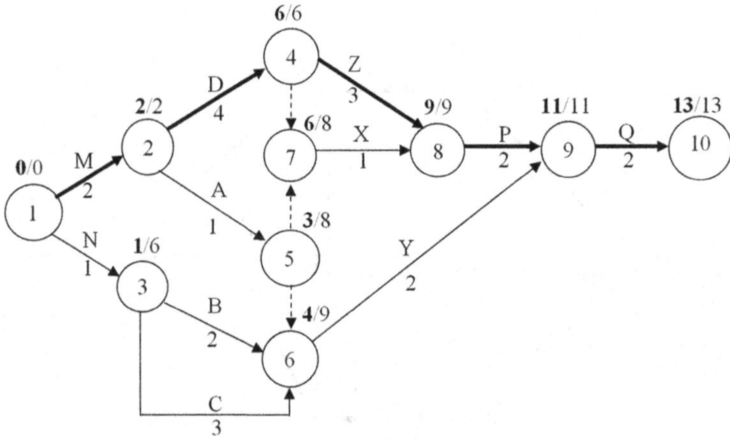

Figure 6.10 A network diagram

Example 6.4: Draw a network diagram for the relationships between the activities given in Table 6.10 and find the critical path length, the four types of time, and the floats of the various activities in the project.

Solution: The network for a given set of logics is shown in Figure 6.11. The critical path is A-B-D-E-H. The critical path length or time duration of the project is 23 days.

Table 6.9 The four types of time and floats of various activities

Activities	*Durations*	*EST*	*EFT*	*LST*	*LFT*	*Total floats*	*Free floats*	*Interfering*	*Independent*
A	1	2	3	7	8	5	0	5	0
B	2	1	3	7	9	6	1	5	0
C	3	1	4	6	9	5	0	5	0
D	4	2	6	2	6	0	0	0	0
M	2	0	2	0	2	0	0	0	0
N	1	0	1	5	6	5	0	5	0
P	2	9	11	9	11	0	0	0	0
Q	2	11	13	11	13	0	0	0	0
X	1	6	7	8	9	2	2	0	0
Y	2	4	6	9	11	5	5	0	0
Z	3	6	9	6	9	0	0	0	0

Table 6.10 Time durations and relationships between various activities

Activities	*Durations (days)*	*Preceding activities*	*Succeeding activities*
A	4	-	B, C
B	4	A	D, F
C	6	A	E, G
D	3	B	E, G
E	5	C, D	H, I
F	6	B	I
G	5	C, D	I
H	7	E	-
I	4	E, F, and G	-

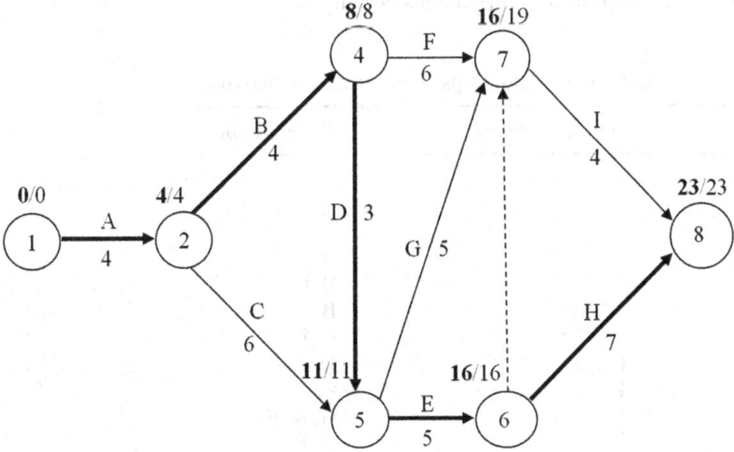

Figure 6.11 A network diagram

Table 6.11 The four types of time and floats of various activities

Activities	*Durations*	*EST*	*EFT*	*LST*	*LFT*	*Total floats*	*Free floats*	*Interfering*	*Independent*
A	4	0	4	0	4	0	0	0	0
B	4	4	8	4	8	0	0	0	0
C	6	4	10	5	11	1	1	0	1
D	3	8	11	8	11	0	0	0	0
E	5	11	16	11	16	0	0	0	0
F	6	8	14	13	19	5	2	3	2
G	5	11	16	14	19	3	0	3	0
H	7	16	23	16	23	0	0	0	0
I	4	16	20	19	23	3	3	0	0

The four types of time and the floats of the various activities in the project are given in Table 6.11.

Exercises

Question 6.1: Draw a network diagram for the relationships between the activities given in Table E6.1 and find the critical path length, the four types of time, and the floats for the various activities in the project.

Solution: The network for a given set of logics is shown in Figure E6.1. The critical paths are C-E-K-J and C-E-H-J. The critical path length or time duration of the project is 10 days. The four types of time, and the floats of the various activities in the project are given in Table E6.2.

Question 6.2: A project consists of eleven activities. The time durations of the various activities therein are given in Table E6.3 and the inter-relationships between them are given as below:

- The project starts when activities A and D start,
- Activity C follows activity A but precedes activity G,
- Activity B follows activity A but precedes activities E and F,
- Activity E follows activities B and C,
- Activity J follows activity F but precedes activity K,

Table E6.1 Time durations and relationships between various activities

Activities	*Durations (days)*	*Depends upon*
A	2	-
B	1	-
C	3	-
D	1	A
E	4	B, C
F	2	B
G	2	C, I
H	1	D, E
I	4	-
J	2	H, K, F, G
K	1	A, E

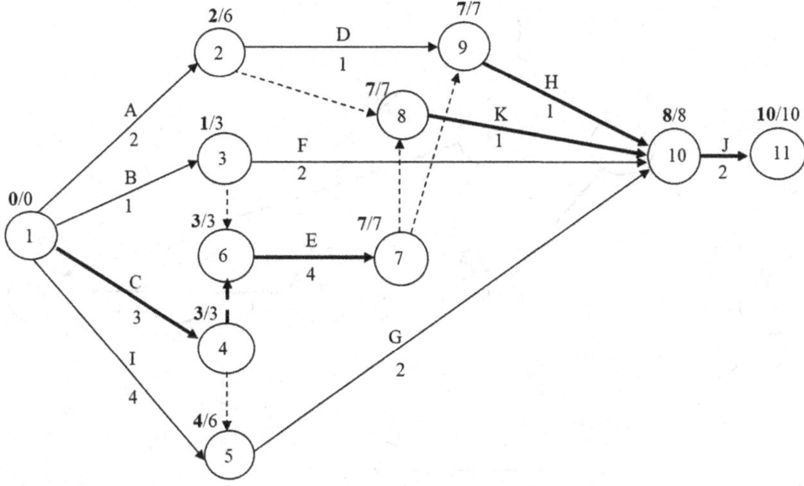

Figure E6.1 Network diagram

Table E6.2 The four types of time and floats of various activities

Activities	*Durations*	*EST*	*EFT*	*LST*	*LFT*	*Total floats*	*Free floats*	*Interfering*	*Independent*
A	2	0	2	4	6	4	0	4	0
B	1	0	1	2	3	2	0	2	0
C	3	0	3	0	3	0	0	0	0
D	1	2	3	6	7	4	4	0	0
E	4	3	7	3	7	0	0	0	0
F	2	1	3	6	8	5	5	0	3
G	2	4	6	6	8	2	2	0	0
H	1	7	8	7	8	0	0	0	0
I	4	0	4	2	6	2	0	2	0
J	2	8	10	8	10	0	0	0	0
K	1	7	8	7	8	0	0	0	0

Table E6.3 Time durations of various activities

Activities	*A*	*B*	*C*	*D*	*E*	*F*	*G*	*H*	*J*	*K*	*L*
Durations (days)	2	1	2	3	4	2	1	2	1	2	1

- Activity H follows activity D but precedes activity L,
- Activity K follows activities G and H, and
- The project ends with the end of activities E, K, and L together.

Draw the network and find the critical path length, the four types of time, and the floats of the various activities in the project.

Solution: The network for a given set of logics is shown in Figure E6.2. The critical paths are A-B-F-J-K and A-C-E. The critical path length or time duration of the project is 8 days. The four types of time, and the floats of the various activities in the project are given in Table E6.4.

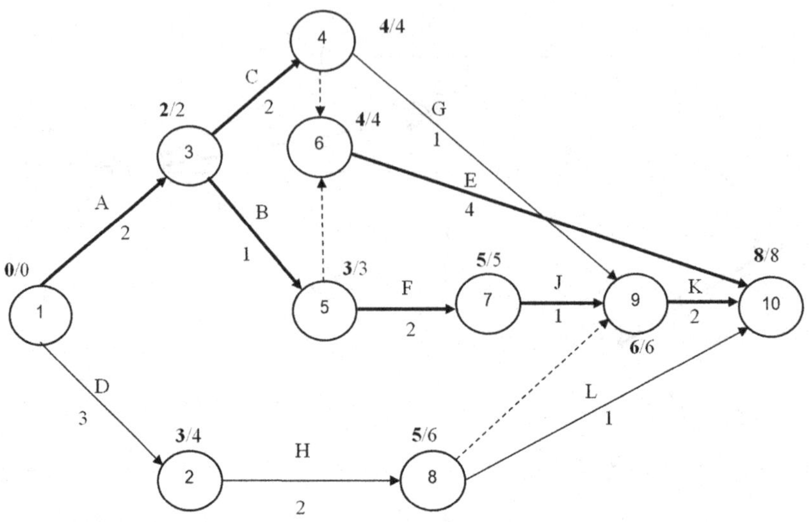

Figure E6.2 Network diagram of 11 activities

Table E6.4 The four types of time and floats of 11 activities

Activities	*Durations*	*EST*	*EFT*	*LST*	*LFT*	*Total floats*	*Free floats*	*Interfering*	*Independent*
A	2	0	2	0	2	0	0	0	0
B	1	2	3	2	3	0	0	0	0
C	2	2	4	2	4	0	0	0	0
D	3	0	3	1	4	1	0	1	0
E	4	4	8	4	8	0	0	0	0
F	2	3	5	3	5	0	0	0	0
G	1	4	5	5	6	1	1	0	1
H	2	3	5	4	6	1	0	1	0
J	1	5	6	5	6	0	0	0	0
K	2	6	8	6	8	0	0	0	0
L	1	5	6	7	8	2	2	0	1

Question 6.3: A project consists of eleven activities. The time durations of the various activities therein are given in Table E6.5 and the inter-relationships between them are given as below:

- The project starts when activities A, B, and C start,
- The project ends when activities Q, M, and R end,
- Activity H precedes activity N but follows activity D,
- Activity E follows activity D but proceeds activity N,
- Activities F and L follow activities K and B,
- Activity F precedes activities M,
- Activity Q follows activities P, L, and N,
- Activity L precedes activity R but follows activity C,
- Activity D precedes activity G,
- Activity K and activity D follow activity A, and
- Activity F precedes activity P but follows activity G.

Table E6.5 Time durations of various activities

Activities	A	B	C	D	E	F	G	H	K	L	N	M	P	Q	R
Durations (days)	2	1	2	3	4	2	1	2	1	2	3	1	3	4	2

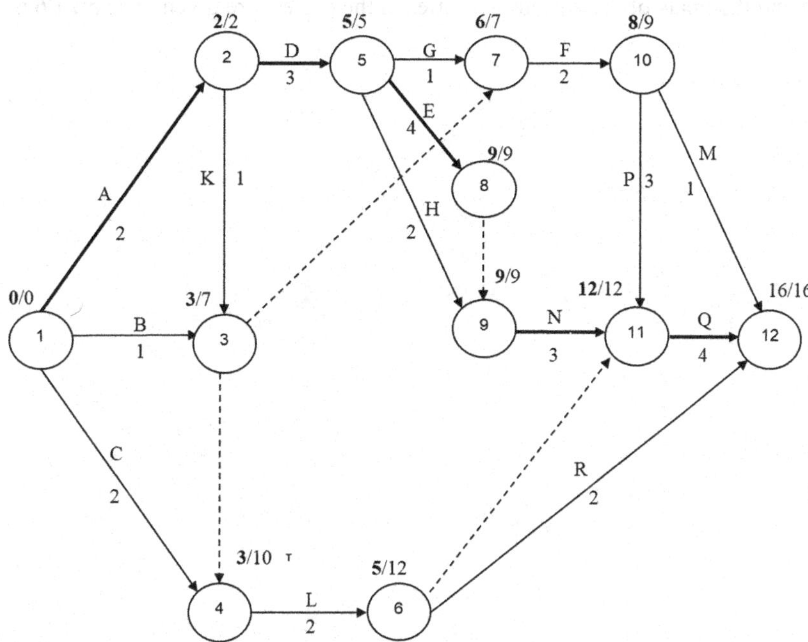

Figure E6.3 Network diagram of 15 activities

Table E6.6 The four types of time and floats

Activities	*Durations*	*EST*	*EFT*	*LST*	*LFT*	*Total floats*	*Free floats*	*Interfering*	*Independent*
A	2	0	2	0	2	0	0	0	0
B	1	0	1	6	7	6	2	4	2
C	2	0	2	8	10	8	1	7	1
D	3	2	5	2	5	0	0	0	0
E	4	5	9	5	9	0	0	0	0
F	2	6	8	7	9	1	0	1	0
G	1	5	6	6	7	1	0	1	0
H	2	5	7	7	9	2	2	0	2
K	1	2	3	6	7	4	0	4	0
L	2	3	5	10	12	7	0	7	0
N	3	9	12	9	12	0	0	0	0
M	1	8	9	15	16	7	7	0	6
P	3	8	11	9	12	1	1	0	0
Q	4	12	16	12	16	0	0	0	0
R	2	5	7	14	16	9	9	0	2

Draw the network and find the critical path length, the four types of time, and the floats of the various activities in the project.

Solution: The network for a given set of logics is shown in Figure E6.3. The critical path is A-D-E-N-Q. The critical path length or time duration of the project is 16 days. The four types of time and the floats of the various activities in the project are given in Table E6.6.

7 Crashing

Time-Cost Tradeoff

7.1 Learning Objectives

After completion of this chapter, readers will be able to:

- Understand the variation of total project cost with the time duration of a project,
- Understand the project cost optimization process, and
- Determine the optimum time duration for a project.

7.2 Introduction

Sometimes it is necessary to complete a project within a time duration shorter than the estimated normal time duration. *Project shortening* is the completion of a project within a time duration shorter than the normal time duration. *Accelerating a project*, *project/schedule compression*, and *project/schedule crashing* are frequently used terms for shortening the normal time duration of a project. The following are a few possible reasons for shortening a project:

- A client wishes to complete a project in a time duration shorter than the normal time duration;
- After the completion of a certain percentage of a project, planned and actual progress is compared, and it is realized that the project is behind schedule and it is accelerated so that it will finish in time;
- Due to some emergency requirements or some political reasons, an executing agency wishes to develop a good reputation by finishing a project before its due date, or an executing agency may get an incentive for finishing a project before its due date;
- An agency is executing has more than one project at the same time. The agency thus accelerates the project under consideration so that it can move resources to some other projects and maximize profit; and
- Sometimes accelerating a project is profitable because of the reduction in the indirect cost of a project.

In the next section, different possible ways to shorten the time duration of a project will be discussed.

7.2.1 Shortening Project Time Durations

The following are a few possible ways to shorten the time duration of a project.

DOI: 10.1201/9781003428992-7

Fast-tracking: This is a technique in which project activities are performed in parallel so as to reduce the project time duration. When activities are performed in a sequence, a finish-to-start relationship is followed. Thus, instead of waiting for the completion of the preceding activity to start the subsequent activity, the largest possible number of activities are performed in parallel, or with the largest possible degree of overlaps. In fast-tracking, good project control is necessary, because multiple activities are performed at the same time. Further, it is also important to constantly monitor project progress and take corrective measures to ensure that project progress is kept on track. Fast-tracking cannot be applied to all activities, as some activities are completely sequential. Thus, if fast-tracking is not possible, project/schedule crashing is used. In project/schedule crashing, additional resources are used to reduce the time durations of project activities. This chapter deals only with project/schedule crashing.

Project/schedule crashing: This technique involves employing more resources or adding additional resources to help finish a project quicker than its normal time duration. This technique is called *project crashing.* In order to provide sufficient workspace for the additional resources employed, sometimes a second or even a third working shift may be planned. The extra cost of putting additional resources must be taken into consideration.

Overtime: his is another way to shorten the time duration of a project, in which more working hours per day and/or more working days per week are used. However, it has been observed in many projects that the productivity of resources declines with more working hours per day or more working days per week.

Others: Depending upon a project, the automation of whatever activities allow for it may also result in project acceleration. Good project control also leads to the completion of a project on time and reduces project delay. Good communication in an organizational structure also helps in the acceleration of a project.

7.3 Project Crashing

Sometimes it is necessary to complete a project in a time duration shorter than its normal time duration. Different techniques which could shorten the normal time duration of a project have been discussed above. Project crashing is a technique used to shorten projects in which additional resources are used to reduce the normal time duration of a project. In the project crashing, the normal time durations of activities that lie on the critical path are shortened by using additional resources to decrease the overall project time duration. Project crashing is started by shortening the normal time duration of an activity lying on the critical path which has a lower crashing cost than other critical activities. This requires an understanding of the time and cost relationship of each activity. The crashing process has been discussed in detail in subsequent sections.

The relationship between the *total project cost* and the time duration of a project must be established. CPM uses the relationship between the total project cost and the time duration of a project to develop a schedule for its minimum cost completion. This chapter revolves around the development of the relationship between the total project cost and time duration of a project. An executing agency or an organization will execute multiple projects simultaneously. It is common practice to shorten the time duration of a specific project so that the saved time may be used for other projects, allowing the organization to earn more profit. In general, the time duration of a project is shortened by shortening the normal time durations of the activities that lie along its critical path. This is achieved by deploying additional resources to encourage the early completion of the activities that lie along the critical path. Shortening the time durations

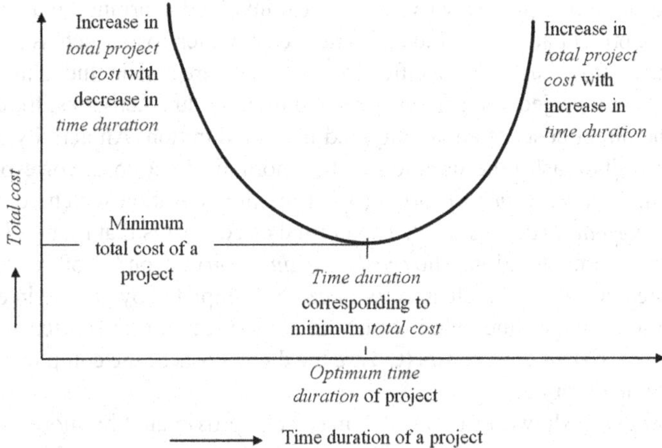

Figure 7.1 Relationship of total project cost to project time duration

of project activities requires the deployment of additional resources, and the use of additional resources increases the cost of a given activity's execution.

A project is divided into various activities; when executing an activity according to its normal time duration, the cost is assumed to be at its minimum. The execution of the same activity within a time duration shorter than the normal time duration increases its execution cost. In CPM, there are two cost estimates associated with each activity: a normal cost estimate and a crash cost estimate. In a normal cost estimate, the focus is on the minimum cost of an activity, and the normal time duration which corresponds to the minimum cost is determined. In a crash cost estimate, the focus is on the minimum time duration and the cost which corresponds to the minimum time duration is determined. The crash cost is the cost corresponding to the minimum time duration required to execute an activity.

In CPM, the relationship between the total project cost and the time duration is taken to be as shown in Figure 7.1. The figure shows that the total project cost increases when the time duration increases beyond an *optimum time duration.* Similarly, the total project cost increases when the time duration shrinks below the *optimum time duration.* The total project cost is at its lowest in a particular time duration. The time duration corresponding to the minimum total project cost of a project is called the *optimum time duration* of a project. The objective is to find the project time duration which corresponds to the minimum total project cost. The optimum time duration is assumed to offer the most economic project execution cost.

7.4 Constituents of Total Project Cost

Total project cost is considered to be the sum of two separate costs: the *direct cost* of project execution and the *indirect cost* related to the project.

7.4.1 Direct Costs

Direct costs of an activity include worker costs, material costs, and equipment costs. Worker costs are the expenses directly involved in the execution of a project activity. Materials costs comprise the costs of the materials required to execute a project activity. Equipment is sometimes needed in the execution of a project activity. The cost involved in arranging the equipment necessary to execute a project activity is the equipment cost. Other costs, such as government fees, legal fees, or consultation fees for a specific project activity are also included in the direct costs.

The direct costs of a project are generally related to individual activities. Figure 7.2 shows a curve between the direct costs of an activity and its time duration. An activity's highest direct costs correspond to its crash time duration and its normal direct costs correspond to its normal time duration. A *normal time duration* (d_n) is the time duration which corresponds to the minimum cost or *normal direct cost* (C_n). The normal direct cost is that required to complete an activity in the normal time duration. The *crash time duration* (d_c), on the other hand, is the minimum possible time duration in which an activity can be completed by putting in extra resources. A crash time duration is the time below which an activity cannot be shortened by any further increase in resources. *Crash direct costs* (C_c) are the direct costs of the completion of an activity within the crash time duration.

The *direct cost curve* shown in Figure 7.2 may be approximated by more than one straight line or by a single straight line depending upon the flatness of the curve. Figure 7.3 (a) shows

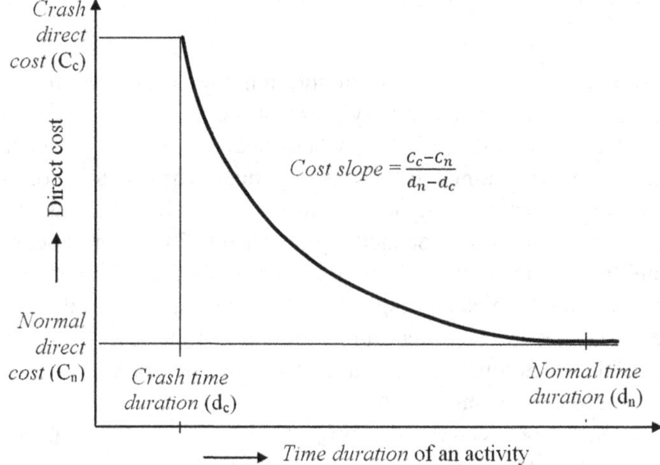

Figure 7.2 Increase in the direct cost of an activity with decrease in time

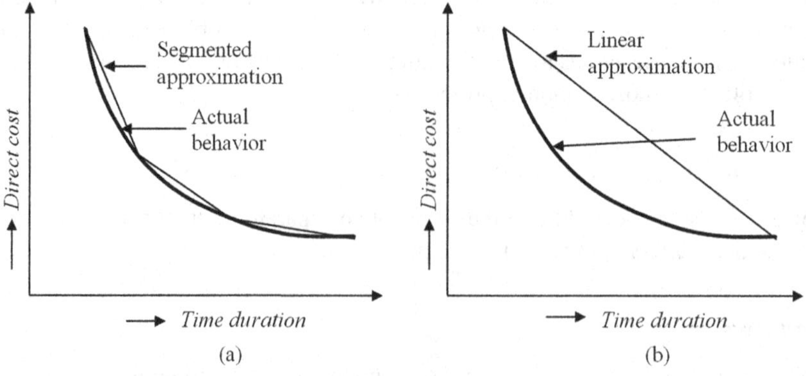

Figure 7.3 Direct cost: (a) segmented approximation of direct cost and (b) linear approximation of the direct cost

a direct cost curve which has been approximated using more than one straight line, and Figure 7.3 (b) shows a direct cost curve which has been approximated using a single straight line. Segmented or straight-line approximation of a direct cost curve is used to carry out a project cost analysis by determining the value of the *cost slope*. The cost slope is the slope of the direct cost curve, approximated as a straight line.

$$Cost\,slope = \frac{C_c - C_n}{d_n - d_c} \qquad 7.1$$

Straight-line approximation of a direct cost curve gives a single value for the cost slope; segmented approximation of a direct cost curve, on the other hand, gives more than one values for the cost slope. The single or multiple value of the cost slope depends upon the non-linearity of the direct cost curve. Segmented approximation of the direct cost curve gives multiple values for the cost slope, resulting in an analysis that is more accurate but which involves more calculations.

7.4.2 Total Indirect Cost

The indirect cost of a project is generally not related to individual activities. Indirect costs are determined for a project as a whole, hence the name *total indirect cost*. This includes expenditures related to project administrative charges, project establishment charges, project overheads, supervision costs, insurance claims, accident costs, other penalties, etc. Project overhead is the cost of the cars and trucks assigned to the project team, the costs of site offices, the construction costs for temporary structures, the costs of office facilities (copying machines, computers, printers, etc.), utility costs (drinking water, telephones, gas, toilets, etc.), etc. The total indirect cost rises along with the time duration of a project. The general relationship between total indirect cost and project duration is shown in Figure 7.4. The relationship of total indirect cost to time duration is not linear, as shown in Figure 7.4. However, in this chapter, the relationship of the total indirect cost to the time duration of a project is approximated as a straight line, with its slope taken as equal to the expenditure per time duration unit of a project. Direct costs fall when the time duration increases, however total indirect costs rise when the time duration of a project increases.

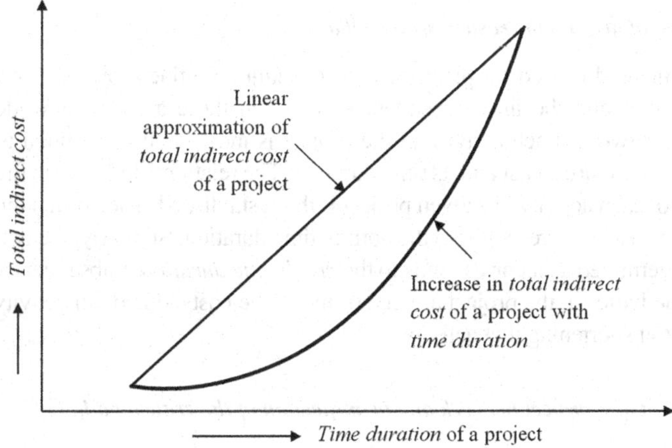

Figure 7.4 Increase in the total indirect cost of a project with an increase in time duration

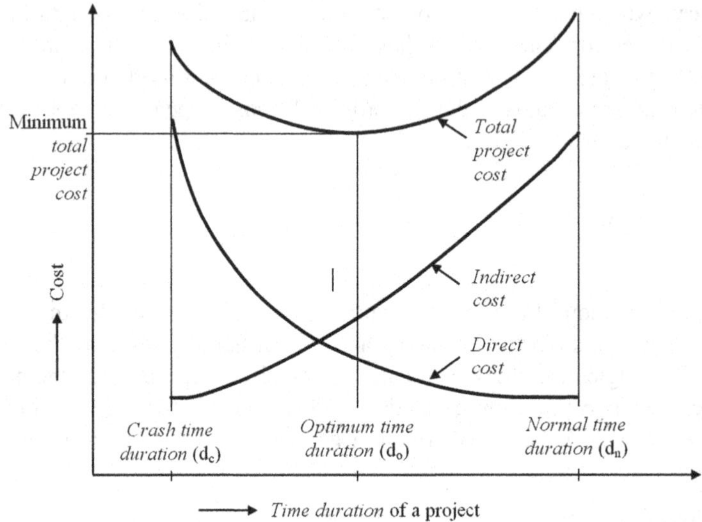

Figure 7.5 Relationship of the *direct cost, indirect cost*, and *total cost* of a project to its time duration

7.5 Total Project Cost and Optimum Time Durations

A total project cost is the sum of the direct costs of all the activities in a project and the indirect cost of a project. Figure 7.5 shows the direct cost curve, indirect cost curve, and total project cost curve of a project. The minimal total project cost is reached in a span of time called the *optimum time duration* (d_o). The project cost corresponding to the *optimum time duration* is the minimum cost for the execution of a project. With a further increase in the project time duration, the total project cost will increase, while with a further decrease in the project time duration, the total project cost will again increase. The decrease in the project time duration to the minimum time duration required to execute a project attains what is called the *project crash time duration* (d_c). At the project crash time duration project costs reach their highest, and this is called the *project crash cost* (C_c).

7.6 The Process of Cost Optimization

Step 1: ***Calculation of direct and crash costs for all activities***

The critical path method is used for planning and scheduling routine projects. A project is divided into various activities and the inter-dependencies between these activities are identified to help develop a project network. Each activity in the project is individually examined to determine the relationship between its direct cost and its time duration. The relationship between the direct cost and the time duration of each activity in a given project is thus established. The normal direct cost of each activity is determined as it corresponds to the normal time duration. Similarly, the crash direct cost of each activity is determined as it corresponds to the *crash time duration.* Subsequently, the cost slope values of all the activities in the project are determined. The cost slope of an activity is the cost per time duration unit of shortening that activity.

Step 2: ***Development of a project network and identification of the critical path***

A logical network is developed using the method previously discussed with regard to CPM. The critical path length of the network is determined using the normal time durations of the various activities therein. The normal time duration that a project takes for to be completed is the sum of the normal time durations of all the activities that lie along the critical path.

Step 3: ***Determination of the normal direct cost, indirect cost, and total cost of a project***

The normal direct costs corresponding to the normal time durations of all the activities in a project are added to determine the normal direct cost of a project. The indirect cost is not estimated for each activity individually. It is estimated for a project as a whole. Thus, indirect cost is estimated per time duration unit. Finally, the total project cost is determined by adding the normal direct costs of all the activities and the indirect cost of the project.

Step 4: ***Production of a time-scaled version of the network***

The time-scaled version of a network is drawn. The length of the arrow representing an activity in the time-scaled version of a network is equal to its normal time duration. The available crash time duration and crashing cost per time duration unity (cost slope) are written on the arrow corresponding to a given activity.

Step 5: ***Crashing of critical activities***

Crashing is assigning additional resources to an activity to finish it earlier than the normal time duration. A project's time duration is shortened by crashing the activities that lie on its critical path. When non-critical activities are crashed, the total direct cost of a project will always increase without any corresponding decrease in its indirect cost. Therefore, a decrease in total project cost is obtained by crashing the project's critical activities. Non-critical activities are not crashed because crashing them would not decrease the project time duration. The process of crashing is started by crashing the activity with the lowest cost slope value. The total project cost of a project for each stage crashed is calculated by adding the *direct cost* of its all activities, its crashing cost, and its *indirect cost*. The crashing of critical activities is continued by selecting critical activities in ascending order by their cost slope values. During the crashing process, non-critical activities can also become critical due to a reduction in the length of the critical path. In such situations, more than one critical activities can be crashed at once. The crashing process is continued until a stage is reached beyond which no further shortening of the time durations of critical activities is possible. The total project cost for each stage crashed is then calculated.

Step 6: ***Determination of the optimum time duration of a project***

Direct cost, indirect cost, and total project cost are calculated for each stage crashed. The curves between the direct cost and the time duration, the indirect cost and the time duration, and the total project cost and the time duration are plotted. The curve between the direct cost and the time duration will show that costs go on increasing as the project time duration is further reduced. The curve between the indirect cost and the time duration shows that costs go on increasing as the time duration of a project is further increased. The total project cost is minimal at the optimum time duration. If the time duration decreases or increases beyond the optimum time duration, the total project cost increases. It may not always be profitable to crash all critical activities to their fullest crashing time duration. The maximum time duration that a project takes for its completion is the sum of the normal time durations of each activity that lies along the critical path. The minimum time duration that a project takes for its completion is greater than or equal to the sum of the crashed time durations of all the activities that lie along the critical path.

7.7 Implementation

The implementation of the project crashing process through the steps described above is discussed below with reference to a very simple project consisting of only two activities.

Step 1: ***Calculation of direct and crash costs of all activities***

Assume a sample project has two activities: activity 1-2 and activity 2-3. The normal direct cost of each activity, corresponding to its normal time duration, and the crash direct cost of each activity, corresponding to its crash time duration, are given in Table 7.1. The *cost slope* value of each activity is determined as below.

The cost slope value of activity 1-2 is

$$Cost\ slope_{1-2} = \frac{13000 - 10000}{10 - 7} = \$1000$$

The cost slope value of activity 2-3 is:

$$Cost\ slope_{2-3} = \frac{9000 - 8000}{8 - 6} = \$500$$

Step 2: ***Development of a network and identification of the critical path***

A logical network for the project is shown in Figure 7.6. The critical path length of the network, determined using the normal time durations of its various activities, is:

Critical path length = 10 + 8 = 18 days

Step 3: ***Determination of the normal direct cost, indirect cost, and total cost of the project***

The normal direct costs corresponding to the normal time durations of all the activities in a project are added to determine the normal direct cost of the project. The normal direct costs of activities 1-2 and 2-3, corresponding to their normal time durations, are \$10000 and \$8000. Therefore, the normal direct cost of the project is:

Normal direct cost of the project = \$10000 + \$8000 = \$18000

Table 7.1 Normal time durations, normal direct costs, crash time durations, and crash direct costs of the various activities in the project

Activities	*Normal time durations (in days)*	*Normal direct costs (in \$)*	*Crash time durations (days)*	*Crash direct Costs (in \$)*
1-2	10	10000	7	13000
2-3	8	8000	6	9000

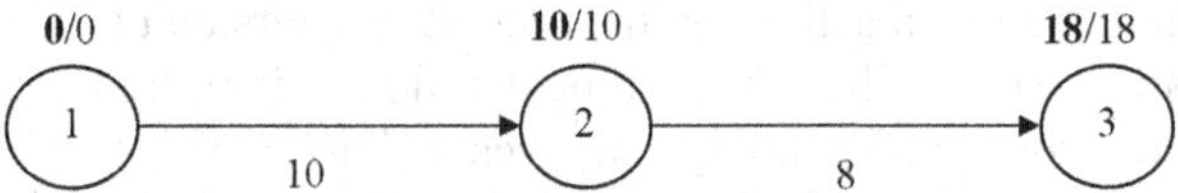

Figure 7.6 Network of the sample project

The indirect cost of the project per time duration unit is then estimated. In the present case, it is assumed to be \$900 per day. The indirect cost of the project over 18 days is:

Indirect cost of the project = \$900 × 18 = \$16200

The total project cost is determined by adding together the normal direct costs of all the activities and the indirect cost of the project. In this case, the total cost of the project is:

Total cost of project = \$18000 + \$16200 = \$34200

Step 4: ***Production of a time-scaled version of the network***

Figure 7.7 shows a time-scaled version of the network in which the normal time duration of each activity have been taken to determine the normal time duration of the project. The length

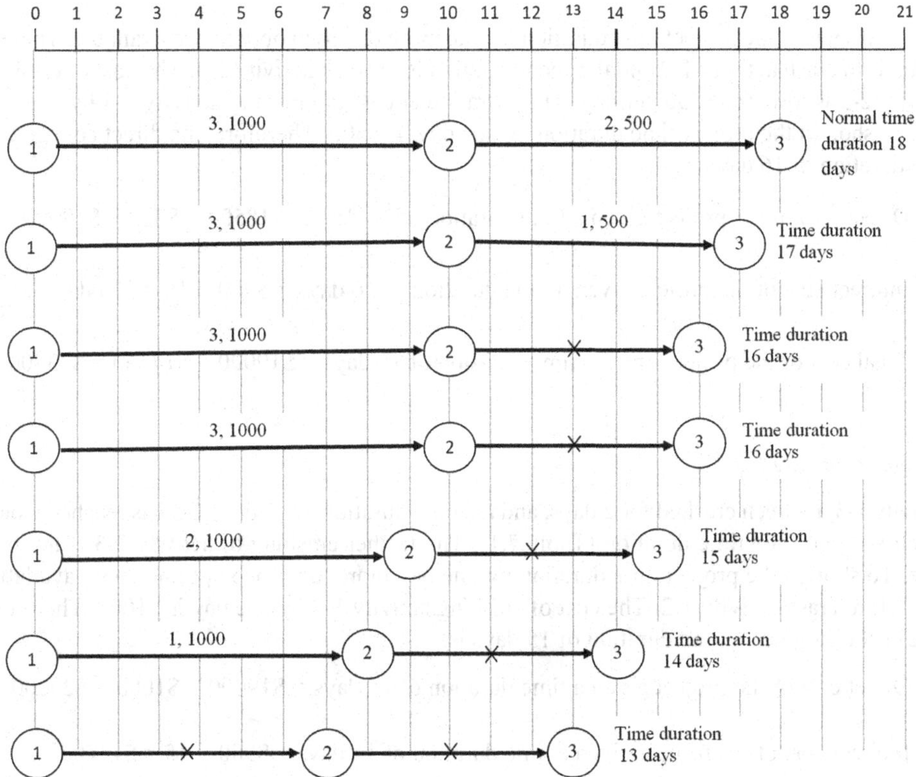

Figure 7.7 Time scaled version of the sample network

of the arrow representing a given activity in the time-scaled version of the network is equal to its normal time duration. The available crash time duration and crashing rate per time duration unit are written on the arrow corresponding to a given activity.

Step 5: *Crashing of critical activities*

Crashing of activity 2-3

A normal project duration is shortened by crashing the activities that lie on the critical path. The process of crashing starts at the activity with the lowest cost slope value. To shorten the time duration of the project by one day, two options are available. The first option is crashing activity 1-2 and the second option is crashing activity 2-3. The cost of crashing activity 1-2 by one day is $1000 and the cost of crashing activity 2-3 by one day is $500. The cost of crashing activity 2-3 is lower at $500 per day. Thus, the extra direct cost of crashing activity 2-3 by one day, so as to shorten the project time duration by one day, is $500. Therefore, the direct cost given a time duration of 17 days is:

Direct cost of the project given a time duration of 17 days = $18000 + $500 = $18500

Indirect cost of the project given a time duration of 17 days = $900 × 17 = $15300

Total cost of the project given a time duration of 17 days = $18500 + $15300 = $33800

Again, to shorten the project time duration by one more day, two options are available. The first option is to crash activity 1-2 and the second option is to crash activity 2-3. The cost of crashing activity 2-3 is lower at $500 per day. The extra direct cost of crashing activity 2-3 by one day, so as to shorten the project time duration by one day, is $500. Therefore, the direct cost given a time duration of 16 days is:

Direct cost of the project given a time duration of 16 days = $18500 + $500 = $19000

Indirect cost of the project given a time duration of 16 days = $900 × 16 = $14400

Total cost of the project given a time duration of 16 days = $19000 + $14400 = $33400

Crashing of activity 1-2

Activity 2-3 has been crashed for 2 days, and the time duration of activity 2-3 has reached 6 days which is its crashed time duration (Table 7.1). The further crashing of activity 2-3 is not possible. To shorten the project time duration by one day more, only one option is now available, which is to crash activity 1-2. The cost of crashing activity 1-2 by one day is $1000. Therefore, the direct cost given a time duration of 15 days is:

Direct cost of the project given a time duration of 15 days = $19000 + $1000 = $20000

Indirect cost of the project given a time duration of 15 days = $900 × 15 = $13500

Total cost of the project given a time duration of 15 days = $20000 + $13500 = $33500

To shorten the project time duration by one more day only one option is available, which is to crash activity 1-2. The cost of crashing activity 1-2 by one day is $1000. Therefore, the direct cost of the project given a time duration of 14 days is:

Direct cost of the project given a time duration of 14 days = $20000 + $1000 = $21000

Indirect cost of the project given a time duration of 14 days = $900 × 14 = $12600

Total cost of the project given a time duration of 14 days = $22000 + $12600 = $33600

To shorten the project time duration by one more day, only one option is available, which is to crash activity 1-2. Therefore, the direct cost of the project given a time duration of 13 days is:

Direct cost of the project given a time duration of 13 days = $21000 + $1000 = $22000

Indirect cost of the project given a time duration of 13 days = $900 × 13 = $11700

Total cost of the project given a time duration of 13 days = $22000 + $11700 = $33700

Activity 1-2 has been crashed for 3 days. The time duration of activity 1-2 has reached 7 days which is its crashed time duration. The further crashing of activity 1-2 is not possible. The critical path length of the network using the crash time durations of its various activities is:

Critical path length = 7 + 6 = 13 days

Step 6: ***Determination of the optimum time duration of a project.***

Figure 7.8 shows the cost-time curves for the direct cost, indirect cost, and total project cost. From Table 7.2 and Figure 7.8, it is clear that the total project cost is at its minimum with a time duration of 16 days. Thus, the optimum time duration of the project is 16 days and the minimum total project cost corresponding to it is $33400.

7.8 Conclusion

Sometimes additional resources are used to shorten the time duration of a project. The shortening of the time duration of a project through the addition of extra resources is called project crashing. Project crashing requires the use of two time durations for each activity. These two time durations are the normal time duration and the shortest possible time duration. The shortest possible time duration is the time beyond which an activity cannot be shortened by increasing its resources. The time duration of a project is shortened by shortening the normal time durations of the activities that lie along its critical path. The objective of project crashing is determining the optimum time duration of a project at which the total project cost is minimal. The schedule finalized in CPM is therefore modified in the project crashing process so as to obtain a schedule corresponding to the minimum total project cost.

Example 7.1: Determine the optimum time duration and cost of a project with the set of data given in Table 7.3; take the indirect cost to be $3000 per day.

Solution:
Step 1: **Calculation of direct and crash costs for all activities**
The sample project has three activities: activities 1-2, 1-3, and 2-3. The normal direct cost of each activity, corresponding to their normal time duration, and the crash direct cost

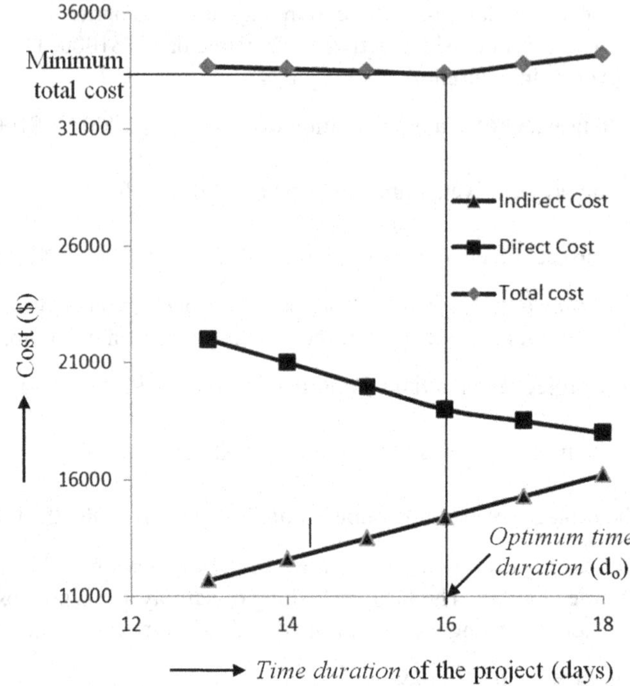

Figure 7.8 Cost-time curves for the direct cost, indirect cost, and total cost of the project

Table 7.2 Direct cost, indirect cost, and total cost of the project

Time duration (in days)	*Direct cost of project (in $)*	*Indirect cost of project (in $)*	*Total cost of project (in $)*
18	18000	16200	34200
17	18500	15300	33800
16	19000	14400	**33400**
15	20000	13500	33500
14	21000	12600	33600
13	22000	11700	33700

for each activity, corresponding to their crash time duration, are given in Table 7.3. The values of the available crash time durations and cost slopes of all activities are listed in the table.

Step 2: ***Development of a network and identification of the critical path***

A logical network for the project is shown in Figure 7.9. The critical path length of the network, using the normal time durations of its various activities, is:

Critical path length = 6 + 4 = 10 days

Step 3: ***Determination of the*** **normal direct cost, indirect cost,** ***and*** **total cost** ***of the project***

The normal direct costs corresponding to the normal time durations of all the activities in the project are added together to determine the normal direct cost of a project.

Table 7.3 Normal time durations, normal direct costs, crash time durations, and crash direct costs of the various activities in the project

Activities	*Normal time durations (in days)*	*Normal direct costs (in $)*	*Crash time durations (days)*	*Crash direct Costs (in $)*	*Available crash time durations (days)*	*Cost slope (in $)*
1-2	6	4000	3	16000	3	4000
1-3	8	5000	5	8000	3	1000
2-3	4	6000	1	12000	3	2000

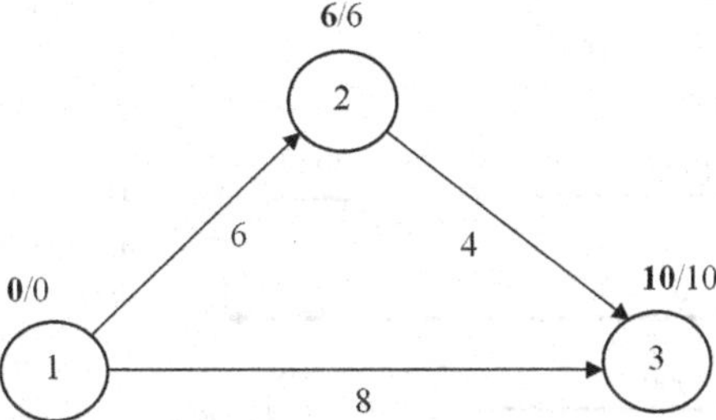

Figure 7.9 Logical network of the sample project

Normal direct cost of the project = \$4000 + \$5000 + \$6000 = \$15000

The indirect cost of the project per time duration unit is \$3000 per day. The total indirect cost of the project over 10 days is:

Indirect cost of the project = \$3000 × 10 = \$30000

The total project cost is determined by adding together the normal direct costs of all the activities and the indirect cost of the project.

Total project cost = \$15000 + \$30000 = \$45000

Step 4: *Production of a time-scaled version of the network*

Figure 7.10 shows a time-scaled version of the network in which the normal time duration of each activity has been taken to determine the normal time duration of the project.

Step 5: *Crashing of critical activities*

Crashing of activity 2-3

To shorten the time duration of the project by one day, two possible options are available. The first option is crashing activity 1-2 and the second option is crashing activity 2-3. The cost of crashing activity 1-2 by one day is \$4000 and the cost of crashing activity 2-3 by one day is \$2000. The cost of crashing activity 2-3 is lower. Thus, the extra direct cost of crashing activity 2-3 by one day, so as to shorten the time duration of the project by one day, is \$2000. Therefore, the direct cost, given a time duration of 9 days, is:

Direct cost of the project given a time duration of 9 days = \$15000 + \$2000 = \$17000

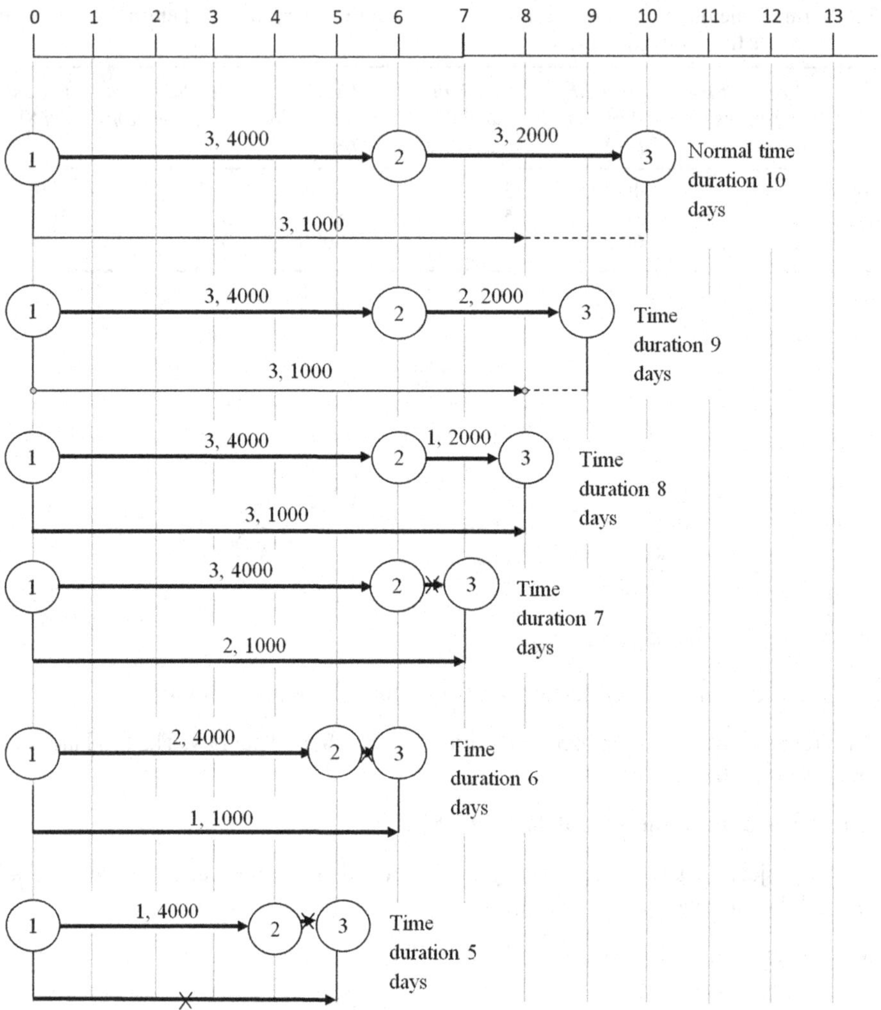

Figure 7.10 Time scaled version of the sample network

Indirect cost of the project given a time duration of 9 days = $3000 × 9 = $27000

Total cost of the project given a time duration of 9 days = $17000 + $27000 = $44000

Again, to shorten the project time duration by one more day, two options are available. The first option is to crash activity 1-2 and the second option is to crash activity 2-3. The cost of crashing activity 2-3 is lower. Therefore, the direct cost given a the time duration of 8 days is:

Direct cost of the project given a time duration of 8 days = $17000 + $2000 = $19000

Indirect cost of the project given a time duration of 8 days = $3000 × 8 = $24000

Total cost of the project given a time duration of 8 days = $19000 + $24000 = $43000

Simultaneous crashing of activity 2-3 and activity 1-3

At this stage of crashing, all activities are critical. To shorten the project time duration by one more day, two possible options are available. The first option is crashing activity 1-2 and activity 1-3 together; the unit cost of this crashing is:

Direct cost of crashing activity 1-2 and activity 1-3 simultaneously = \$4000 + \$1000 = \$5000

The second alternative is crashing activity 2-3 and activity 1-3 simultaneously; the unit cost of this crashing is:

Direct cost of crashing of activity 2-3 and activity 1-3 simultaneously = \$2000 + \$1000 = \$3000

The cost of crashing of activity 2-3 and activity 1-3 together is lower. Therefore, the direct cost given a time duration of 7 days is:

Direct cost of the project given a time duration of 7 days = \$19000 + \$3000 = \$22000

Indirect cost of the project given a time duration of 7 days = \$3000 × 7 = \$21000

Total cost of the project given a time duration of 7 days = \$22000 + \$21000 = \$43000

Crashing of activity 1-2 and activity 1-3 together

Activity 2-3 has been crashed for 3 days. The time duration of activity 2-3 has reached one day, this being the *crashed time duration* of activity 2-3. The further crashing of activity 2-3 is not possible. All activities are critical; to shorten the project time duration by one more day, only one option is available – crashing activity 1-2 and activity 1-3 together. The unit cost of this crashing is:

Direct cost of crashing activity 1-2 and activity 1-3 simultaneously = \$4000 + \$1000 = \$5000

Direct cost of the project given a time duration of 6 days = \$22000 + \$5000 = \$27000

Indirect cost of the project given a time duration of 6 days = \$3000 × 6 = \$18000

Total cost of the project given a time duration of 6 days = \$27000 + \$18000 = \$45000

To shorten the project time duration by one more day, only one option is available. This option is crashing activity 1-2 and activity 1-3 simultaneously; the unit cost of this crashing is:

Direct cost of crashing activity 1-2 and activity 1-3 simultaneously = \$5000

Direct cost of the project given a time duration of 5 days = \$27000 + \$5000 = \$32000

Indirect cost of the project given a time duration of 5 days = \$3000 ×5 = \$15000

Total cost of the project given a time duration of 5 days = \$32000 + \$15000 = \$47000

Activity 1-3 has been crashed for 3 days. The time duration of activity 1-3 has reached 5 days which is its crashed time duration. Activity 1-2 has been crashed by 2 days; activity 1-2 can still be crashed for one day, but its time duration has reached 5 days, which is its *crashed time duration*. Therefore, crashing activity 1-2 by one more day is not possible. The critical path length of the network, calculated using crash time durations, is 5 days.

Step 6: ***Determination of the optimum time duration of the project.***

From Table 7.4 and Figure 7.11, it is clear that the total cost is at its minimum with project duration of 7 days. The optimum time duration of the project is 7 days and the minimum total project cost corresponding to it is \$43000.

Example 7.2: Determine the optimum time duration and cost of a project with the set of data given in Table 7.5. Take the indirect cost to be \$2500 per day.

Solution:
Step 1: Calculation of direct and crash costs for all activities

The sample project has five activities, as shown in Figure 7.12. The normal direct cost, normal time duration, crash direct cost, and crash time duration of each activity are given in Table 7.6. The available crash time duration and cost slope values of all the activities are listed in the same table.

Step 2: ***Development of a network and identification of the critical path***

A logical network for the project is shown in Figure 7.12; the critical path length of the network, calculated using the normal time durations of its various activities, is 14 days.

Step 3: ***Determination of normal direct cost, indirect cost, and total project cost***

The normal direct costs, corresponding to the normal time durations of all the activities in the project are added together to determine the normal direct cost of the project.

The normal direct cost of the project = \$28000

Table 7.4 Direct cost, indirect cost, and total cost of the project

Time duration (in days)	*Direct cost of project (in \$)*	*Indirect cost of project (in \$)*	*Total cost of project (in \$)*
10	15000	30000	45000
9	17000	27000	44000
8	19000	24000	43000
7	22000	21000	**43000**
6	27000	18000	45000
5	32000	15000	47000

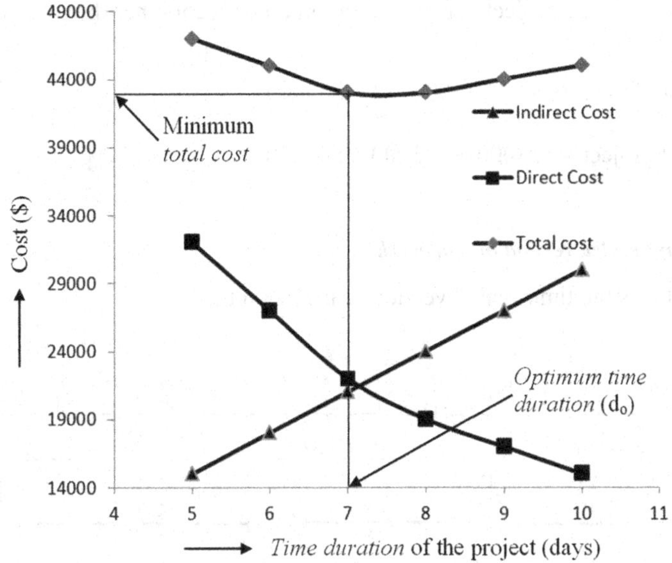

Figure 7.11 Cost-time curves for the direct cost, indirect cost, and total cost of the project

Table 7.5 Normal time duration, normal direct cost, crash time duration, and crash direct cost of the various activities in the project

Activities	*Normal time durations (in days)*	*Normal direct costs (in $)*	*Crash time durations (days)*	*Crash direct Costs (in $)*	*Available crash time durations (days)*	*Cost slope (in $)*
1-2	5	8000	2	14000	3	2000
1-3	7	5000	4	8000	3	1000
2-3	4	6000	2	7000	2	500
2-4	6	3000	3	12000	3	3000
3-4	5	6000	3	14000	2	4000

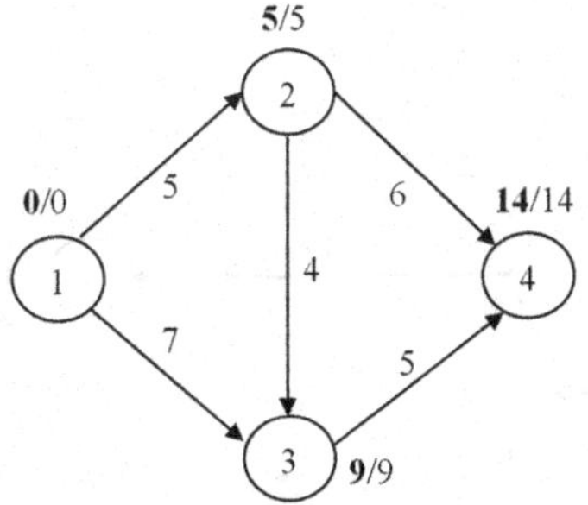

Figure 7.12 Logical network of the sample project

The indirect cost of the project per time duration unit is \$2500 per day. The indirect cost of the project over 14 days is:

Indirect cost of the project = \$2500 × 14 = \$35000

Total cost of project = \$28000 + \$35000 = \$63000

Step 4: *Draw a time scaled version of a network*

Figure 7.13(a) shows the time-scaled version of the network.

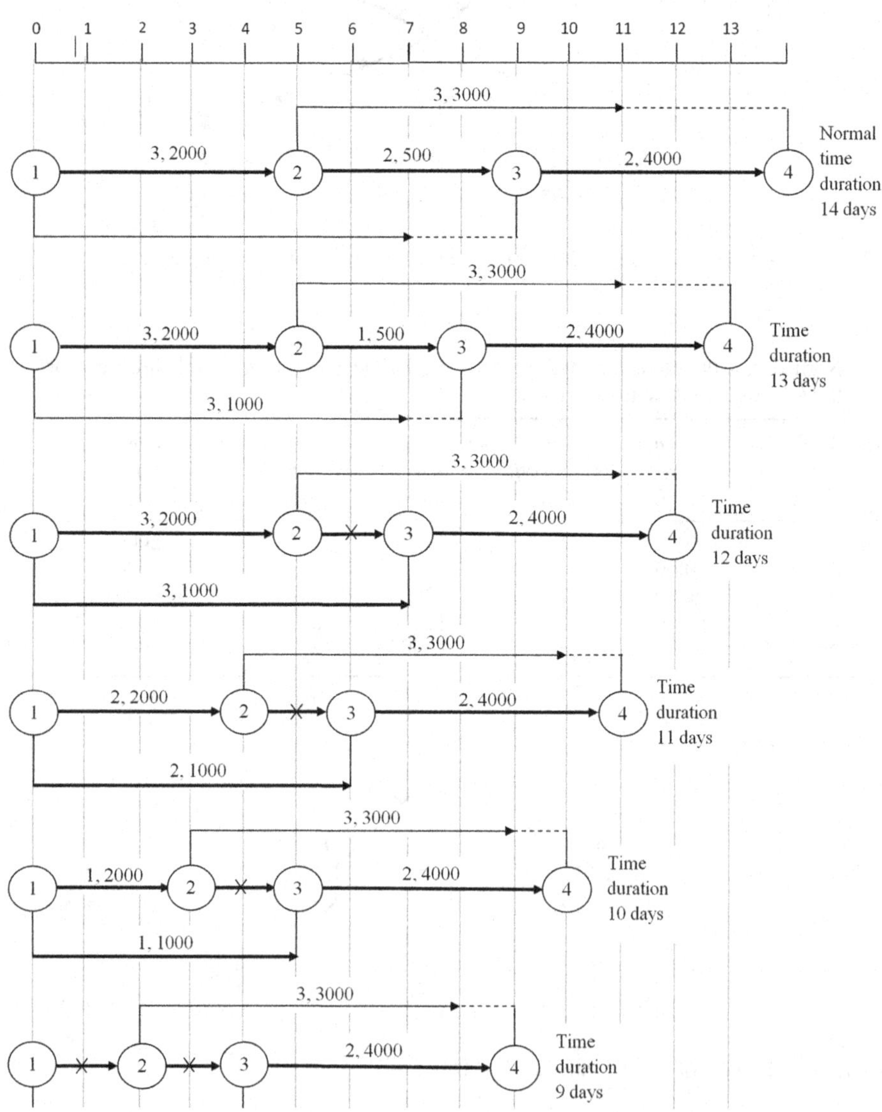

Figure 7.13 (a) Time scaled version of the sample network

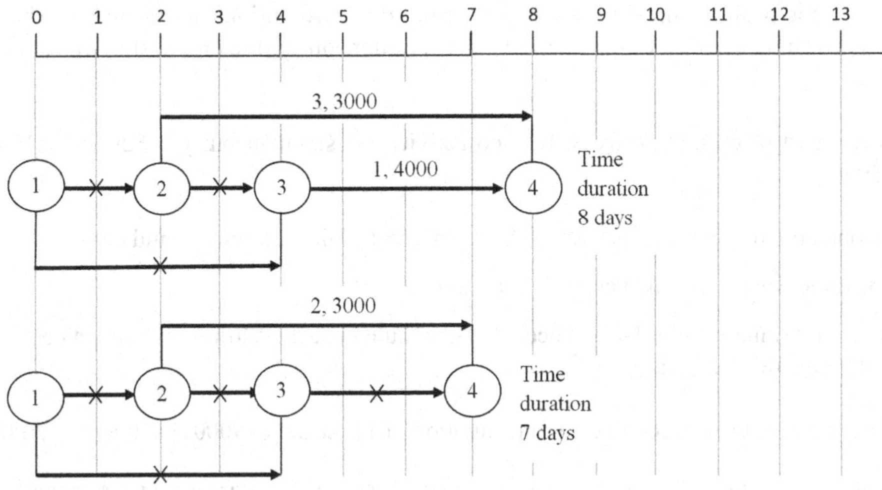

Figure 7.13 (b) Time scaled version of the sample network

Step 5: *Crashing of critical activities*

Crashing of activity 2-3

To shorten the time duration of the project by one day three possible options are available. The first option is crashing activity 1-2, the second is crashing activity 2-3, and the third is crashing activity 3-4. The cost of crashing activity 1-2 by one day is $2000, the cost of crashing activity 2-3 by one day is $500, and the cost of crashing activity 3-4 by one day is $4000. The cost of crashing activity 2-3 is lower. Thus, the extra direct cost of crashing activity 2-3 by one day, so as to shorten the time duration of the project by one day, is $500. The direct cost given a the time duration of 13 days is:

Direct cost of the project given a time duration of 13 days = $28000 + $500 = $28500

Indirect cost of the project given a time duration of 13 days = $2500 × 13 = $32500

Total cost of the project given a time duration of 13 days = $28500 + $32500 = $61000

Again, to shorten the project time duration by one more day three options alternatives are available. Between these options, the cost of crashing of activity 2-3 is the lowest. Thus, the direct cost given a time duration of 12 days is:

Direct cost of the project given a time duration of 12 days = $28500 + $500 = $29000

Indirect cost of the project given a time duration of 12 days = $2500 × 12 = $30000

Total cost of the project given a time duration of 12 days = $29000 + $30000 = $59000

Simultaneous crashing of activity 1-2 and activity 1-3

Activity 2-3 has been crashed for 2 days. The time duration of activity 2-3 has reached 2 days which is its *crashed time duration.* The further crashing of activity 2-3 is not possible.

To shorten the project time duration by one more day, two options are available. The first option is crashing activity 1-2 and activity 1-3 simultaneously; the unit cost of this crashing would be:

Direct cost of crashing activity 1-2 and activity 1-3 simultaneously = \$2000 + \$1000 = \$3000

The second option is crashing activity 3-4, the unit cost of this crashing would be:

The direct cost of crashing activity 3-4 = \$4000

The cost of crashing activity 1-2 and activity 1-3 simultaneously is lower. The direct cost given a time duration of 11 days is:

Direct cost of the project given a time duration of 11 days = \$29000 + \$3000 = \$32000

Indirect cost of the project given a time duration of 11 days = \$2500 × 11 = \$27500

Total cost of the project given a time duration of 11 days = \$32000 + \$27500 = \$59500

To shorten the project time duration again by one more day, two options are available. The first option is crashing activity 1-2 and activity 1-3 simultaneously; the unit cost of this crashing would be \$3000. The second alternative is crashing activity 3-4; the unit cost of this crashing would be \$4000. The cost of crashing activity 1-2 and activity 1-3 simultaneously is lower. The direct cost given a time duration of 10 days is:

Direct cost of the project given a time duration of 10 days = \$32000 + \$3000 = \$35000

Indirect cost of the project given a time duration of 10 days = \$2500 × 10 = \$25000

Total cost of the project given a time duration of 10 days = \$35000 + \$25000 = \$60000

To shorten the project time duration again by one more day, two options are available. The cost of crashing activity 1-2 and activity 1-3 simultaneously is lower. Thus, the direct cost given a time duration of 9 days is:

Direct cost of the project given a time duration of 9 days = \$35000 + \$3000 = \$38000

Indirect cost of the project given a time duration of 9 days = \$2500 × 9 = \$22500

Total cost of the project given a time duration of 9 days = \$38000 + \$22500 = \$60500

Crashing of activity 3-4

Activity 1-2 has been crashed for 3 days. The time duration of activity 1-2 has reached 2 days, which is its crashed time duration. The further crashing of activity 1-2 is not possible. Similarly, activity 2-3 has been crashed for 3 days. The time duration of activity 2-3 has reached 4 days, which is its crashed time duration. The further crashing of activity 2-3 is not possible. To shorten the project time duration by one more day only one option is available, the crashing of activity 3-4. The unit cost of this crashing would be:

Direct cost of crashing activity 3-4 = \$4000

Direct cost of the project given a time duration of 8 days = \$38000 + \$4000 = \$42000

Indirect cost of the project given a time duration of 8 days = \$2500 × 8 = \$20000

Total cost of the project given a time duration of 8 days = \$42000 + \$20000 = \$62000

Simultaneous crashing of activity 3-4 and activity 2-4

At this stage of crashing, all the activities in the project are critical. To shorten the project time duration by one more day, only one option is available. This option is crashing activity 3-4 and activity 2-4 simultaneously; the unit cost of this crashing would be:

Direct cost of crashing activity 3-4 and activity 2-4 simultaneously = \$4000 + \$3000 = \$7000

Direct cost of the project given a time duration of 7 days = \$42000 + \$7000 = \$49000

Indirect cost of the project given a time duration of 7 days = \$2500 × 7 = \$17500

Total cost of the project given a time duration of 7 days = \$49000 + \$17500 = \$66500

Activity 3-4 has been crashed for 2 days. The time duration of activity 3-4 has reached 3 days which is its crashed time duration. Further, activity 2-4 has been crashed by 1 day; activity 2-4 can still be crashed for two days, but the time duration of activity 2-4 has reached 3 days, which is its *crashed time duration*. Therefore, crashing activity 2-4 by two more days is not possible. The minimum critical path length of the network, calculated using crash time durations, is 7 days.

Step 6: ***Determination of the optimum time duration of the project.***

Figure 7.14 shows the cost-time curves of the direct cost, indirect cost, and total project cost. The total project cost is minimal at a project time duration of 12 days. *The optimum time duration* of the project is 12 days and the minimum total cost corresponding to it is \$59000.

Exercises

Question 7.1: Determine the optimum time duration and the corresponding cost for the set of data given in Table E7.1; take the indirect cost to be \$600 per day.

Answer: The optimum time duration of the project is 18 days and the minimum total project cost corresponding to it is \$283800. Detail has been provided in Table E7.2.

Question 7.2: Determine the optimum time duration and the corresponding cost for the set of data given in Table E7.3; take the indirect cost to be \$1200 per day.

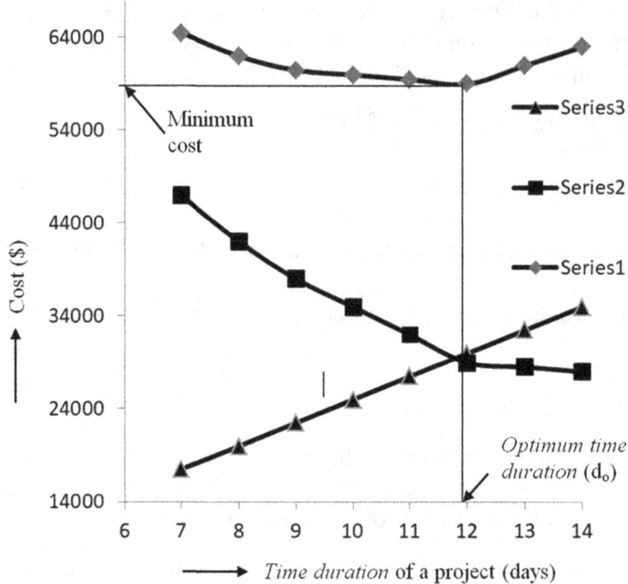

Figure 7.14 Cost-time curves for the direct cost, indirect cost, and total cost of the project

Table 7.6 Direct cost, indirect cost, and total cost of the project

Time duration (in days)	*Direct cost of project (in \$)*	*Indirect cost of project (in \$)*	*Total cost of project (in \$)*
14	28000	35000	63000
13	28500	32500	61000
12	29000	30000	**59000**
11	32000	27500	59500
10	35000	25000	60000
9	38000	22500	60500
8	42000	20000	62000
7	49000	17500	66500

Table E7.1 Normal time durations, normal direct costs, crash time durations, and crash direct costs of the various activities in the project

Activities	*Normal time durations (in days)*	*Normal direct costs (in \$)*	*Crash time durations (days)*	*Crash direct Costs (in \$)*
1-2	12	8000	8	12000
2-3	8	4000	6	5000

Answer: The optimum time duration of the project is 10 days and the minimum total cost corresponding to it is \$40000. Details are provided in Table E7.4.

Question 7.3: Determine the optimum time duration and cost of the project with the set of data given in Table E7.5; taking the indirect cost to be \$325 per day.

Table E7.2 Direct cost, indirect cost, and total project cost of the project

Time duration (in days)	*Direct cost of project (in $)*	*Indirect cost of project (in $)*	*Total cost of project (in $)*
20	12000	12000	24000
19	12500	11400	23900
18	**13000**	**10800**	**23800**
17	14000	10200	24200
16	15000	9600	24600
15	16000	9000	25000
14	17000	8400	25400

Table E7.3 Normal time durations, normal direct costs, crash time durations, and crash direct costs of the various activities in the project

Activities	*Normal time durations (in days)*	*Normal direct costs (in $)*	*Crash time durations (days)*	*Crash direct Costs (in $)*
1-2	8	12000	4	16000
1-3	10	6000	8	7000
2-3	4	8000	1	14000

Table E7.4 Direct cost, indirect cost, and total cost of the project

Time duration (in days)	*Direct cost of project (in $)*	*Indirect cost of project (in $)*	*Total cost of project (in $)*
12	26000	14400	40400
11	27000	13200	40200
10	**28000**	**12000**	**40000**
9	29500	10800	40300
8	31000	9600	40600

Table E7.5 Normal time duration, normal direct cost, crash time duration, and crash direct cost of the various activities in the project

Activities	*Normal time durations (in days)*	*Normal direct costs (in $)*	*Crash time durations (days)*	*Crash direct Costs (in $)*
1-2	5	600	4	800
2-3	3	400	1	600
2-4	8	900	5	1200
2-5	4	600	2	1200
3-5	4	500	3	700
4-5	2	300	1	500
5-6	3	300	2	600

Table E7.6 Direct cost, indirect cost, and total cost of the project

Time duration (in days)	*Indirect cost of project (in $)*	*Direct cost of project (in $)*	*Total cost of project (in $)*
18	5850	3600	9900
17	5525	3700	9225
16	5200	3800	9000
15	4875	3900	8775
14	4550	4100	8650
13	4225	4400	8625
12	**3900**	**4700**	**8600**

Table E7.7 Normal time duration, normal direct cost, crash time duration, and crash direct cost of the various activities in the project

Activities	*Normal time durations (in days)*	*Normal direct costs (in $)*	*Crash time durations (days)*	*Crash direct Costs (in $)*
1-2	3	4000	2	5000
1-3	4	6000	2	12000
2-4	4	4000	2	8000
3-4	5	1000	1	5000
3-5	6	5000	4	7000
4-5	4	12000	1	24000

Table E7.8 Direct cost, indirect cost, and total cost of the project

Time duration (in days)	*Direct cost of project (in $)*	*Indirect cost of project (in $)*	*Total cost of project (in $)*
13	32000	16250	48250
12	33000	15000	48000
11	**34000**	**13750**	**47750**
10	36000	12500	48500
9	40000	11250	51250
8	45000	10000	55000
7	50000	8750	58750

Answer: The optimum time duration of the project is 12 days and the minimum total project cost corresponding to it is $8600. Detail has been provided in Table E7.6.

Question 7.4: Determine the optimum time duration and cost of the project with the set of data given in Table E7.7; taking the indirect cost to be $1250 per day.

Answer: The optimum time duration of the project is 11 days and the minimum total project cost corresponding to it is $47750. Detail has been provided in Table E7.8.

8 Limited Time Resource Scheduling

8.1 Learning Objectives

After the completion of this chapter, readers will be able to:

- Draw histograms for different resource requirements,
- Determine the total usage of a resource over the time duration of a project, and
- Smooth a resource requirement histogram.

8.2 Introduction

The resources used to execute of the various activities in a project can include manpower, materials, money, machines, etc. *Resource allocation* is the term used for the assignment of the required resources to a project activity to help ensure that it is finished within the allocated time duration. During the planning stage of a project, one's available resources are assumed to be either unlimited or limited. When a project is planned with the assumption that one's resources are unlimited, it is assumed that the different types of necessary resource are available as and when they are required. In reality, one's available resources are not unlimited and are not available as and when they are required. Normally resources are not at rest; their availability is not always a simple matter. As a result, the various activities in a project are planned or scheduled according to the availability of the various types of resource required. In this planning model, one's available resources are considered limited. In such cases, projects are scheduled so as to ensure the availability of the various types of resources required. Projects are scheduled in such a way that their various resource requirements are kept almost uniform throughout the time duration of a project, because in practice, variations in the resource requirements become difficult to manage. In such cases, one's various resource requirements must not exceed one's available resources. The concepts of unlimited or limited resource availability are applied to resources that are generally hired, like laborers, equipment, etc. In general, materials requirements are not made uniform.

8.3 Resource Profiles

A project is divided into its various activities, an execution sequence is decided on, and a project network is prepared. The starting and finish times of all the activities are calculated on an early or late start basis. Each activity in a project requires a different type and different number of resources for its execution, depending upon its time duration. A time-scaled version of the project network is developed, and daily requirements for different types of resources are generated.

DOI: 10.1201/9781003428992-8

Different resource requirements do not normally remain uniform throughout the time duration of a project. A graph showing the relationship between a particular resource requirement and the time duration of a project is thus prepared. This graphical representation of the relationship between a particular resource requirement and the time duration of a project is called a resource profile. A resource profile is generated for each type of resource necessary to execute a project. The variations in the resource profiles of different types of resources are different.

Consider a sample project network, as shown in Figure 8.1. The project consists of 10 activities. Table 8.1 lists the requirements regarding a particular type of resource (manpower) needed for the execution of multiple activities in the project. The four types of time corresponding to the project's different activities are also listed in Table 8.1. A time-scaled version of the network is shown in Figure 8.2. Figure 8.3 shows the resource profile of a particular type of required resource. It is clear from the resource profile that the resource requirement is not uniform across the time duration of the project. The requirement is as high as 11 hands on days 5, 6, 11, and 12. On the other hand, the requirement is as low as 2 hands on the first four and last seven days of the project. It is highly uneconomical to keep permanently employed resources standing by to fulfill such a fluctuating requirement. When resources are employed on a temporary basis, however, the availability of resources on the required date and for the required time duration may not be a simple matter, as these resources may not be available when required.

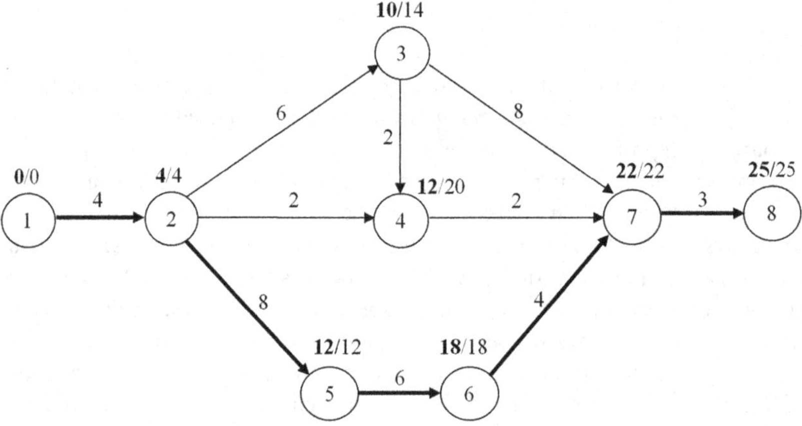

Figure 8.1 A sample project network

Table 8.1 Activities in a project and their time durations, resource requirements, times, and total floats

Activities	*Time durations*	*Resource requirements*	*EST*	*EFT*	*LST*	*LFT*	*Total floats*
1-2	4	2	0	4	0	4	0
2-3	6	3	4	10	8	14	4
2-4	2	4	4	6	18	20	14
2-5	8	4	4	12	4	12	0
3-4	2	3	10	12	18	20	8
3-7	8	4	10	18	14	22	4
4-7	2	2	12	14	20	22	8
5-6	6	2	12	18	12	18	0
6-7	4	2	18	22	18	22	0
7-8	3	2	22	25	22	25	0

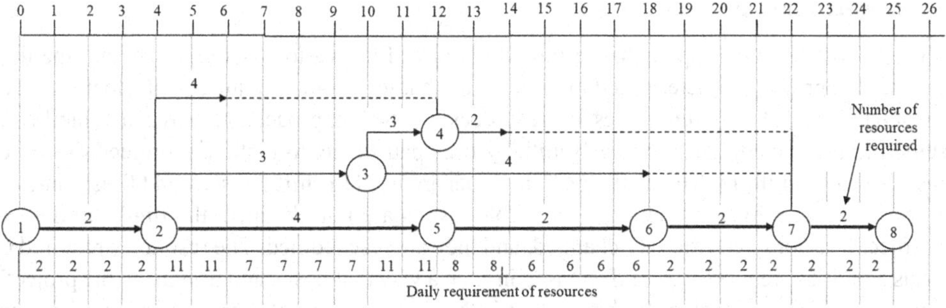

Figure 8.2 Calculation of daily resource requirements from the time scaled version of the network

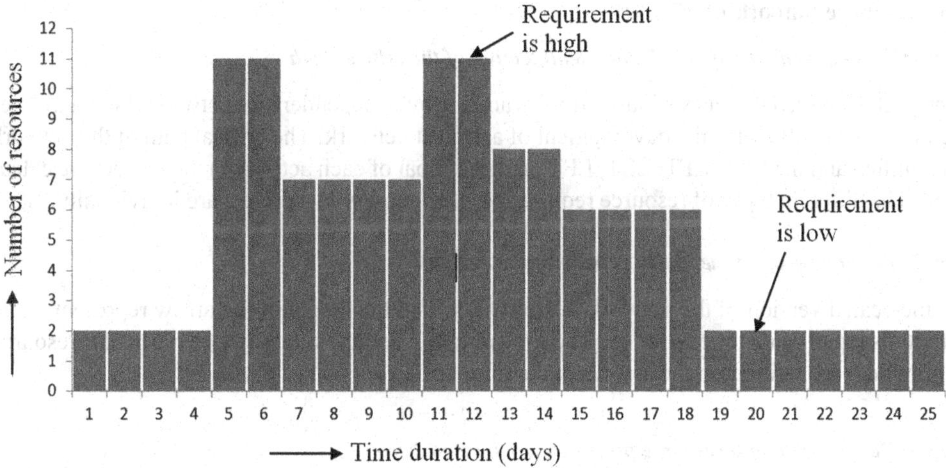

Figure 8.3 Resource profile of a particular type of resource

8.4 Resource Planning

In the planning stage, a planner plans a project in such a way that all their resource requirements remain almost uniform across a project. To keep resource demands almost uniform across a project, two types of resource planning are carried out: *resource smoothing* and *resource leveling.*

8.4.1 Resource Smoothing (Limited Time Resource Scheduling)

In the resource smoothing approach, the time duration of a project is not changed, but the start or finish times of its non-critical activities are adjusted (according to path float availability) in such a way that resource requirements remain nearly uniform. Resource availability, in this case, is assumed to be unlimited. It is assumed that resources are available as and when they are required. This is also known as *limited-time resource scheduling.* It is used a project's the time duration is limited. The objective is to complete a project within its allocated time duration while avoiding peaks and troughs in its resource profiles. It is the minimization of fluctuations in daily resource demands throughout the time duration of a project by shifting the start/finish times of its non-critical activities within their available float limits.

8.4.2 Resource Leveling

In the resource leveling approach, the time durations of the various activities in, and the time duration of, a project are determined in such a way that its resource requirements do not exceed the availabilities of different types of resources. In this approach, resource availability is assumed to be limited. This approach initially uses path floats to make the project's resource demands nearly uniform. The available path floats are used to shift the start and finish times of the project's non-critical activities; if desirable results are not obtained, the time durations of the project's various activities are changed, and they are rescheduled. The resources required by various activities are increased or decreased in such a way that peaks and troughs in the project's resources profiles are leveled. In this case, it is the time duration of a project or time durations of its various activities that are changed. The activities in a project are scheduled in such a way that demand for different resource types do not exceed their availabilities.

8.5 Resource Smoothing Process

Step 1: ***Development of a network and identification of the critical path***

A project is divided into its various activities and the inter-dependencies between these activities are identified to allow for the development of a project network. The critical path of the network is identified and the EST, EFT, LST, LFT, and path float of each activity in the project are calculated. The different types of resource required by each activity in a project are individually listed.

Step 2: ***Production of a time scaled version of the network***

A time-scaled version of the network is drawn in which the length of the arrow representing an activity is equal to its time duration. The time duration and the quantity of a particular resource requirement are written on the arrows in the time-scaled version of a network.

Step 3: ***Development of a resource profile for each resource***

A graph showing the relationship between a particular resource and the time duration of a project is prepared. The graphical representation of a particular resource requirement is called a resource profile. A resource profile is generated for each type of resource required to execute the project. The variations in the resource profiles of different types of resource are different.

Step 4: ***Resource smoothing***

Resource profiles corresponding to each type of resource required to execute a project are generated. Resource smoothing is done separately for each type of resource required to execute a project. The starting and finish times of the various activities in a project are adjusted (according to the availability of path floats) in such a way that resource requirements remain almost uniform across the project. It is assumed that resources are available as and when they are required. The objective is to complete the project within the allocated time duration while avoiding peaks and troughs in its resource profiles.

8.6 Understanding the Process

The network shown in Figure 8.4 is used to demonstrate the concept of resource smoothing. The resource requirements of the various activities in the project are given in Table 8.2. It also

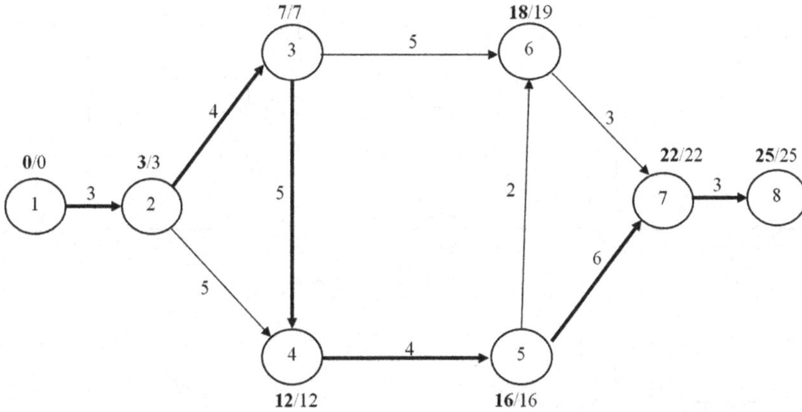

Figure 8.4 A sample project network

Table 8.2 Activities in a project and their time durations, resource requirements, times, and total floats

Activities	*Time durations*	*Resource requirements*	*EST*	*EFT*	*LST*	*LFT*	*Total floats*
1-2	3	3	0	3	0	3	0
2-3	4	2	3	7	3	7	0
2-4	5	2	3	8	7	12	4
3-4	5	2	7	12	7	12	0
3-6	5	4	7	12	14	19	7
4-5	4	2	12	16	12	16	0
5-6	2	1	16	18	17	19	1
5-7	6	1	16	22	16	22	0
6-7	3	1	18	21	19	22	1
7-8	3	1	22	25	22	25	0

gives the EST, EFT, LST, LFT, and total float values of each activity in the project. A time-scaled version of the network is shown in Figure 8.5. A time duration and a particular resource requirement are written on the arrows representing activities. It also provides a resource profile in which the resource requirements of an individual activity is shown. The resource requirement of an activity is shown using a rectangle in the resource profile. The length along the x-axis of a rectangle is equal to the time duration of an activity and the width along the y-axis is equal to the quantity of the resource required. The name of an activity and the quantity of a particular type of resource required are written in the rectangle representing that activity in the resource profile.

In the network under consideration, activities 1-2, 2-3, 3-4, 4-5, 5-7, and 7-8 are critical. These activities have path float values of zero, therefore the start and finish times of these activities must remain unchanged in the resource smoothing process. The resource requirements of these critical activities are represented in the form of rectangles in the first layer of the resource profile as shown in Figure 8.5. Activity 1-2 has a duration of 3 days, thus the length of the rectangle corresponding to activity 1-2 is 3 units. It requires a resource quantity of 3 units, therefore the width of the rectangle is 3 units. Activity 2-3 has a time duration of 4 days, therefore the length of the rectangle corresponding to activity 2-3 is 4 units. It requires a resource quantity of 2 units, therefore the width of the rectangle is 2 units. The particular resource requirements of all the project's critical and non-critical activities are represented in the same way, as shown in Figure 8.5. The daily resource requirement reaches as high as 8 units for day 8, but the requirement is as low as 1 unit for the last four days of the project.

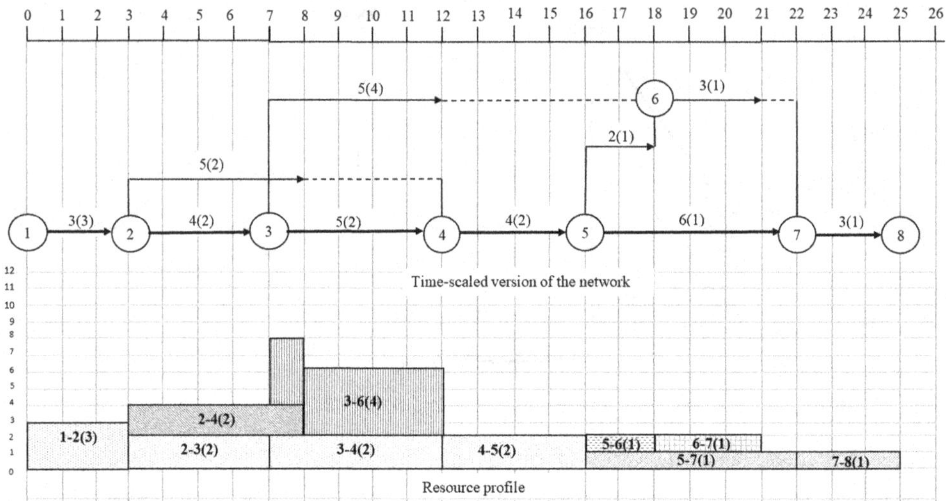

Figure 8.5 Time scaled version of the network and resource profile

The starting and finish times of the project's non-critical activities are adjusted (depending upon total float availability) in such a way that the day-to-day requirement of a particular type of resource is almost uniform across the project. This is done by shifting the rectangle representing the resource requirement of an activity in the resource profile graph within its total float limit. It is assumed that resources are available as and when they are required. The objective is to complete the project within the allocated time duration while avoiding peaks and troughs in the resource profile.

To reduce the peak resource demand by two days, activity 3-6 is rescheduled. It starts with at start of day 9 and finishes at the end of day 13. The rectangle corresponding to activity 3-6, with a length of 5 of units and a width of 4 units, is shifted between day 9 and day 13 as shown in Figure 8.6. The peak requirement of the resource has reduced from 8 units to 6. Similarly, the

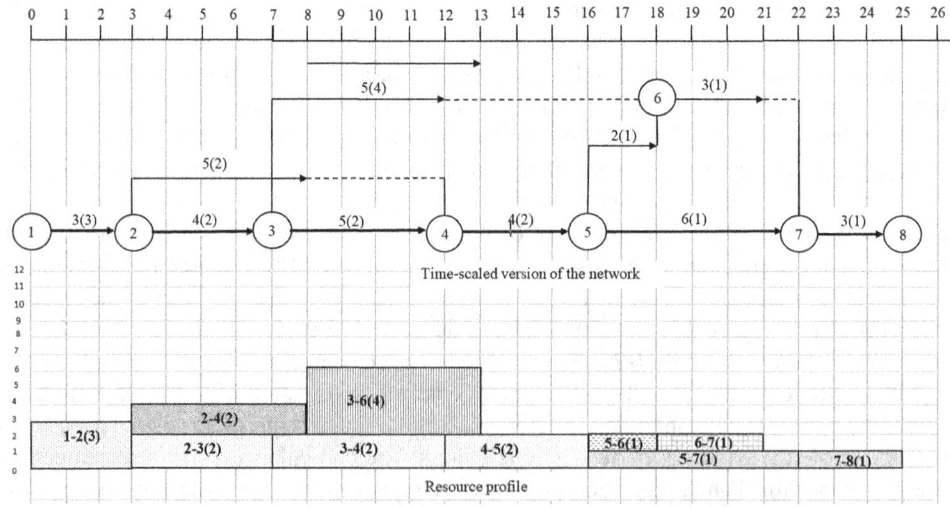

Figure 8.6 Time scaled version of the network and resource smoothing

starting and finish times of the project's non-critical activities are determined so that peaks and troughs in the resource profile are minimized.

8.7 Illustrative Example 1

Consider the network shown in Figure 8.1. The resources required for the various activities in the project are given in Table 8.1. The table also gives the EST, EFT, LST, LFT, and total float values corresponding to each activity in the project. The time-scaled version of the network and its resource profile are shown in Figure 8.7. The resource requirements of individual activities are shown using a rectangle in the resource profile. In the network under consideration, activities 1-2, 2-5, 5-6, 6-7, and 7-8 are critical. These activities have path and total float values of zero, therefore, the start and finish times of these activities remain unchanged during the resource smoothing process. The resource requirements of the project's critical activities are represented in the form of rectangles in the first layer of the resource profile in Figure 8.7. Activity 1-2 has a time duration of 4 days, thus the length of the rectangle corresponding to activity 1-2 is 4 units. It has a resource requirement of 2 units, thus the width of the rectangle is 2 units. Activity 2-5 has a time duration of 8 days, thus the length of the rectangle corresponding to activity 2-5 is 8 units. It has a resource requirement of 4 units, thus the width of the rectangle is 4 units. The resource requirements of all the critical and non-critical activities are represented in the same way. Resource requirements are as high as 11 units on days 5, 6, 11, and 12, and as low as 2 units on the first four and last 7 days of the project.

Non-critical activities 2-3 and 3-4 are not disturbed; these activities retain their original start and finish times. Activity 2-3 starts at the start of day 5 and ends at the end of day 10. Activity 3-4 starts at the start of day 11 and ends at the end of day 12. Event number 4 is shifted from the end of day 12 to the end of day 14, within its available float limit. To reduce the resource demand, activity 2-4 is rescheduled. It starts at the start of day 13 and ends at the end of day 14. The rectangle corresponding to activity 2-4, which has a length of 2 units and a width of 4 units, is shifted between days 13 and 14, as shown in Figure 8.8.

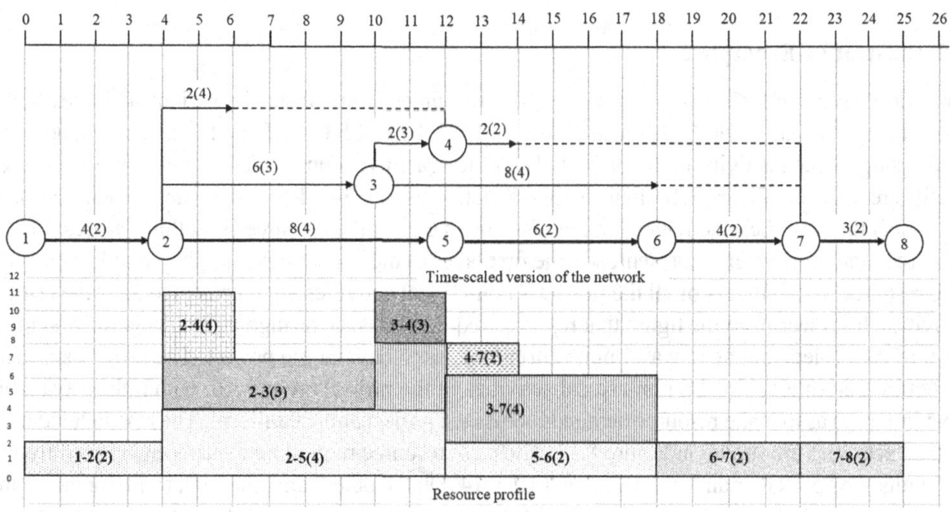

Figure 8.7 Time scaled version of the network and resource profile

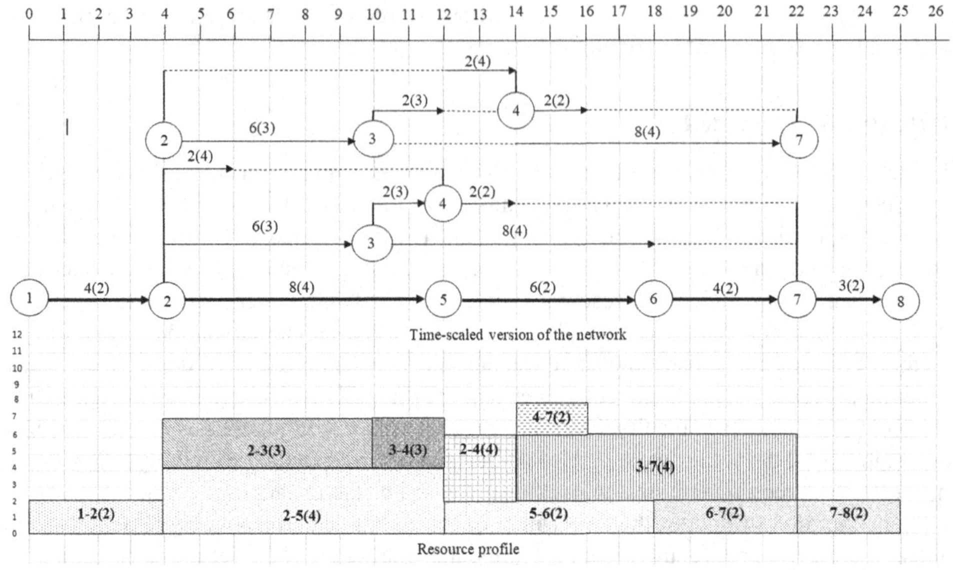

Figure 8.8 Time scaled version of the network and resource smoothing

Activity 4-7 is also rescheduled, starting at the start of day 15 and ending at the end of day 16. The rectangle corresponding to activity 4-7, which has a length of 2 units and a width of 2 units, is shifted between days 15 and 16. To reduce peak resource demand, activity 3-7 is also rescheduled. It starts at the start of day 15 and ends at the end of day 22. The rectangle corresponding to activity 3-7, which has a length of 8 units and a width of 4 units, is shifted between days 15 and 22. The project's peak resource demand has been reduced from 11 to 8 units, and the peaks and troughs in the resource profile have been reduced in comparison to those of the resource profile drawn on the earliest start time basis. The start and finish times of the non-critical activities in a project are decided in such a way that peaks and troughs in the resource profile are minimized.

8.8 Illustrative Example 2

Consider the network shown in Figure 8.9. The resources required for the various activities in the project are given in Table 8.3. It also gives the EST, EFT, LST, LFT, and total float values corresponding to each activity in the project. The time-scaled version of the network and its resource profile are shown in Figure 8.10. In the network, activities 1-3, 3-4, 4-6, and 6-8 are critical. The start and finish times of these activities remain unchanged during the resource smoothing process. Their resource requirements are represented in the first layer of the resource profile shown in Figure 8.10. The resource requirements of all the project's non-critical activities are represented after the critical activities. It is clear from the figure that resource requirements are as high as 8 units in the first three days of the project, but as low as 3 units during the last 6 days of the project. The project's critical activities are immovable. All non-critical activities in the project are moved, within their available float limits, so as to make resource demands across the project almost uniform. The rescheduled non-critical activities are shown in Figure 8.11. The peak resource requirement has been reduced from 8 to 5 units. The peak demand is uniform up to day 14. On the other hand, the peaks and troughs in the resource profile have also been reduced in comparison to the resource profile drawn on the earliest start time basis.

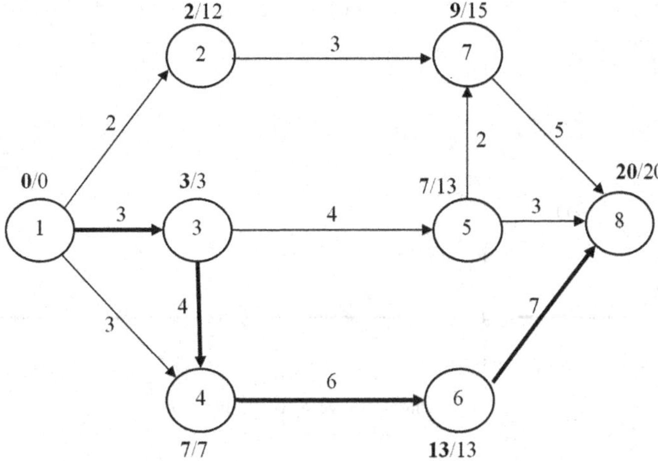

Figure 8.9 A sample project network

Table 8.3 Activities in a project and their time durations, resource requirements, times, and total floats

Activities	*Time durations*	*Resource requirements*	*EST*	*EFT*	*LST*	*LFT*	*Total floats*
1-2	2	1	0	2	10	12	10
1-3	3	5	0	3	0	3	0
1-4	3	2	0	3	4	7	4
2-7	3	1	2	5	12	15	10
3-4	4	3	3	7	3	7	0
3-5	4	1	3	7	9	13	6
4-6	6	3	7	13	7	13	0
5-7	2	1	7	9	13	15	6
5-8	3	1	7	10	17	20	10
6-8	7	3	13	20	13	20	0
7-8	5	1	9	14	15	20	6

8.9 Resource Leveling vs. Resource smoothing

Although the objectives of resource leveling and resource smoothing are the same, both approaches differ in the following ways.

1. In resource leveling the time durations of the various activities in a project and the time duration of the project change, while in the case of resource smoothing, the time durations of the various activities in a project and the time duration of the project are not changed.
2. In resource leveling, the project's critical path length changes because of changes to the time durations of activities and to the project time duration while in the case of resource smoothing the critical path length does not change because of the time durations of the various activities in a project and the project time duration are not changed,
3. Resource leveling is used when the quantity of resources allocated to a project is greater than or lower than the quantity required and project re-scheduling is thus needed to help re-allocate resources within their respective availabilities. Resource smoothing is used when resources are unevenly allocated between the various activities in a project and resource

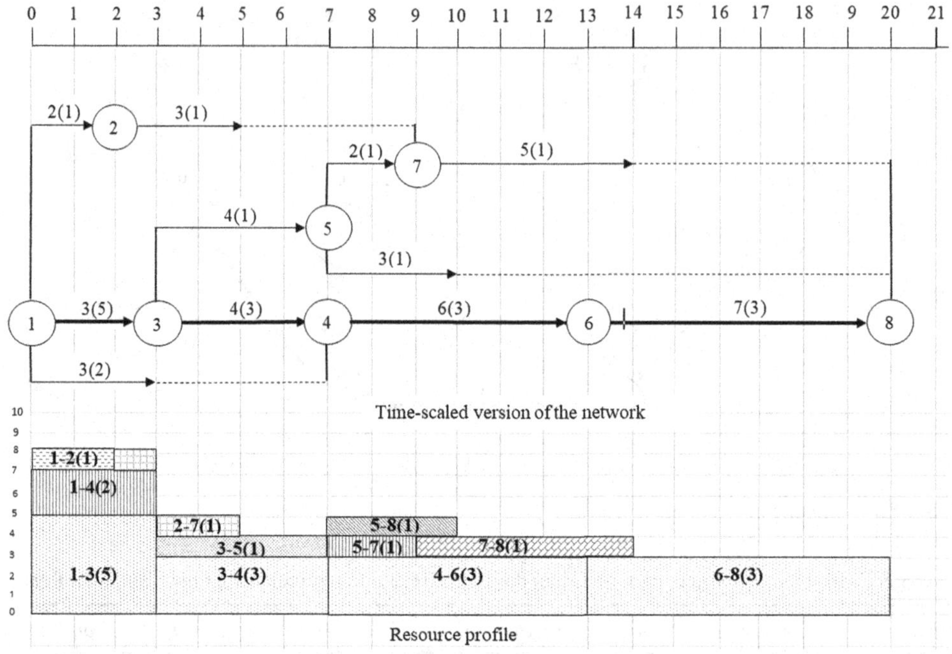

Figure 8.10 Time scaled version of the network and resource profile

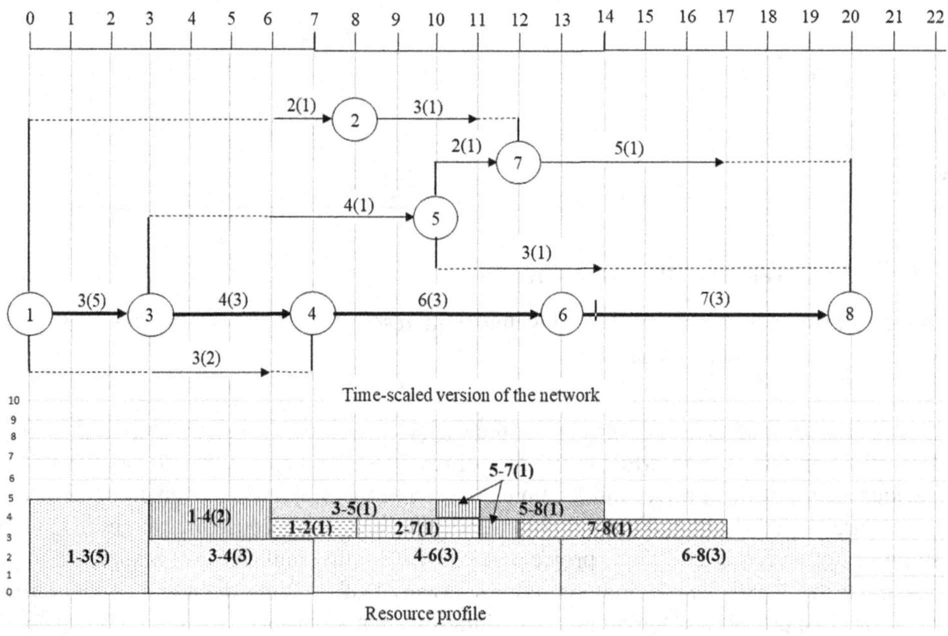

Figure 8.11 Time scaled version of the network and resource smoothing

demands ought to be made as uniform as possible across the project, depending upon the total float availability of the various activities in the project.

4. In resource leveling, resource availability is the main constraint, thus project re-scheduling is performed with resource availability in mind when aiming towards the execution of the various activities in a project, while in resource smoothing, project duration is the main constraint, aiming to finish a project within the allocated time duration,
5. Resource leveling is applied to a project's critical as well as non-critical activities, whereas resource smoothing is applied to only the non-critical activities in a project. The activities which lie on the critical path are not disturbed in resource smoothing,
6. Resource leveling involves the adjustment of a project's start and finish times, depending upon resource constraints. Resource smoothing involves the adjustment of the start and finish times of a project's non-critical activities to help avoid peaks and valleys in the resource profiles.
7. Resource leveling is also called *resource-constrained scheduling* and resource smoothing is also called *time-constrained scheduling*.

8.10 Conclusion

During the planning stage of a project, available resources are assumed to be either unlimited or limited. In reality, available resources are not unlimited and are not available as and when they are required. When resources are assumed to be unlimited, the starting or end times of a project's non-critical activities project are adjusted (depending on the path float availability) in such a way that resource requirements become nearly uniform throughout a project's time duration because variations in resource requirements are difficult to manage in practice. Sometimes the various activities in a project are planned and scheduled according to the availability of the various types of resource required. In this case, available resources are considered limited. Requirements for various types of resource are thus not increased beyond their availability.

Exercises

Question 8.1: Consider the network shown in Figure E8.1. The time durations of and resources required for the various activities in the project are given in Table E8.1. Draw a resource profile for and perform resource smoothing on this project.

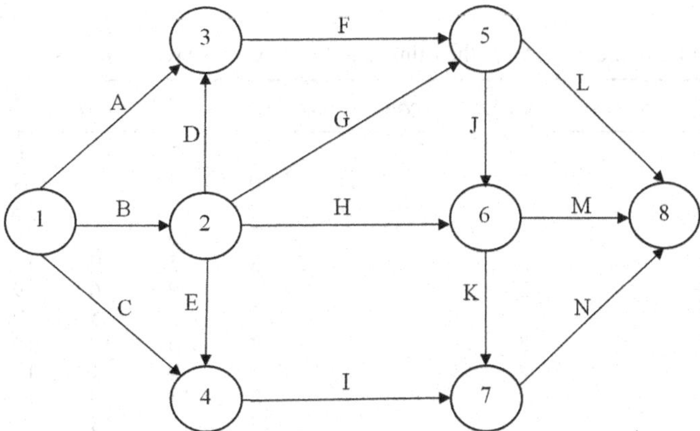

Figure E8.1 Network of a sample project

Table E8.1 Time durations of and resources required for various activities

Activities	*Time durations*	*Resource requirements*
A	2	3
B	3	2
C	2	1
D	3	1
E	2	2
F	3	2
G	4	3
H	3	3
I	4	2
J	4	2
K	2	2
L	2	4
M	2	2
N	3	2

Solution: The network calculations are shown in Figure E8.2 and their values are given in Table E8.2. The time-scaled version of the network and resource profile are shown in Figure E8.3. The revised time-scaled version of the network and its resource smoothing are shown in Figure E8.4.

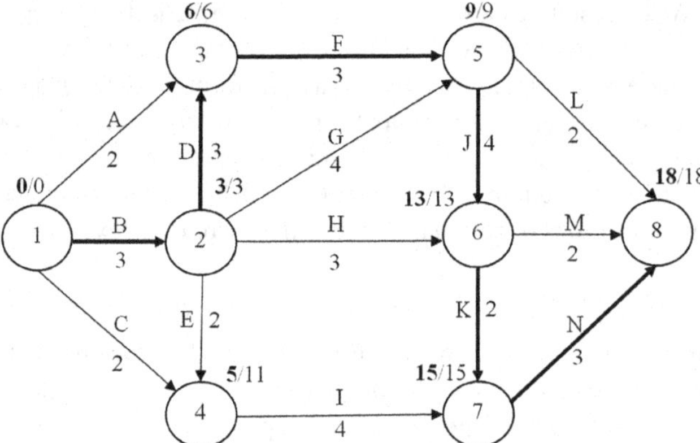

Figure E8.2 Network calculations

Table E8.2 Activities in the project and their time durations, resource requirements, times, and total floats

Activities	*Time durations*	*Resource requirements*	*EST*	*EFT*	*LST*	*LFT*	*Total floats*
A	2	3	0	2	4	6	4
B	3	2	0	3	0	3	0
C	2	1	0	2	9	11	9
D	3	1	3	6	3	6	0
E	2	2	3	5	9	11	6
F	3	2	6	9	6	9	0
G	4	3	3	7	5	9	2
H	3	3	3	6	10	13	7
I	4	2	5	9	11	15	6
J	4	2	9	13	9	13	0
K	2	2	13	15	13	15	0
L	2	4	9	11	16	18	7
M	2	2	13	15	16	18	3
N	3	2	15	18	15	18	0

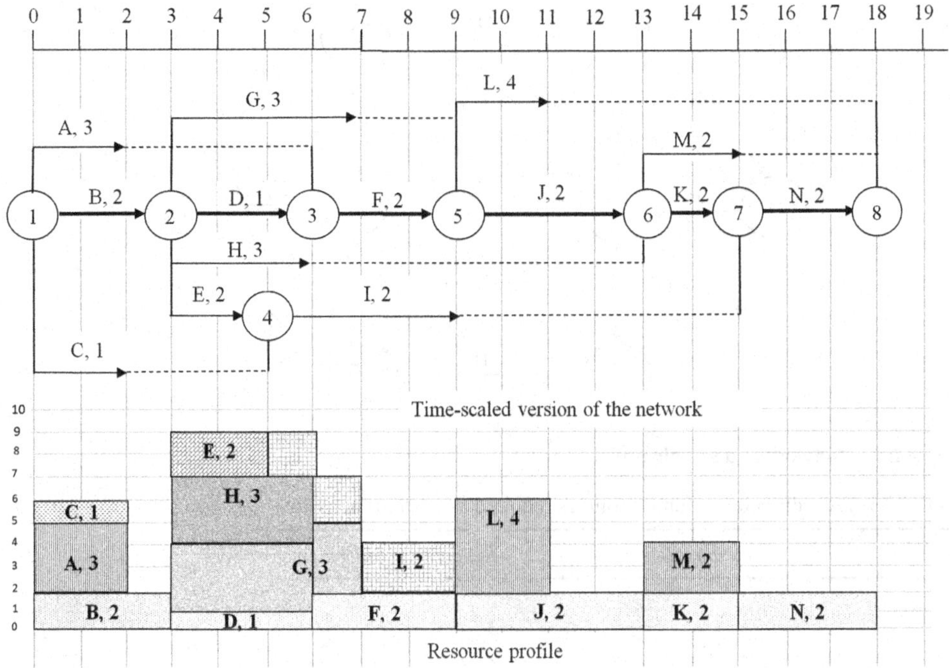

Figure E8.3 Time scaled version of the network and resource profile

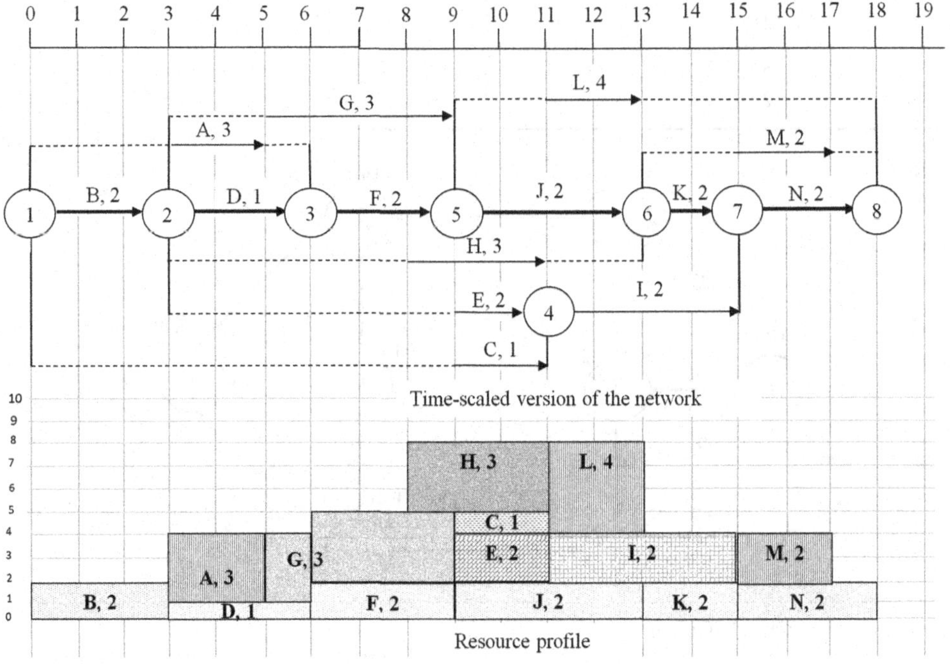

Figure E8.4 Revised time scaled version of the network and resource smoothing

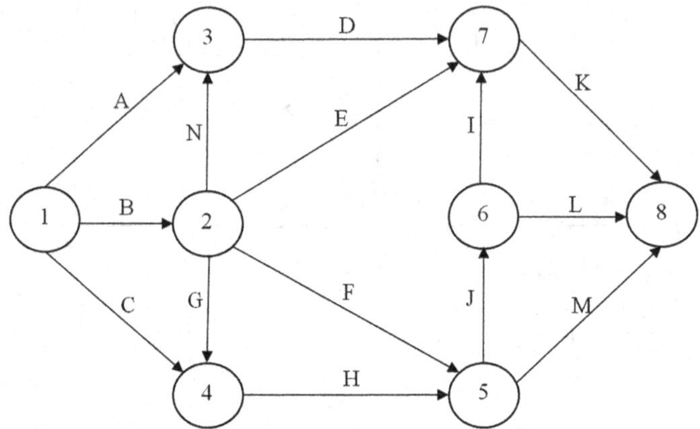

Figure E8.5 Network of a sample project

Table E8.3 Time durations of and resources required for various activities

Activities	*Time durations*	*Resource requirements*
A	2	2
B	3	3
C	2	2
D	2	2
E	3	4
F	2	2
G	4	2
H	3	2
I	3	2
J	3	4
K	4	2
L	3	2
M	2	2
N	2	2

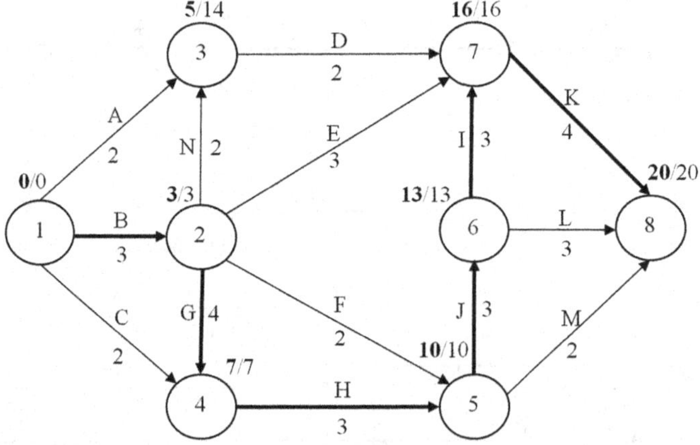

Figure E8.6 Network calculations

Table E8.4 Activities in the project and their time durations, resource requirements, times, and total floats

Activities	*Time durations*	*Resource requirements*	*EST*	*EFT*	*LST*	*LFT*	*Total floats*
A	2	2	0	2	12	14	12
B	3	3	0	3	0	3	0
C	2	2	0	2	5	7	5
D	2	2	5	7	14	16	9
E	3	4	3	6	13	16	10
F	2	2	3	5	8	10	5
G	4	2	3	7	3	7	0
H	3	2	7	10	7	10	0
I	3	2	13	16	13	16	0
J	3	4	10	13	10	13	0
K	4	2	16	20	16	20	0
L	3	2	13	16	17	20	4
M	2	2	10	12	18	20	8
N	2	2	3	5	12	14	9

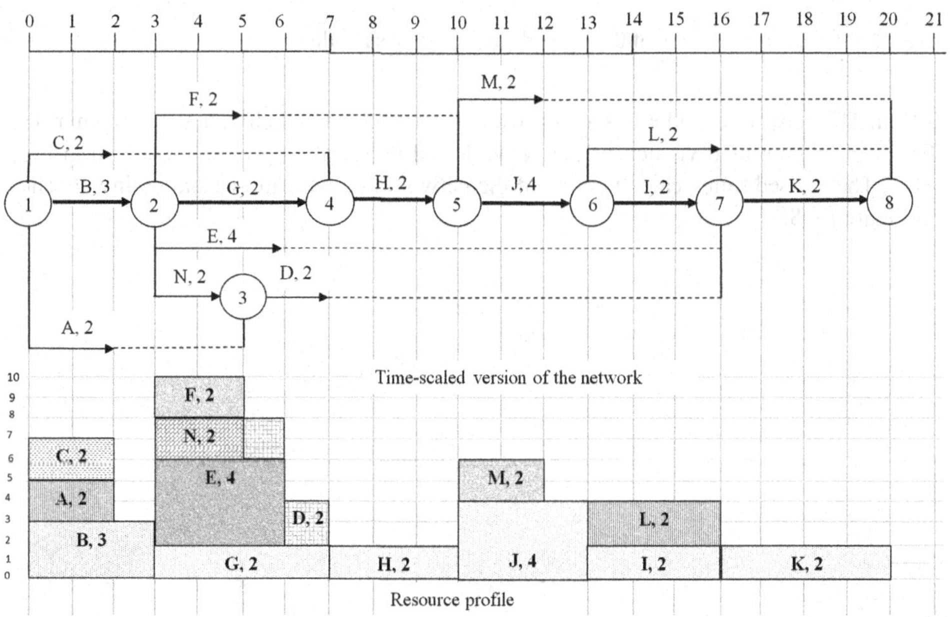

Figure E8.7 Time scaled version of the network and resource profile

Question 8.2: Consider the network shown in Figure E8.5. The time durations of and resources required for the various activities in the project are given in Table E8.3. Draw a resource profile for and perform resource smoothing on this project.

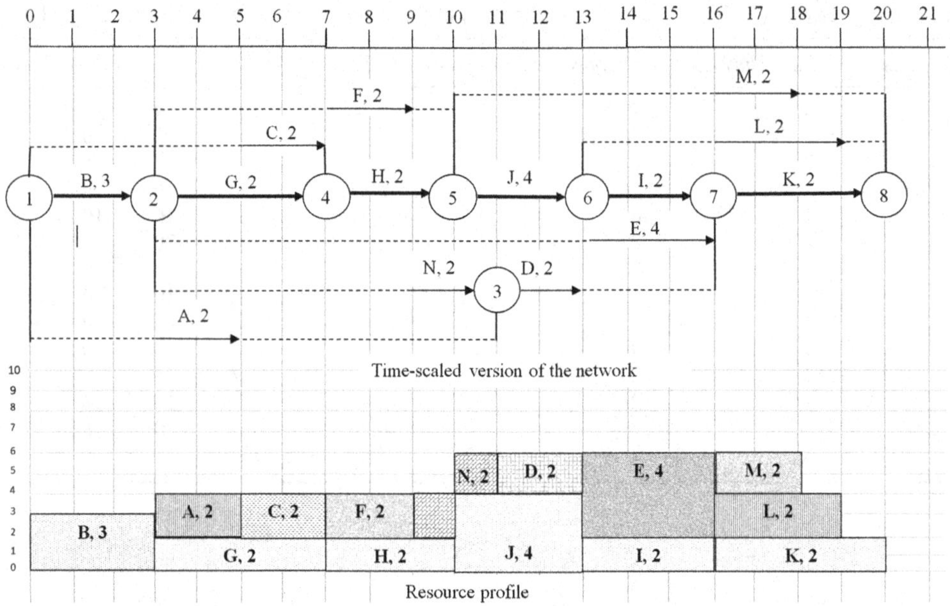

Figure E8.8 Time scaled version of the network and resource smoothing

Solution: The network calculations are shown in Figure E8.6 and their values are given in Table E8.4. The time-scaled version of the network and its resource profile are shown in Figure E8.7. The revised time-scaled version of the network and its resource smoothing are shown in Figure E8.8.

9 Project Control and Updating

9.1 Learning Objectives

After the completion of this chapter, readers will be able to:

- Understand project control as a step in project management,
- Update a project schedule and understand its requirements, and
- Draw an S-curve to depict the cumulative fund flow associated with a project.

9.2 Introduction

A project plan looks to the future, but it is developed based on the skills, knowledge, and experience gained from the present and past. A developed plan may not be always perfect or accurate. Deviations from a developed plan may occur during the execution of a project. A developed project plan enforces the performance outcomes that the resources of a project are expected to attain. The attainment of performance outcomes is needed for a project to be completed within its planned time duration, allocated budget, and agreed quality. Project control is aimed at the development of a mechanism for monitoring and controlling the project to ensure the achievement of performance outcomes. Effective project control helps in the detection of deviations from the initial plan and the formulation of a remedial strategy. Project control is also required to help provide a safe working environment, manage risks, decide on measures for efficient resource use, identify issues and rectify them, and make adjustments to the organizational structure. However, the main purpose of project control is to ensure the successful accomplishment of project objectives. Project control starts with project execution, but project execution, project control, and schedule refinements run parallel and make a closed loop, as discussed in chapter 1. Without suitable project control a project may sometimes go off track and become unmanageable and expensive. The purpose of this chapter is to understand project control from the points of view of scheduling and project cost only.

9.3 Project Control

As discussed in Chapter 1, project control is a step in project management. Project control comprises three functions: *monitoring*, *comparing*, and *correcting*. Monitoring is the continuous tracking of a project's progress during its execution. Comparing is the time-to-time comparison of the actual progress obtained during a project's execution with its planned progress. This comparison helps identify the activities in a project that are following the plan and the activities which have deviated from the plan. Correcting involves determining and measuring

DOI: 10.1201/9781003428992-9

deviations and deciding on corrective measures so that project objectives can be achieved within the planned project time duration, allocated budget, and required quality. The comparison of actual and planned progress, as well as current project information and knowledge of ongoing trends, help in making informed decisions during project execution. It mainly helps in taking appropriate corrective measures to bring project activities back on schedule.

Project control involves the collection of information related to actual progress and planned progress, the assessment of current trends, the projection of future outcomes, and the application of this learning in deciding on corrective measures for keeping the project on schedule or bringing the project back on schedule. Continuous monitoring and control help in keeping a project on track. Project monitoring and control ensure the seamless execution of project activities and improve their productivity. A good project control system mainly focuses on the following tasks.

Data Collection: This involves gathering project information including the project's plan, execution schedule, finance schedule, resource schedules, and the outputs of different types of resource as assumed in the planning stage, etc. It also includes the documentation of newly acquired information and knowledge which may affect the project under consideration.

Product Quality Assessment: The quality of products produced or services rendered must meet their previously agreed upon specifications. Thus, a mechanism for evaluating the quality of products or services is essential for an effective project control system.

On-going Progress Trends: The assessment of ongoing progress trends, the projection of future project outcomes as based on ongoing trends, and finally the application of this learning in deciding on corrective measures are parts of an effective project control system. It is possible to find the causes of delays that may sometimes have come about because of an inadequate allocation of resources to the delayed activities.

Performance Reporting: Project execution and control run parallel; thus, a project performance reporting mechanism is essential to ensure the efficient use of resources and project team members and the timely completion of project activities. Accurate information of regarding project performance with respect to the actual time taken and maintaining its record help planners make suitable decisions in time.

Comparison of Planned and Actual Performance: Planned performance at a particular point in time is compared with the actual performance achieved during project execution. Comparison with regard to a particular point in time helps in determining which activities have been completed, which activities are in progress, and which activities are yet to be started. This comparison helps in identifying deviations, the causes of these deviations, and the strategies which will help bring the project back to its planned level of performance.

Updating Schedules: A planner needs to revise the time durations of ongoing activities and those of activities yet to be started, and to enrich the project plan using the experience gained during the execution of a certain portion of the project. The time durations and resource requirements of the activities which are in progress and which are not yet started are re-adjusted based on on-going progress trends. Finally, the project schedule is updated based on the revised time durations of its activities, ongoing progress trends, and the correction measures decided on for the improvement of project performance. The output obtained during the execution of a project is used to calculate revised time durations for its ongoing and remaining activities.

Revision of Expenditures and Other Information: Project expenditure is generally expressed in terms of an S-curve, which is revised based on current expenditure trends. The comparison of the actual cost incurred with the planned cost helps in the identification of deviations. Resources schedules and other forms of project-related information are also revised or updated based on current performance. A project control system should be developed and implemented as part of a continuous process. This process should be started immediately upon the project execution.

9.4 Schedule Updating

Schedule updating is a part of the project control process. Project execution rarely goes according to the plan or the schedule; deviations from the original plan and schedule always appear. The process of revising the plan and schedule of a project based on the experience gained during the execution of a certain portion of a project is called *project updating*. If a project is progressing according to its planned progress, its updating is not essential. However, if a project deviates from its initial plan and schedule during its execution, project updating becomes essential. The new skills, knowledge, and related information obtained during the execution of a certain portion of a project help in revising its plan and schedule. During project execution, the actual progress achieved at a particular point in time is compared with the planned progress. This comparison helps to categorize the project activities being executed into three classes: activities that are on schedule, activities that are ahead of schedule, and activities that are behind schedule.

Activities that are ahead of or behind schedule require revisions to their time durations, and schedule calculations are revised accordingly. Further, changes to the time durations of a project's critical necessitates an update. The updating process results in changes to the project time duration, and, accordingly, changes to the start and finish times of various activities in a project. The identification of any error in the estimation of an activity's time duration during the planning stage also necessitates an update. Sometimes updating becomes essential due to unpredictable circumstances like delays in the availability of the workforce or of other resources, natural calamities, pandemics, etc. If all the activities in a project are proceeding according to the plan and the schedule, project updating is not necessary.

A project plan needs to be revised when activities are behind or ahead of the planned progress. Depending on updated information and the experience gained during the execution of a certain portion of a project, revisions to the time durations of ongoing activities and activities that are yet to be started and schedule calculations become necessary. After being updated, the time duration of a project changes because of changes to the time durations of project activities. If the time duration of a project increases, the time durations of activities that lie on the critical path may be decreased through the deployment of more resources in an effort to decrease the project time duration.

9.5 Schedule Updating Process

Project updating starts with the collection of the necessary information. This collected information and the knowledge acquired during project execution are used to revise the time duration of project activities and re-perform the scheduling calculations.

9.5.1 Information Required

During the execution of a project the following pieces of information are collected to help update a project:

a. The initial/previous project plan, execution schedule, resource schedules, scheduling calculations for the project in question, etc. are collected to start the project updating process.

b. The point in time or date at which a project is to be updated is decided. It plays a major role in the updating process. It is the date at which the actual project progress is compared with the planned progress.
c. The actual project progress at the point of time or date decided for the project update is measured. This helps in developing the list of activities that have been finished, are in progress, or are yet to be started. It includes all works completed up to the project update time or date.
d. Information about activities that have been added or deleted along with their time durations, interdependencies, resource requirements, cost estimates, etc. are also collected. Information about when an activity actually started and finished is also documented.
e. Information about all activities which have undergone changes to their interdependencies is collected. These changes affect logic within a project network,
f. If any change to the number of working days in a week or any holiday is reported, the schedule will need revised accordingly.
g. The latest information that may affect the time duration of project activities is also reported. This includes assessments of on-going progress trends and projections of future outcomes based on those on-going trends.

This is not a complete list of information to be collected; this varies from one project to another, from time to time, according to the number of updates planned, etc.

9.5.2 Updating Process

The simple updating process has the following steps:

Step 1: The point in time or date of the update is decided. The point in time of the update is taken as the earliest event time (EET) of the starting event of the project network that needs to be updated. In the case of calendar date scheduling, the update date is taken as the EET of the starting event. Generally, for the sake of simplification, network calculations are carried out assuming an EET of zero for the starting event of a project, and start and finish dates are calculated afterward.

Step 2: The actual project progress at the point in time or date decided for the project update is measured. The comparison of the actual project progress with the progress planned for the same point in time or date provides a list of activities that have been finished, that are in progress, and that are yet to be started.

Step 3: The time durations of activities that have been completed are assigned values of zero. The time durations of activities that are in progress at the time of the update are assigned revised values based on ongoing trends. The time durations of activities that are yet to be started after the update are also assigned revised values based on ongoing trends. These revised time durations are based on the experience of executing a portion of the project with a different environment and under different conditions or constraints.

Step 4: The network calculations are carried out again based on the revised time durations of the various activities in the project.

9.5.3 Frequency of Updates

Updates are made frequently for projects with small time durations by taking into account new knowledge and the latest information obtained during the execution of a certain portion of the project. For projects with long time durations, on the other hand, updates are made after relatively long intervals when a project starts, but as a project progresses updates are mad more frequently, at smaller intervals. The frequency of updates is increased as a project progresses. The time remaining to a project goes on decreasing as a project progresses, thus a project with little time left to go is treated as a small project. Its update frequency is therefore increased according to the decrease in time duration.

Knowledge, information, and experience increase as a project progresses, hence the need for an increased update frequency. A major change in the time duration of an activity immediately necessitates an update. An update also becomes essential if any change is observed in the time duration of any critical activity. Remedial measures are necessary if the time duration of a critical activity increases or decreases. If it is not acceptable to delay a project, activities on the new critical path of its revised network are accelerated through the allocation of extra resources. Updating enables us to take corrective measures in time, and to make managerial decisions in problem areas. The opportunity to take corrective action cannot exist unless actual progress is measured. Updating a plan or schedule after a long interval may decrease the update's effectiveness.

9.6 Fund Flow Forecasting: Inflow and Outflow

Projects are awarded by clients (an owner or a public sector or government organization) through a bidding process in an effort to bring down project costs. An executing agency (an organization, firm, or individual) makes every effort to complete the project at its minimum cost. It is always the tendency of an executing agency to complete a project at a cost below the awarded amount, so as to make that project profitable. To survive in a competitive market, an executing agency must keep its project on track in terms of its cost, time duration, and the quality of its project deliverables. Therefore, the plan and schedule of a project must be physically and financially feasible.

To determine the physical feasibility of a project, the resources required for the execution of a project, their availability, and other physical constraints are checked. Financial feasibility is checked by forecasting the inflow and outflow of funds required for the physical execution of a project. The payments received by an executing agency from a client constitute their fund inflow. The payments made by an executing agency to get the work done constitute their fund outflow, or expenditure. The inflow and outflow of funds takes place in all kinds of project. The inflow and outflow of funds taken together determine the net funds required to execute a project.

The planning of fund inflow and outflow across the time duration of a project and their correct monitoring are two of the most important tasks in project management. A comprehensive examination of the (accurately estimated) inflow and outflow of a project's funds is necessary to determine whether a project is financially feasible or not. A positive fund flow indicates that the inflow is greater than the outflow and a negative fund flow indicates the opposite. The majority of projects have a negative fund flow until the final payment is received after the completion of a project. A negative fund flow indicates a need for net funds during the project execution. This necessitates the use of the executing agency's working capital or the arrangement of funds from external funding agencies. It is the tendency of an executing agency to minimize the need for external funding by keeping the sum of their fund inflow as close as possible to the sum of their fund outflow.

9.6.1 S-Curve

An S-curve is a simple graph used to represent the cumulative funds associated with a project as plotted against the project's time duration. Figure 9.1 shows the daily expenditures of a

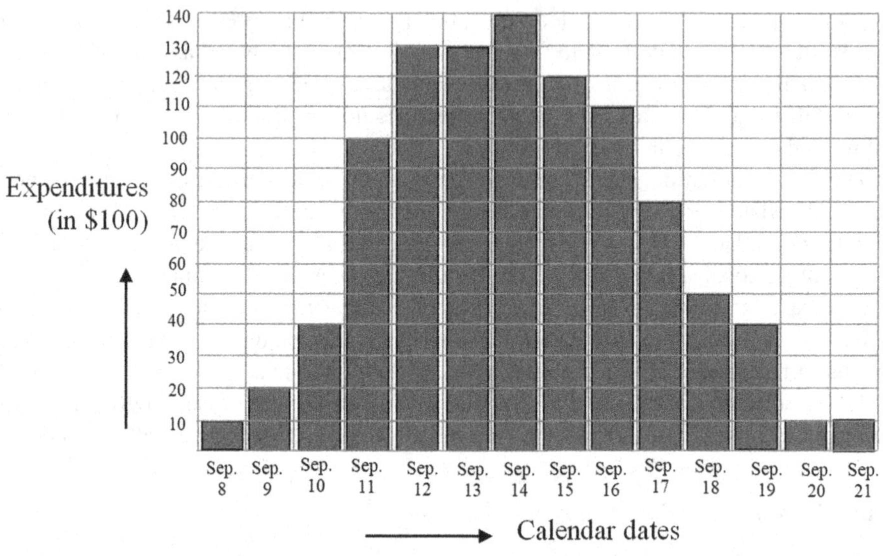

Figure 9.1 Daily expenditure of a sample project

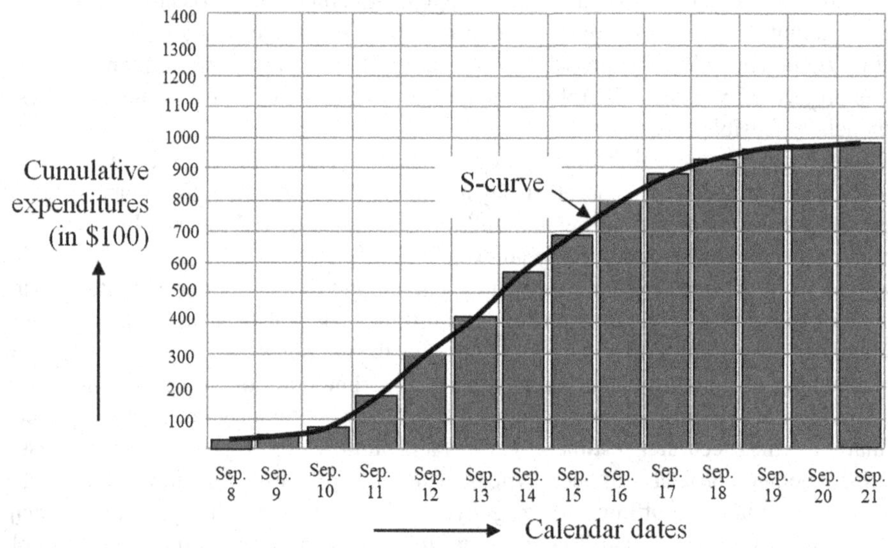

Figure 9.2 Daily cumulative expenditure/S-curve of the sample project

sample project. Figure 9.2 shows the daily cumulative expenditure, or S-curve, of the same project. Project expenditure is most commonly depicted using an S-curve. As shown in the figure, expenditure usually starts slowly, is at its highest in the middle of a project, and then tapers off near the project's completion. The shape of the S-curve looks like the letter 'S', hence, called the S-curve.

In general, the S-curve is flatter at the start and end of a project, and steeper in the middle of a project. A project starts with a slow progress rate because of an initial inertia. This slow progress rate explains the gradual upward rise of the S-curve. With time, the progress rate will accelerate rapidly and the S-curve will suddenly rise upward in the middle of a project. Team members work with maximum efficiency in the middle of a project. Toward the end of a project, the majority of the work has already been completed, thus the progress rate is slow, forming the upper part of the curve.

9.6.2 Financial Planning

In the beginning of a project, the sum of the outflow or expenditure of an executing agency are usually higher than the sum of the inflow or income received for work done. Outflow or expenditure, in the beginning, is higher when activities start at their EST, and lower in comparison when activities start at their LST. This helps determine the funds required to carry out a project. In Figure 9.3, the upper curve depicts the cumulative expenditure or outflow given activities starting as early as possible, taken as the upper limit. The lower curve depicts the cumulative expenditure or outflow given activities starting as late as possible, taken as the lower limit. In Figure 9.4, a closer set of curves is shown; the distance between the two curves is indicative of

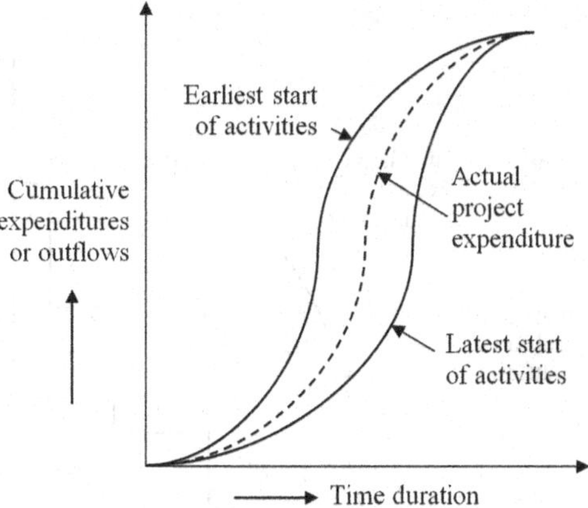

Figure 9.3 S-curve representing cumulative expenditure or outflow

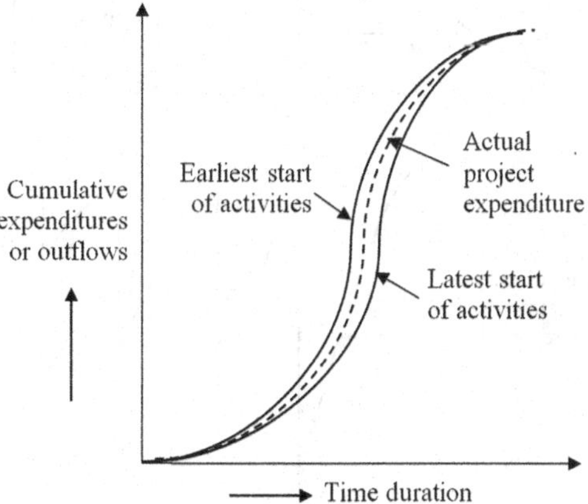

Figure 9.4 S-curve representing cumulative expenditure; a smaller distance between the curves indicates lower flexibility

the funding flexibility of a given project. is the smaller the distance between the two curves, the lower the flexibility; the higher the distance between the two curves, is the higher the flexibility.

The actual project expenditure curve must be in between the upper and lower curves which represent the upper and lower limits of the project's outflow. Thus, as long as the actual project expenditure curve falls between the two outflow curves, the project is financially feasible. The nearer the actual project expenditure curve comes to the latest start outflow curve, the higher the financial risk. The actual project expenditure curve may be brought closer to the earliest start outflow curve by using short-term loans or by using the working capital of the executing agency. Financial risk is reduced as the actual project expenditure curve moves closer to the earliest start fund outflow curve.

As a project progresses, part payments or a single payment after the completion of a project are made to the executing agency by their client. Part payments are those made for a partial amount of work. Part payments are also called progress payments, the receipt of which from a client represents fund inflow. However, progress payments always lag behind the total value of the work completed by an executing agency, as shown in Figure 9.5. The curve helps in determining the amount of additional funds required besides the progress payments that are expected from a client as the project progresses, as shown in Figure 9.5. Sometimes funds are required

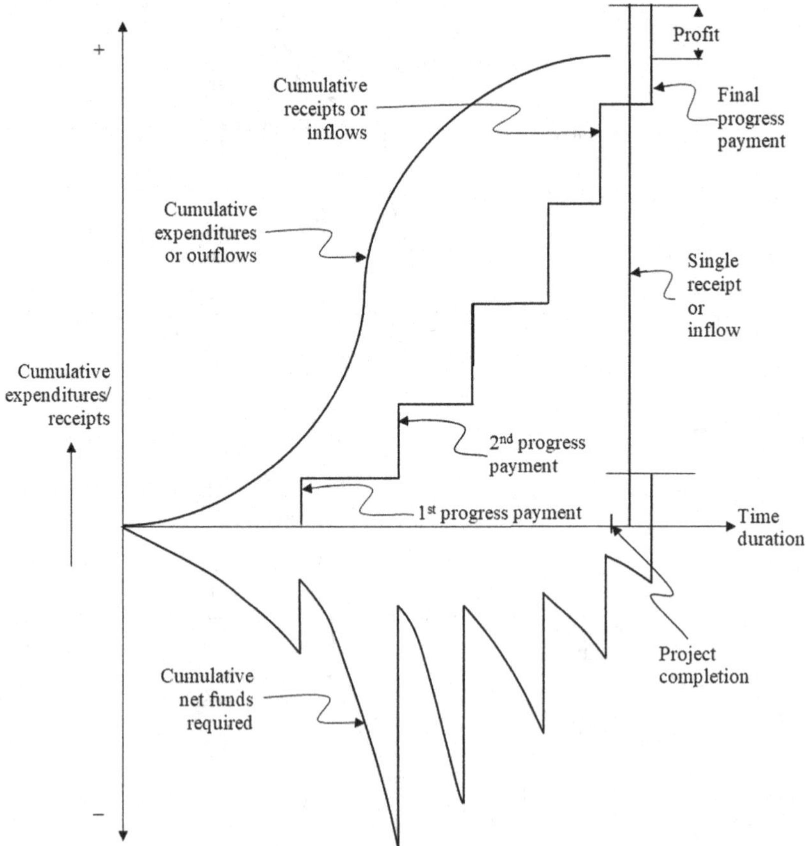

Figure 9.5 S-curve representing cumulative expenditure, cumulative receipts, and net cumulative funds required

beyond that offered by the working capital of an executing agency. The majority of projects have a negative fund flow, as shown in the figure, until the final payment is received upon completion of a project. A negative funds flow indicates a need for funds during the project execution. To minimize the interest on borrowings, it is recommended to balance the outflow of funds against the inflow to ensure the financial feasibility of a project – therefore, financial planning is always essential.

9.7 Conclusion

Project control is the process of analyzing the planned and actual performance of a project to keep its plan, schedule and expenditures on track. Project control is an iterative process for measuring the current status of a project, forecasting likely outcomes based on its current status, and improving project performance when forecasted outcomes are unacceptable. The actual progress obtained at a given point is compared with the planned progress at regular intervals and necessary corrective measures are taken to keep the progress in concordance with the plan. To keep project progress on track requires a certain level of control, thus the plan and schedule are sometimes revised. Revision of a schedule is called schedule updating. Therefore, during the execution of a project, suitable control is to ensure its completion within the allocated time duration. Project control involves taking necessary corrective measures regularly to keep project progress in concordance with planned progress.

Example 9.1: Figure 9.6 shows a project plan in the form of a network. The project is to be updated at the end of day 20. The following situations have arisen by the end of day 20.

- Activity 1-2 was completed, in concordance with the initial plan, in 8 days.
- Activity 1-3 was completed more rapidly than the initial plan of the project and took 7 days.
- Activity 2-3 took longer than the initial plan suggested; the revised time duration was 9 days.
- Activity 2-5 is in progress and requires 2 days more to complete.
- Activity 3-4 is in progress and requires 14 days more to complete.

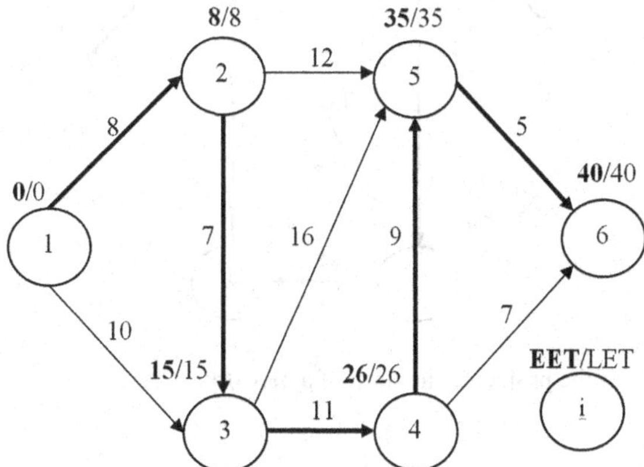

Figure 9.6 Plan of the project in the form of a network

- Activity 3-5 is also in progress and requires 8 days more to complete.
- The revised time duration of activity 4-5 is 10 days.
- The revised time duration of activity 4-6 is 8 days.
- The revised time duration of activity 5-6 is 4 days.

Update the project plan and determine its revised time duration.

Solution: Figure 9.6 shows the project plan in the form of a network; the time duration of the project is 40 days. Table 9.1 summarizes the situation of the project at the time of the update.

The updated plan of the project in the form of a network is shown in Figure 9.7. The project has a time duration of 42 days, instead of the 40 days planned initially.

Table 9.1 Situation of the project at the time of update

Activities	*Planned time durations*	*Status at the time of update*	*Revised remaining time durations*
1-2	8	Completed	0
1-3	10	Completed	0
2-3	7	Completed	0
2-5	12	In-progress	2
3-4	11	In-progress	8
3-5	16	In-progress	14
4-5	9	Yet to start	10
4-6	7	Yet to start	8
5-6	5	Yet to start	4

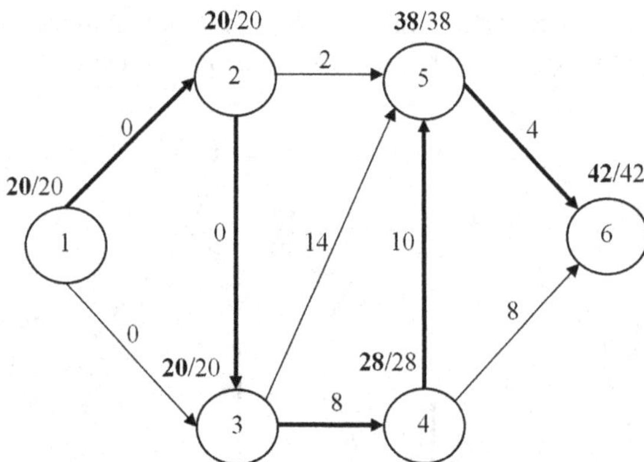

Figure 9.7 Updated plan of the project in the form of a network

Exercises

Question 9.1: What is project control, and why it is essential in project management?

Question 9.2: What is schedule updating, and why it is done?

Question 9.3: What are the different types of information required in schedule updating?

Question 9.4: Figure E9.1 shows the plan of a project in the form of a network. The project is to be updated at the end of day 7. The following situations have arisen by the end of day 7:

- Activity 1-2 was completed more rapidly than the plan suggested and took 1 day.
- Activity 1-3 was completed as per the initial plan in 4 days.
- Activity 1-4 took longer than the initial plan suggested; its revised time duration was 4 days.
- Activity 2-5 took longer than the initial plan suggested; the revised time duration was 4 days.
- Activity 3-6 is in progress and requires 2 days more to finish.
- Activity 4-7 was completed, as per the initial plan, in 1 day.
- The revised time duration of activity 6-8 is 1 day.
- The revised time duration of activity 7-9 is 4 days.
- The revised time duration of activity 8-9 is 5 days.

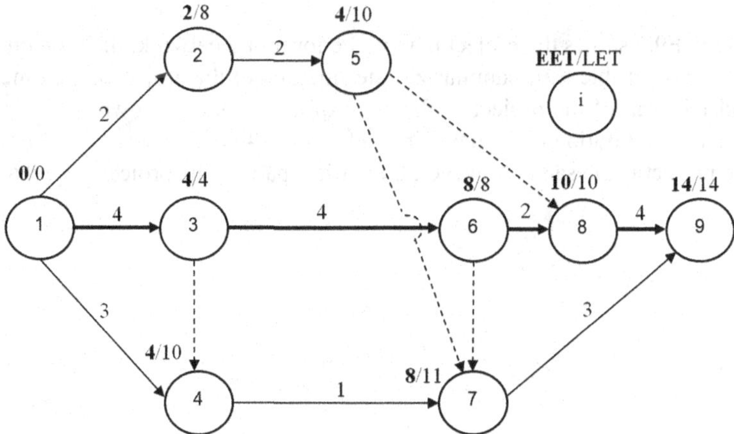

Figure E9.1 Plan of the project in the form of a network

Table E9.1 Situation of the project at the time of update

Activities	*Planned time durations*	*Status at the time of update*	*Revised remaining time durations*
1-2	2	Completed	0
1-3	4	Completed	0
1-4	3	Completed	0
2-5	2	Completed	0
3-6	4	In-progress	2
4-7	1	Completed	0
6-8	2	Yet to start	1
7-9	3	Yet to start	4
8-9	4	Yet to start	5

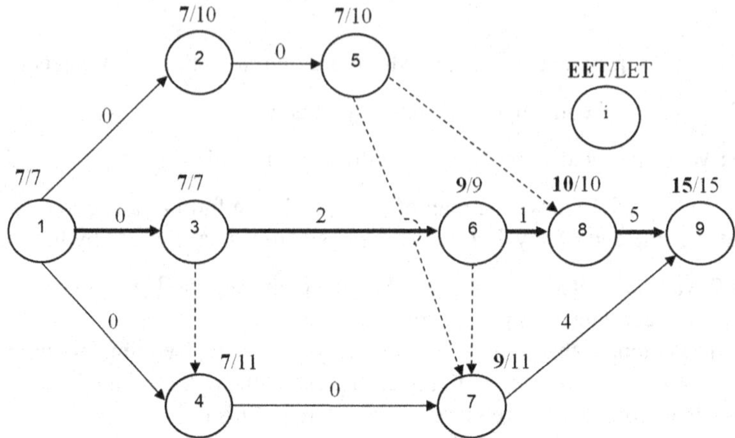

Figure E9.2 Updated plan of the project in the form of a network

Update the project plan and determine its revised time duration.

Solution: Figure E9.1 shows the project plan in the form of a network; the time duration of the project is 14 days. Table E9.1 summarizes the situation of the project at the time of update.

The updated plan of the project in the form of a network is shown in Figure E9.2. The project has a time duration of 15 days, instead of the 14 days planned initially. The critical path of the project is the same; however, the critical path of the project may change.

10 Programme Evaluation and Review Technique – I

Event Times

10.1 Learning Objectives

After the completion of this chapter, readers will be able to:

- Determine the time durations of project activities,
- Calculate earliest and the latest event times from a time-scaled version of a network, and
- Calculate earliest and the latest event times using forward and backward pass calculations.

10.2 Introduction

In 1958, the U.S. Navy's Special Projects Office, together with the aerospace company Lockheed Missile Systems and the consultancy firm Booz Allen & Hamilton, developed the *program evaluation and review technique* (PERT), for the Polaris missile program. Research and development for the technique was carried out simultaneously with, but independently of, Remington and DuPont's CPM technique. PERT is a road map for identifying the major activities in a project and their inter-dependencies. Unlike Remington and DuPont's CPM technique, PERT only revolves around time constraints and does not deal with the cost and quality of a product or project.

PERT is used for projects which are not routine. Thus, the time duration estimates required for the execution of a project's activities lack a fair degree of accuracy. When CPM is used, the project in question either consists of activities with which a planner is thoroughly familiar, is routine, or the planner has sufficient experience in handling such projects. Thus when using CPM, time duration estimates have a fair degree of accuracy. In the case of PERT, however, because of uncertainties present due to the unfamiliarity of the planner with the project under consideration, accurate time duration estimates for the completion of various activities are difficult to produce. Therefore, PERT is used when planning projects that are non-routine, or for projects in which accurate time duration estimation is not possible. Such projects have a lot of uncertainty regarding estimated time durations. These are projects of their own nature. The time duration estimation required for the completion of a project's activities involves uncertainties; thus, an activity's time duration may not be expressed by a single value.

10.3 Activity Time Durations

Generally, two methods are used for estimating of time duration of the completion of an activity. The first is called the *deterministic method.* In the deterministic method, it is assumed that a planner is familiar with the activity and a single time duration value is thus assigned to each

DOI: 10.1201/9781003428992-10

activity in a project. Further, this single time duration value is considered accurate. This method is used in CPM. The second method is the *non-deterministic* or *probabilistic method*, in which estimates of an activity's time duration lack a fair degree of accuracy. In the non-deterministic method, a planner may assign more than a single value for an activity's time duration. In the non-deterministic method, a planner assigns lower and upper limits within which the time duration of an activity will lie. This method is used in PERT.

In PERT it is difficult to determine an accurate single time duration value for the completion of an activity. Therefore, in PERT, one determines lower and upper time duration limits for the completion of each activity. The uncertainties involved are incorporated by these lower and upper time duration limits. The lower limit is an optimistic estimate of the time duration of an activity. The upper limit is a pessimistic estimate of the time duration of an activity. PERT also uses a most likely time duration, which lies between an activity's lower and upper time duration limits. Thus, to consider the uncertainties of time involved, in PERT the following three time estimates are determined for each activity in a project.

- An *optimistic time duration.*
- A *pessimistic time duration.*
- A *most likely time duration.*

10.3.1 Optimistic Time Durations

This is the minimum possible time duration in which an activity can be completed. The minimum possible time duration for an activity's completion occurs in ideal or near to ideal conditions. The optimistic time duration is the shortest possible time in which an activity can be completed when everything is favorable, everything goes perfectly well, no difficulty takes place, and no adverse conditions appear. In general, an activity can be completed within the optimistic time duration when provided with better than normal conditions. An optimistic time duration is denoted by d_O.

10.3.2 Pessimistic Time Durations

This is the maximum possible time duration that an activity may take. It is the maximum possible time duration in which an activity can be completed given the worst possible conditions. *The pessimistic time duration* is the maximum possible time in which an activity can be completed when everything is unfavorable, things are not well, problems take place, or adverse conditions appear. However, *pessimistic time durations* do not include possible delays because of catastrophes such as strikes, fires, earthquakes, floods, etc. A *pessimistic time duration* is denoted by d_P.

10.3.3 Most Likely Time durations

The most likely time duration is also called the *most probable time.* It is the estimate of an activity's time duration under normal conditions. The value of the most likely time duration lies between those of the optimistic and pessimistic time durations of an activity. The most likely time duration reflects the time required in normal situations where neither the most favorable conditions nor the worst conditions appear. The most likely time duration is denoted by d_L. The uncertainties involved in estimating the time duration of an activity in PERT analysis are incorporated by the optimistic time duration and pessimistic time duration.

10.4 Time Duration Estimation

The estimation of three-time durations for the completion of each activity in a project is not a simple task. The projects in which the PERT technique is used are not routine, and so require three-time durations for each activity. The lower and upper time duration limits corresponding to an activity incorporate the uncertainties involved in this estimation. The range of the lower and upper limits of an activity also provides information about the uncertainties involved.

In CPM, a planner divides a project into various activities, relying on their experience in handling such projects. However, in the case of PERT, a planner is not likely to be familiar with the project under consideration. In PERT, a planner divides a project into activities in such a way that information about each activity in a project may be collected or arranged from different sources. In PERT, experience in executing projects like that under consideration is not available, however, experience of executing a few activities or a part of the project under consideration may be available. Thus, a project is divided into various activities in such a way that information about each activity or a part of the project may be collected from the different sources or parties concerned.

PERT's three time durations are difficult to estimate without any information from the different sources or parties concerned. The information from the different sources or parties concerned, and the experience of a planner, are the bases that determine the lower and upper time duration limits time duration in PERT. Further, the time duration required to complete an activity under different conditions is the main determinant of these time limits. The time duration is minimized when better than normal conditions appear. The time duration is maximized when adverse conditions appear. However, the likelihood of finishing an activity in its minimum or maximum time duration are very low.

10.4.1 Frequency Distribution Curve

The information collected from the different sources and parties concerned, and the experience of a planner, is used to determine the lower and upper time duration limits of an activity. The information collected concerns the time durations required by different parties to perform an activity under varying conditions. A curve is plotted, showing the relationship between a time duration and the number of times a given activity has been completed within that time duration. This curve is called a *frequency distribution curve.* It is clear from the frequency distribution curve, as shown in Figure 10.1, that in the majority of cases, an activity will be completed within its most likely time duration. Point 'b' corresponds to the majority of cases in which an activity is completed within its most likely time duration. Point 'a' corresponds to a very small number of cases in which an activity is completed within its minimum time duration, which is referred to as its optimistic time duration. Point 'c' corresponds to the very small number of cases in which an activity is completed within its maximum time duration, which is referred to as its pessimistic time duration.

The curve shown in Figure 10.1 is symmetrical o both sides of its peak, 'b'. The curves that are symmetrical to both sides of their peaks are known as *normal distribution curves*. However, frequency distribution curves may not be symmetrical to both sides of their peaks; such curves are known as the *β distribution curves*. β distribution curves may be *skewed left* or *skewed right*, as shown in Figures 10.2 and 10.3. Figure 10.2 shows a frequency distribution curve that is skewed left, in which the difference between d_L and d_O is 4 – 3 = 1 day, while the difference between d_P and d_L is 8 – 4 = 4 days. This skewed-left distribution is followed by optimistic time estimators. Figure 10.3 shows a frequency distribution curve that is skewed right, in which the

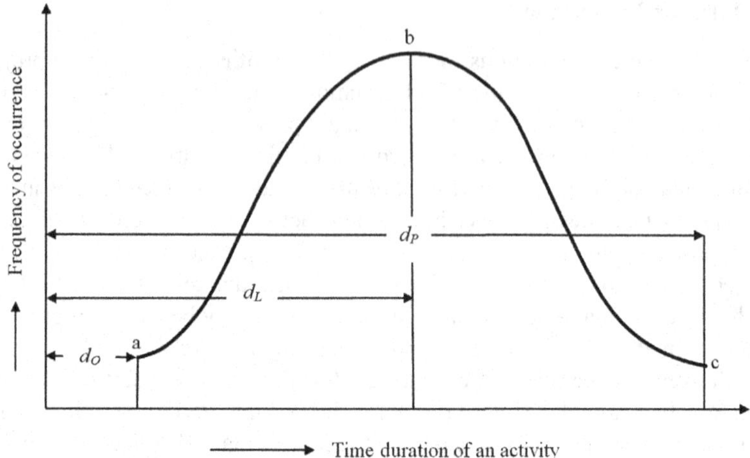

Figure 10.1 Frequency distribution curve with a normal distribution

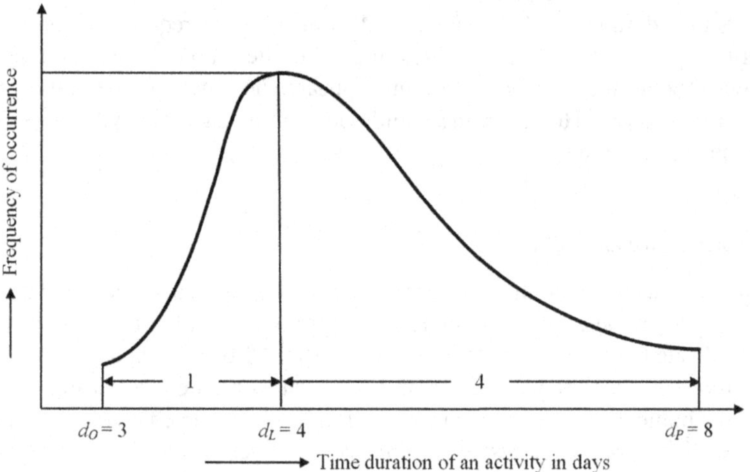

Figure 10.2 Frequency distribution curve with a β distribution: skewed left

difference between d_L and d_O is 6 – 2 = 4 days, while the difference between d_P and d_L is 8 – 6 = 2 days. This skewed-right distribution is followed by pessimistic estimators.

Consider the three frequency distribution curves for activities A, B, and C shown in Figure 10.4. These curves are symmetrical about their peaks, having normal frequency distributions. In Figure 10.4, the curve corresponding to activity 'C' has a wider range between d_P and d_O. This wider range expresses more uncertainty regarding the time duration estimates. The curve corresponding to activity 'A' has a smaller range between d_P and d_O. This smaller range expresses less uncertainty regarding the time duration estimates, thus suggesting a more reliable time estimate for activity 'A' than that of activity 'C'. However, the range of activity 'B' lies between that of activity 'A' and that of activity 'C'. A wider time duration estimate range time duration expresses more uncertainty and a narrow range expresses less.

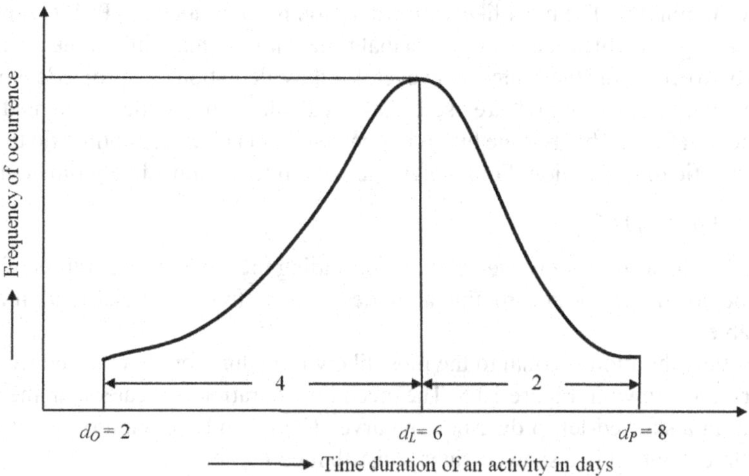

Figure 10.3 Frequency distribution curve with a β distribution: skewed right

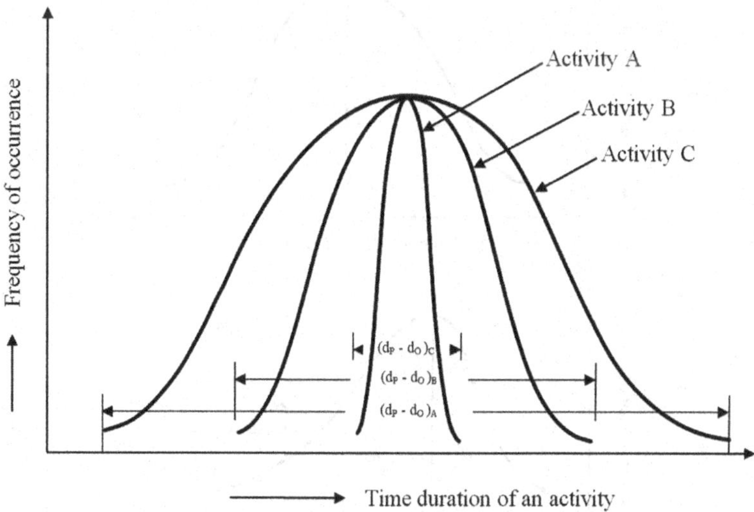

Figure 10.4 Frequency distribution curves: a wider range in the time duration estimates expresses more uncertainty

10.5 Mean Time Duration

In PERT, it is difficult to assign a single time duration value to each activity. Thus, as discussed earlier, an activity is assigned three time durations. A frequency distribution curve is drawn for each activity to determine those three-time durations. For example, consider an activity that involves traveling through a crowded city from station A to station B. The actual time taken will depend upon the traffic, the weather, and the surface condition of the road. If everything proceeds favorably, the travel time may correspond to an optimistic time duration, say 30 minutes. If everything proceeds unfavorably, the travel time may correspond to a pessimistic time duration, say 50 minutes. The most likely time duration is 40 minutes. In PERT, time durations are given as a range in which the activity's actual time duration may fall; in the given example, this range is between 30 to 50 minutes. However, for the calculation of a project's time duration, activities' *mean time durations* (*d*) are used. This is calculated by taking a weighted average of the three time durations. This is done by giving one weight each to the optimistic time duration and the pessimistic time duration. Four weights are given to the most likely time duration:

$$d = (d_O + 4\, d_L + d_P) / 6 \qquad 10.1$$

This is called the *weighted average method* for finding the mean time duration of each activity. It must be noted that for non-routine activities, precise frequency distribution curves may not be available.

The mean time duration is equal to the most likely time duration in a normal frequency distribution curve, as shown in Figure 10.5. The mean time duration is greater than the most likely time duration in a skewed-left β distribution curve. The mean time duration is lower than the most likely time duration in a skewed-right β distribution curve.

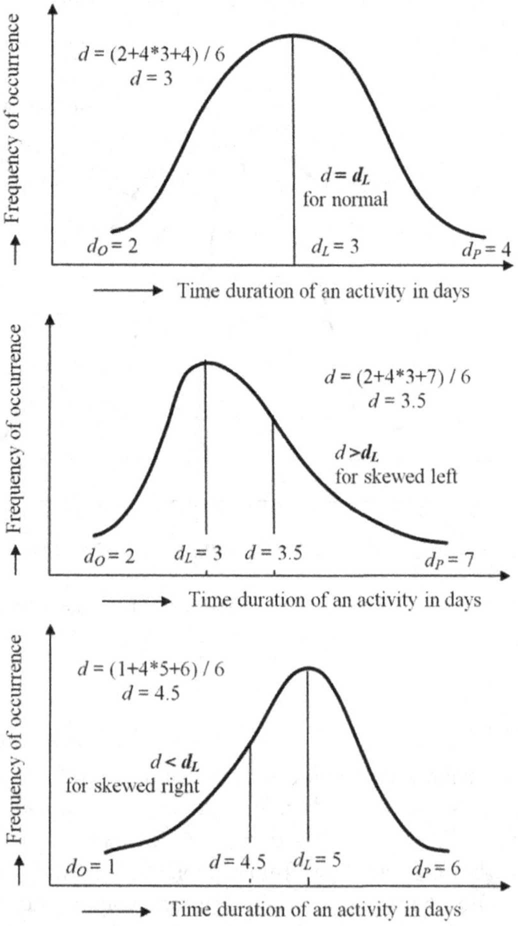

Figure 10.5 Mean time duration and most likely time duration

10.6 Event Times

An event time is the point in time when all the activities preceding an event have just been completed and all the activities following it are just beginning. Consider the execution of an activity. The activity is a trip from station A to station B and its time duration is 15 hours, as shown on the time scale in Figure 10.6. However, the actual available time to perform the journey is 20 hours. The available time runs from 12.00 midnight to 08.00 PM, making the total available time duration equal to 20 hours. At its earliest, the activity can be started at 12.00 noon; this is the earliest event time of the start event of the activity in question. The earliest event time is the earliest possible time at which an event can occur.

i. ***Earliest Event Time (EET):*** The earliest point in time at which all the activities preceding an event have just been completed and all the activities following it have just begun. The earliest event time is the earliest possible time at which an event in a network can occur. It is calculated by moving from the start to the end event of a network.

   In Figure 10.6, the available time duration is 20 hours and the estimated time duration of the activity is 15 hours. If the activity starts at its earliest, at 12.00 midnight, it will end at 3.00 PM. There is a free time interval of 5 hours from 3.00 to 8.00 PM after the activity's early end. If the start time of the activity in question is delayed by 5 hours from 12.00 midnight to 5.00 AM, the activity will start at 5.00 AM and end at 8.00 PM without going beyond its available time duration. The activity can be started at its latest at 5.00 AM, bearing in mind the available time; this is the latest event time of the start event of the activity in question.
ii. ***Latest Event Time (LET):*** The latest possible time by which an event can occur without causing any delay to the completion of a project. It is the latest time at which activities preceding the event in question may be allowed to finish. It is the latest time before which all activities preceding the event in question may be completed (as late as possible) so as to allow the subsequent activities to start (as late as possible). The latest event time is the latest allowable time for an event in a network to occur. It is calculated by moving from the end to the start event of a network.

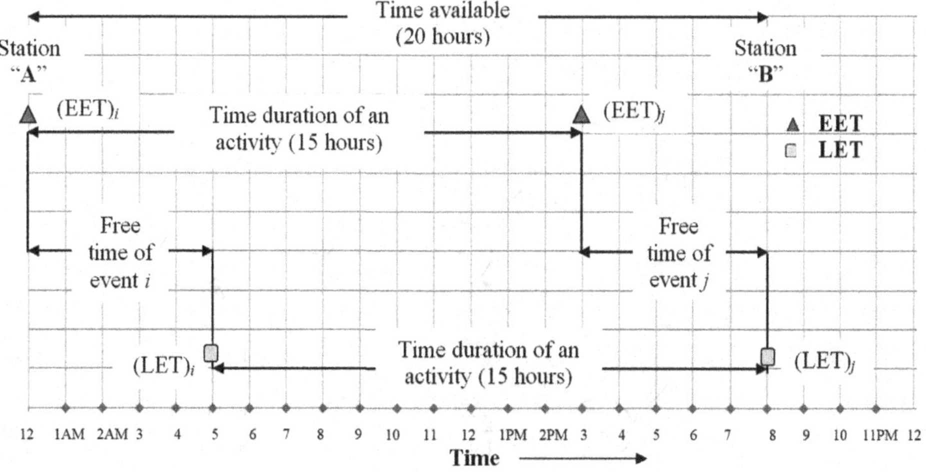

Figure 10.6 Earliest and latest event times according to the time scale

As shown in Figure 10.6, the start event of the activity in question can be delayed by a maximum of up to 5 hours – that is, 12.00 midnight to 5.00 AM – without going beyond its available time duration.

10.7 Time-Scaled Networks: Calculations Of Earliest Event Times

The network used to demonstrate earliest event time calculations is shown in Figure 10.7. This network has been redrawn according to the time scale shown in Figure 10.8. The time-scaled version of a network is the most convenient way to understand the concept of EET. The x-axis is used as the time axis and on the y-axis activities are represented by arrows. The anticipated start and end times of each activity in a network are depicted as the start and end of the arrow corresponding to the activity under consideration. The length of an arrow represents the time duration of an activity. The number of arrows in a network is equal to the number of activities in a project. Dummy activities are shown using dotted lined arrows.

The simplest method for drawing a time-scaled network is to first draw the activities that lie on its longest path. In Figure 10.8, activities B, D, and G make up the longest path, with a time duration of 11 (3 + 5 + 3) days. All these activities are represented initially as shown in the figure. The length of the time-scaled version of a network is equal to the project's time duration or the sum of the time durations of all the activities which lie on its longest path. The path just smaller than the longest path is drawn after the longest path is completed. In this way, all the paths/activities in a network are drawn according to a time scale. It must be noted that when activity A is drawn between nodes 1 and 3, the time available between nodes 1 and 3 is 3 days, however activity A has a time duration of 2 days. Activity A starts with the start of the project and finishes at the end of day 2. The remaining day between nodes 1 and 3 is shown using a dotted line. The starting event, numbered as 1, starts at time zero. In other words, the start time of the project and the start time of event number 1 is assumed as zero.

At event 1, the project starts along with the simultaneous start of activities A and B. The starting event, numbered as 1, starts at time zero. Therefore, the EET of event 1 is zero. Activities A and B have time durations of 2 and 3 days respectively, as shown in Figure 10.8. The arrow corresponding to activity A ends at the end of day 2. The arrow corresponding to activity B ends at the end of day 3. It is assumed that activity A will start as soon as possible.

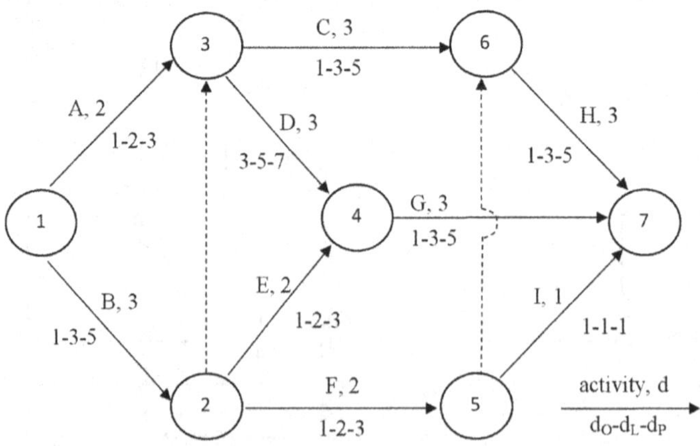

Figure 10.7 A network used to demonstrate the calculation of earliest event times

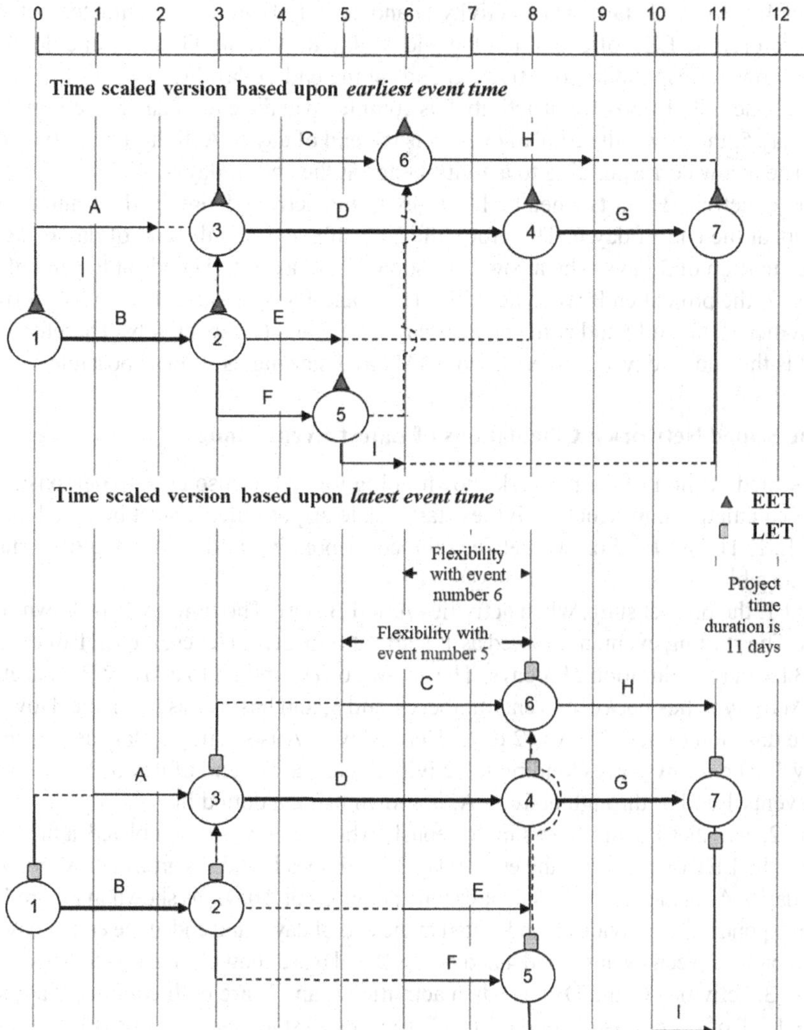

Figure 10.8 Time-scaled version of the network shown in Figure 10.7

At event 2, activities E and F start simultaneously when activity B is completed, at its earliest, at the end of day 3. Activities E and F start at the end of day 3, therefore the EET of event 2 is at the end of day 3. Activities E and F both have time durations of 2 days. The arrows corresponding to activities E and F end at the end of day 5. Activities E and F will start as soon as possible. Activity E starts at the end of day 3 and ends at the end of day 5. The 3 days on the path between events 2 and 4, and 1 day on the path between events 5 and 6, are shown using dotted lines.

At event 3, activities C and D start when activities A and B are both completed at the end of day 3. The EET of event 3 is the end of day 3. Activity C takes 3 days to complete. The arrow corresponding to activity C ends at the end of day 6. Activity D has a 5-day time duration. The arrow corresponding to activity D ends at the end of day 8.

At event 4, activity G starts when activity D and activity E are both completed at the end of day 8. Therefore, the EET of event 4 is the end of day 8. Activity G has a time duration of 3 days. The arrow corresponding to activity G ends at the end of day 11.

At event 5, activity I starts when activity F is completed at the end of day 5. Activity I starts at the end of day 5, therefore, the EET of event 5 is the end of day 5. Activity I has a time duration of 1 day. The arrow corresponding to activity I ends at the end of day 6.

At event 6, activity H starts when both activity C and activity F are finished, and event number 5 occurs at the end of day 6. Therefore, the EET of event 6 is the end of day 6. Activity H has a time duration of 3 days. The arrow corresponding to activity H ends at the end of day 9.

At event 7, the project ends when activities H, G, and I are finished. The arrows corresponding to activities H, G, and I end at the ends of days 9, 11, and 6 respectively. Therefore the EET of event 7 is the end of day 11, given activities H and I starting as soon as possible.

10.8 Time Scaled Networks: Calculations of Latest Event Times

The time-scaled version of the network shown in Figure 10.8 is also used to demonstrate latest event time calculations, in which activities start as late as possible. It must be noted that activities A, C, E, F, H, and I are drawn between the corresponding nodes, keeping their start times as late as possible.

At event 1, the project starts when activities A and B start. The time scale is drawn on a late-start basis. The starting event, numbered as 1, starts at time zero. Hence, the LET of event 1 is 0. Activity B has a time duration of 3 days. The arrow corresponding to activity B ends at the end of day 3. Activity A lies between event numbers 1 and 3, and has 3 days available. However, the actual time duration of activity A is 2 days. Here activity A is starting as late as possible at the end of day 1. The arrow corresponding to activity A ends at the end of day 3. 1 day on the path between events 1 and 3, through activity A, is shown using a dotted line.

At event 2, activities E and F start simultaneously when activity is B completed at the end of day 3, therefore, the LET of event 2 is the end of day 3. Activities E and F start as late as possible with the end of day 6. Activities E and F both have time durations of 2 days, as shown in Figure 10.8. The arrows corresponding to activities E and F start at the end of day 6 and end at the end of day 8. The 3 days on the path between events 2 and 4 and events 2 and 6 are shown using dotted lines.

At event 3, activities C and D start when activities A and B are both completed at the end of day 3. The LET of event 3 is the end of day 3. Activity C starts as late as possible at the end of day 5 and takes 3 days to complete. The arrow corresponding to activity C ends at the end of day 8. Activity D has a 5-day time duration. The arrow corresponding to activity D also ends at the end of day 8.

At event 4, activity G starts when activity D and activity E are both completed at the end of day 8. Hence, the LET of event 4 is at the end of day 8. Activity G has a time duration of 3 days. The arrow corresponding to activity G ends at the end of day 11.

At event 5, activity F finishes as late as possible at the end of day 8. Thus, the LET of event 5 is at the end of day 8. Activity I starts as late as possible at the end of day 10 and the arrow corresponding to it ends at the end of day 11.

At **event 6**, activity H starts when activity C finishes as late as possible and event 5 occurs as late as possible at the end of day 8. Thus, the LET of event 6 is at the end of day 8. Activity H has a time duration of 3 days. The arrow corresponding to activity H ends at the end of day 11.

At event 7, the project ends when activities H, G, and I end. The arrows corresponding to activities H, G, and I end at the end of day 11. Therefore, the LET of event 7 is the end of day 1, given activities H and I starting as late as possible.

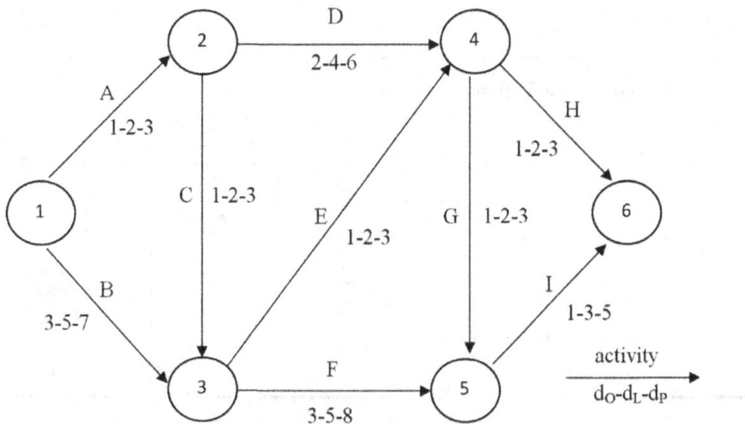

Figure 10.9 A sample network

Example 10.1: For the network shown in Figure 10.9, calculate the earliest event time and latest event time of each event using a time-scaled version of the network.

Solution: The earliest and latest event times of each event in the network have been calculated by drawing the time-scaled version of the network, as shown in Figure 10.10.

10.9 Forward Pass: Calculations of EET

The concepts of earliest event times and latest event times have been explained in the previous sections with the help of the time-scaled version of a network. However, a project may consist of hundreds of activities; in such cases, calculating event times using the time-scaled version of a network is difficult. In such cases, event time calculations are performed using forward and backward pass calculations. Event time calculations involve first a forward and then a backward pass calculation through a network.

Forward pass calculations proceed sequentially from the starting event to the terminal event of a network moving along with the direction of its arrows. Forward pass calculations give the earliest event time of each event. Forward pass calculations are started by assigning an arbitrary earliest start time to the starting event of a project. A time value of zero is usually assigned to the starting event as its earliest start time. It is assumed that a given project begins at time zero and has only one initial and one terminal event. The computations proceed on the assumption that each activity will start as soon as possible, that is, as soon as all of its preceding activities are completed. Forward pass computations proceed sequentially from the starting event to the terminal event of a network following the steps given below.

Step 1: Assume that the starting event of a project occurs at time zero. The earliest event time of the starting event is assumed to be zero.

Earliest event time of the starting event = 0

Step 2: Assume that all the activities in a network start as soon as possible. In other words, as soon as all the activities preceding the event under consideration are completed, all

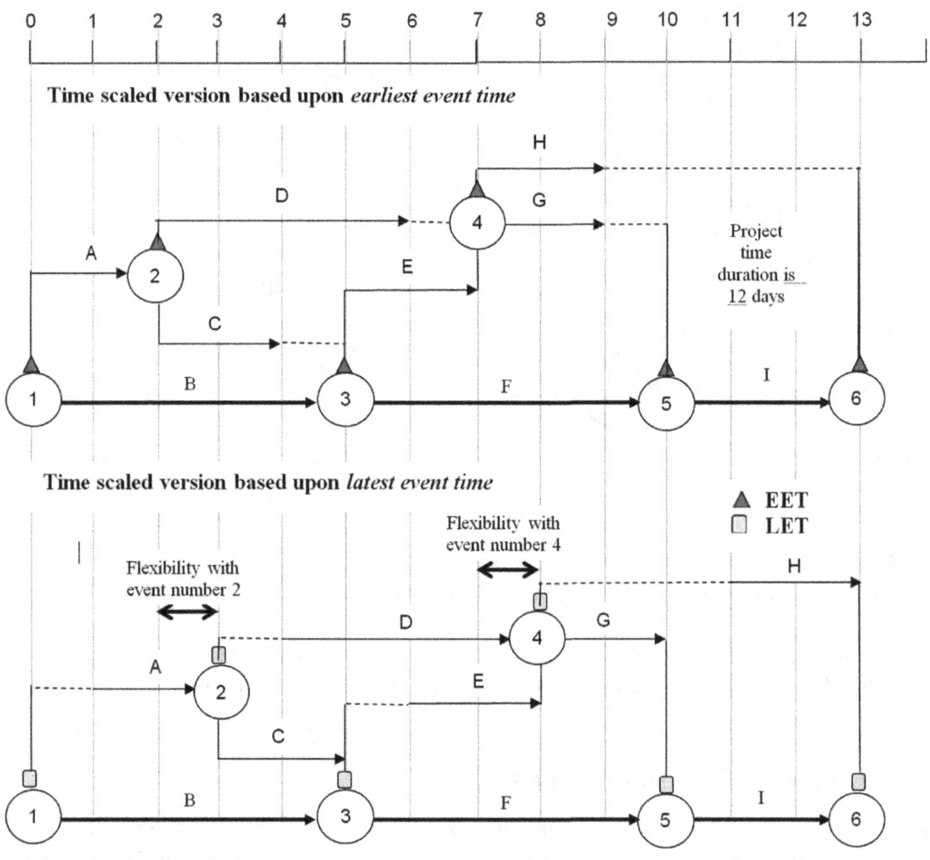

Figure 10.10 Time scaled version of the network shown in Figure 10.9

activities succeeding it will start as early as possible, without any delay. In this case, an event occurs at its earliest event time, thus the EET of the event (i) under consideration is given by:

EET_i = Maximum earliest finish time of the activities immediately preceding (all activities ending at node i) the event under consideration

Step 3: The earliest finish time (EFT_{ij}) of activity a_{ij} is the algebraic sum of the earliest event time (EET_i) of the starting event (i) and the time duration of the activity under consideration (d_{ij}). The earliest finish time of activity a_{ij} is given by:

$$EFT_{ij} = EET_i + d_{ij} \qquad 10.2$$

The computations proceed sequentially from the starting event to the terminal event of a network along the direction of the arrows. Forward pass calculations give *the earliest event time* for each event of a network. The EET computed for each event is generally written on the event itself.

10.9.1 Calculation of Earliest Event Times Using a Forward Pass

Consider the network shown in Figure 10.7 for the demonstration of the use of forward pass calculations to find the EETs of all events. The calculations are shown in Figure 10.11. In forward pass calculations, the starting event, numbered as 1, starts at time zero. The start time of the project is zero (as discussed in step 1).

At event 1, the project starts when activities A and B start simultaneously. The starting event, numbered as 1, starts at time zero. Therefore, the EET of event number 1 is 0. The EFT of an activity is calculated by adding its time duration to its EET. Therefore, the EFTs of activities A and B are 2 (0 + 2) and 3 (0 + 3) respectively.

At event 2, activities E and F start simultaneously when activity B is completed at the end of day 3. Therefore, the EET of event number 2 is at the end of day 3. Activities E and F start at the end of day 3. The EFTs of activities E and F are 5 (3 + 2) and 5 (3 + 2) respectively.

At event 3, activities C and D start simultaneously when activities A and B are complete. Activity A takes 2 days to complete and activity B takes 3 days to complete. Activities C and D start simultaneously when both activities A and B are completed at the end of day 3. This is the maximum EFT of the activities immediately preceding event number 3, when both activities A and B are finished (discussed in step 2). The EFTs of activities C and D are at the ends of days 6 (3 + 3) and 8 (3 + 5) respectively.

The crux of a forward pass calculation occurs at a merge event, where the maximum EFT of its preceding activities is held to determine the EET of the event under consideration.

At event 4, activity G starts when activities D and E are completed. Activity D takes 8 days to complete and activity E takes 5 days to complete. Activities D and E are both completed at the end of day 8. This is the maximum EFT of the activities immediately preceding event number 4. Hence, the EET of event number 4 is at the end of day 8. The EFT of activity G is the end of day 11 (8 + 3).

At event 5, activity I starts when activity F is completed. Activity F takes 5 days to complete. The EET of event number 5 is at the end of day 5. The EFT of activity I is the end of day 6 (5 + 1).

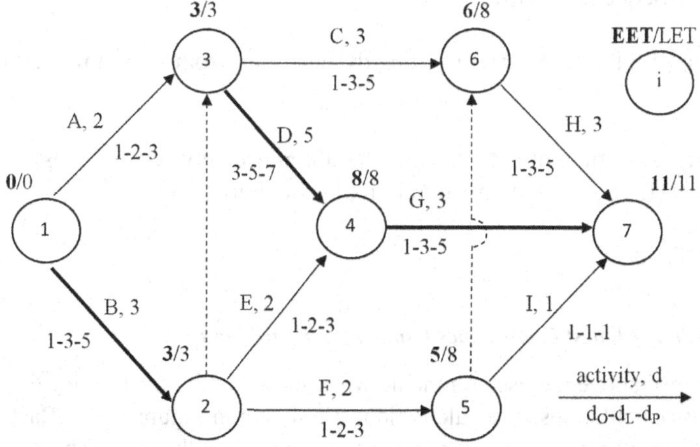

Figure 10.11 Earliest and latest event times of each event, found using forward and backward pass calculations

At event 6, activity H starts when activities C and F are completed. Activity C takes 6 days to complete and activity F takes 5 days to complete. Activities C and F are both completed at the end of day 6. This is the maximum EFT of the activities immediately preceding event number 6. Hence, the EET of event number 6 is at the end of day 6. The EFT of activity H is the end of day 9 (6 + 3).

At event 7 the project ends when activities H, G, and I end. H, G, and I finish at the ends of days 9, 11, and 6 respectively. The maximum EFT of the activities immediately preceding event number 7 is the end of day 11. Therefore, the EET for event number 7 is the end of day 11, and the time duration of the project is 11 days.

10.10 Backward Pass: Calculations of LET

Backward pass calculations are done sequentially from the terminal event to the starting event of a project by moving against the direction of the arrows. Backward pass calculations provide the latest event times of each event in a network. Backward pass calculations start by deciding on a value for the scheduled completion time duration of a project. The value of the scheduled completion time duration is assigned to the terminal event as its latest event time. The project's terminal event must occur on or before the scheduled completion time duration. If the scheduled completion time duration of a project is not specified, the latest event time of the terminal event is taken to be equal to its earliest event time as calculated in the forward pass calculations. Backward pass calculations proceed sequentially from the terminal event to the starting event of a network following the steps given below.

Step 1: The latest event time of the terminal event of a project is taken to be equal to the project's scheduled completion time duration (d_S) or its earliest event time as computed in the forward pass computations.

Latest event time of the terminal event = Scheduled completion time duration (d_S) of the project or earliest event time of the terminal event

Step 2: The latest event time of an event is equal to the smallest, or earliest, of the latest start times of its subsequent activities.

LET = Minimum LST of the activities directly emerging from or following the event under consideration

Step 3: The latest start time of activity a_{ij} is the difference between the latest event time of its terminal event (LET_j) and the time duration of an activity (d_{ij}).

$$LST_{ij} = LET_j - d_{ij} \qquad 10.3$$

10.10.1 Calculation of Latest Event Times Using a Backward Pass

The steps discussed above are used on the network shown in Figure 10.7 for the demonstration of backward pass calculations. The calculations are shown in Figure 10.11. Backward pass calculations start from the terminal event of a network, event 7 in this case. The earliest event time of event 7, as computed in the forward pass calculations, is taken to be its latest event time, that is, the end of day 11, as discussed in step 1.

At event 7, the project finishes when activities G, H, and I end at the end of day 11. Hence, the LET of event 7 is at the end of day 11. The LST of activity H is the end of day 8 (11–3), for activity G it is the end of day 8 (11–3), and for the activity I, it is the end of day 10 (11–1).

At event 6, the LET is the minimum LST of the activities directly following the event (step 2). The LST of activity H, directly following event 6, is the end of day 8. Thus, the LET of event 6 is the end of day 8. The LST of activity C is the end of day 5 (8–3).

At event 5, its LET is equal to the smallest or earliest LST of its subsequent activities, H and I. The LST of activity H is at the end of day 8 and the LST of the activity I is at the end of day 10. The smallest or earliest LST of its subsequent activities is the end of day 8. The LET of event 5 is the end of day 8. The LST of activity F is the end of day 6 (8–2).

At event 4, its LET is equal to the LST of its subsequent activity G. The LST of activity G is the end of day 8. Thus, the LET of event 4 is the end of day 8. The LST of activity D is the end of day 3 (8-5) and for activity E it is the end of day 6 (8–2).

At event 3, its LET is equal to the smallest or earliest LST of its subsequent activities C and D. The LST of activity C is the end of day 5 and for activity D it is the end of day 3. The smallest or earliest LST of the subsequent activities is the end of day 3. Thus, the LET of event 3 is at the end of day 3. The LST of activity A is the end of day 1 (3–2).

At event 2, its LET is equal to the smallest or earliest LST of its subsequent activities C, D, E, and F. The LSTs of activities C, D, E, and F are the ends of days 5, 3, 6, and 6 respectively. The smallest or earliest LST of the subsequent activities is the end of day 3. Thus, the LET of event 2 is at the end of day 3. The LST of activity B is 0 (3–3).

At event 1, its LET is the smallest or earliest LST of its subsequent activities A and B. The LST of activity A is the end of day 1 and that of activity B is 0. The smallest or earliest LST of the subsequent activities is 0. Thus, the LET of event 1 is 0.

Example 10.2: For the network shown in Figure 10.9, calculate the earliest and latest event times of each event using forward and backward pass calculations.

Solution: The earliest and latest event times of each event in the network, found using forward and backward pass calculations, are shown in Figure 10.12.

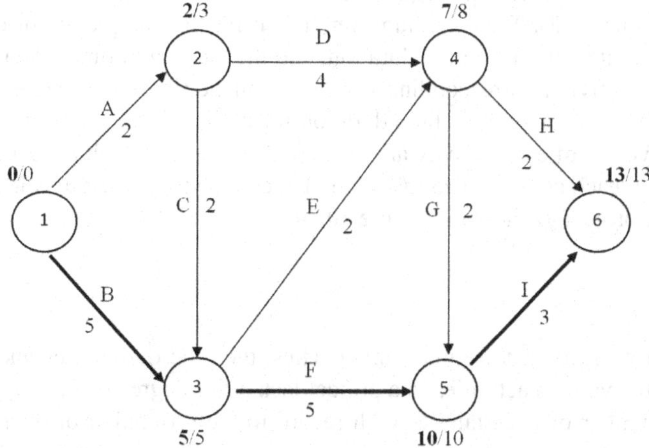

Figure 10.12 Earliest and latest event times of each event, found using forward and backward pass calculations

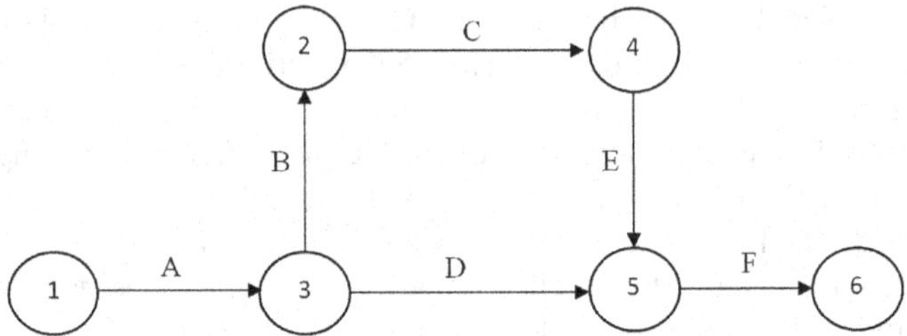

Figure 10.13 The project network

Table 10.1 Time duration used to execute various activities by different parties

Activities	*Time durations in hours*
A	3.5, 3, 1, 3, 1.5, 4.5, 3, 2, 2, 2, 4, 3, 2.5, 2.5, 4, 2.5, 3.5, 3.5, 3, 2.5, 1.5, 3, 4, 3, 3, 3.5, 4.5, 5, 3
B	5, 4, 2, 4, 6, 4, 4.5, 4, 2.5, 5.5, 4, 3, 4, 5, 3, 4, 4.5, 3, 5, 3.5, 3.5, 4, 3.5, 3.5, 4.5, 4.5, 4, 4.5, 4.5, 2.5, 3, 5.5, 3.5, 4.5
C	2, 2, 1.75, 2, 2, 1, 3, 2, 1.5, 2, 2, 1.5, 2, 1.75, 2, 2.75, 2, 1.5, 2, 1.75, 1.75, 2.75, 2.75, 3
D	2, 2, 1, 3, 2, 2, 1.75, 1.75, 2.75, 2.75, 3,1.5, 2, 2, 1.5, 2, 1.75, 2, 2.75, 2, 1.5, 2, 2, 1.75
E	2, 2, 1, 2, 3, 2, 1.5, 2, 1.5, 1.5, 2, 1.75, 2, 1.75, 2, 2.75, 2, 2.75, 2, 1.75, 1.75
F	3.5, 3, 2, 4, 3, 2.5, 2.5, 4, 2.5, 3.5, 1, 3, 1.5, 4.5, 3, 2, 2, 4, 3.5, 3, 2.5, 1.5, 3, 3, 3, 3.5, 4.5, 5, 3

Example 10.3: A project consists of six activities A, B, C, D, E, and F. The relationships between the different activities are represented in the form of a network given in Figure 10.13. The time taken (in hours) to execute the various activities by the different parties is given in Table 10.1. Each party executed an activity by employing a group that consists of the same quantities and types of resources. Calculate the time duration of the project.

Solution: From Table 10.1, the frequencies of occurrence of the different activities are determined (Table 10.2). The minimum time taken for the completion of each activity, corresponding to its optimistic time duration, and the maximum time taken for the completion of each activity, corresponding to its pessimistic time duration, are determined. From Table 10.2, the most likely time duration for each activity is also determined. The d_O, d_P, and d_L values of each activity are given in Table 10.3, which are used to calculate the '*d*' values of each activity. The '*d*' value of each activity is used to calculate the time duration of the project as shown in Figure 10.14.

10.11 Conclusion

PERT is used for projects that are not routine. Thus, the time duration estimates required for the execution of the various activities in a project lack a fair degree of accuracy. Such projects include a large number of uncertainties with regard to the estimation of their activities' time durations, therefore the minimum, maximum, and most likely time durations for each activity are decided. These three time durations are converted to a single time duration value and used to determine the project's time duration. PERT is an event-based technique in which event times

Table 10.2 Frequency distribution of the various activities in the project

Activity A (records = 29)		*Activity B (records = 34)*		*Activity C (records = 24)*		*Activity D (records = 24)*		*Activity E (records = 21)*		*Activity F (records = 29)*	
d	fr.	d	fr.	d	fr.	d	fr.	d	fr.	d	fr.
1	1	2	1	1	1	1	1	1	1	1	1
1.5	2	2.5	2	1.5	3	1.5	3	1.5	3	1.5	2
2	3	3	4	1.75	4	1.75	4	1.75	4	2	3
2.5	4	3.5	5	2	11	2	11	2	10	2.5	4
3	9	4	9	2.75	3	2.75	3	2.75	2	3	9
3.5	4	4.5	7	3	2	3	2	3	1	3.5	4
4	3	5	3	-	-	-	-	-	-	4	3
4.5	2	5.5	2	-		-	-	-	-	4.5	2
5	1	6	1	-	-	-	-	-	-	5	1

Table 10.3 d_O, d_L, d_P and d for each activity of the project

Activities	d_O	d_L	d_P	*d*
A	1	3	5	3
B	2	4	6	4
C	1	2	3	2
D	1	2	3	2
E	1	2	3	2
F	1	3	5	3

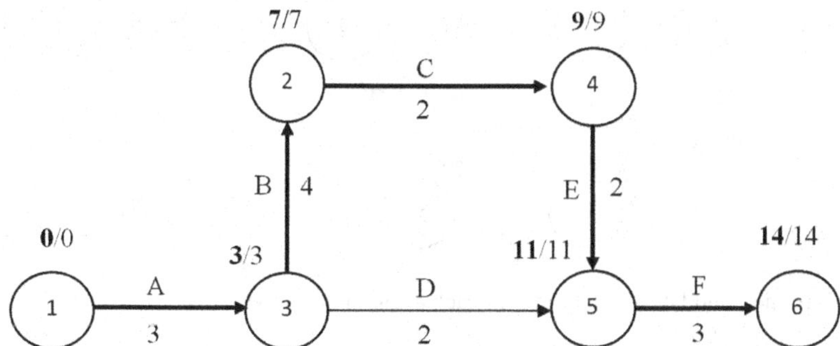

Figure 10.14 Earliest and latest event times of each event, found using forward and backward pass calculations

are calculated. This chapter has covered the calculation of earliest and latest event times using a time-scaled version of a network that also demonstrates the physical significance of those two times. Further, the calculation of earliest and latest event times has also been demonstrated with the help of forward and backward pass calculations.

Exercises

Question 10.1: The inter-dependencies between the various activities in a network and their mean time durations are given in Table E10.1. Calculate the earliest and latest event times of each event using forward and backward pass calculations.

Table E10.1 Inter-dependencies between various activities and their mean time durations

Activities	*Depends upon*	*Mean time durations (days)*
N	-	2
O	N	3
P	O, S	2
Q	P, Y	1
R	Q, U	3
S	N	4
T	S	2
U	T, Y	2
X	N	3
Y	S, X	1
Z	R	2

Solution: The earliest and latest event times of each event, found using forward and backward pass calculations, are shown in Figure E10.1.

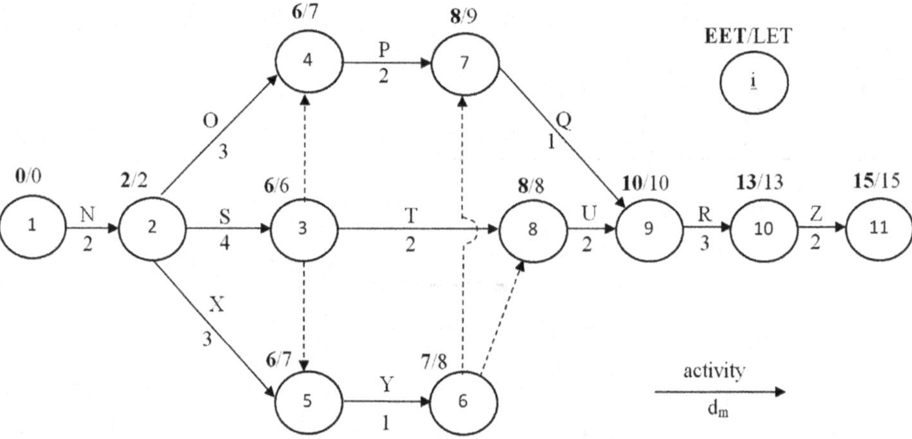

Figure E10.1 Earliest and latest event times of each event in the network

Table E10.2 Inter-dependencies between various activities and their mean time durations

Activities	*Depends upon*	*Mean time durations (days)*
A	-	2
B	-	4
C	-	3
D	A	2
E	B, C	1
F	B	4
G	D, E, F	3
H	F	2
I	D, H	4
J	I, G	3

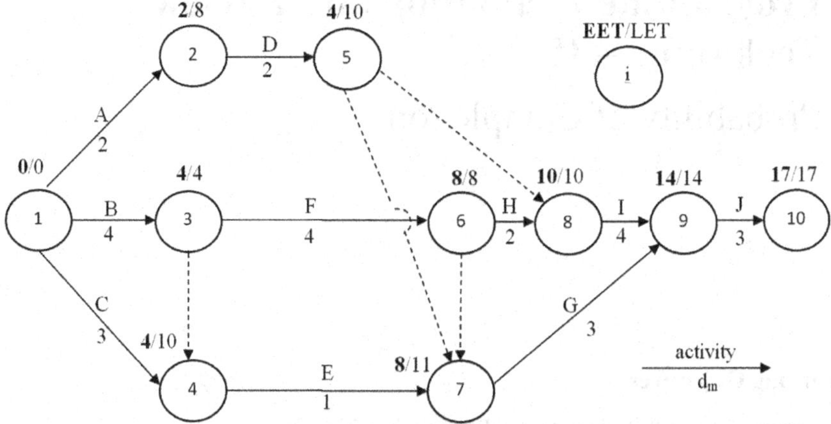

Figure E10.2 Earliest and latest event times of each event in the network

Question 10.2: The inter-dependencies between the various activities in a network and their mean time durations are given in Table E10.2. Calculate the earliest and latest event times of each event using forward and backward pass calculations.

Solution: The earliest and latest event times of each event, found using forward and backward pass calculations, are shown in Figure E10.2.

11 Programme Evaluation and Review Technique – II

Probability of Completion

11.1 Learning Objectives

After the completion of this chapter, readers will be able to:

- Determine the critical path and critical activities of a project,
- Determine the probability of completing of a project within a time duration longer or shorter than the mean time duration, and
- Understand the difference between CPM and PERT.

11.2 Introduction

The programme evaluation and review technique (PERT) is an event-oriented technique. It is used in research and development (R&D), one-time, complex, and non-routine projects. In such projects, time duration estimates for their various activities are highly uncertain. Also, the past records of the estimate of time durations of various activities of the PERT network are very less. Therefore, PERT uses minimum and maximum time duration limits for each activity. A probability is associated with the calculated project time duration. This probability depends upon uncertainties in the time durations of the project's various activities. PERT is used for projects in which the time duration is the primary factor, rather than the project's cost.

11.3 Critical Path

A network consists of many paths running from its starting event to its terminal event. The different possible paths in a network have different path lengths. The length of a path is the time duration required to complete it. This is calculated by adding together the time durations of all the activities that lie on the path. Out of all the possible paths in a network, the critical path is the longest path that connects the start event and the end event of a network. A critical path is usually denoted on a network using thick-lined arrows. Out of all the possible paths in a network, at least one path is critical. A network may have more than one critical path, however, and in extreme conditions all the paths in a network may become critical paths. However, the lengths of all critical paths are equal.

The network shown in Figure 11.1 has seven possible paths: A-C-H (path length = 8), A-D-G (path length = 10), B-C-H (path length = 9), B-D-G (path length = 11), B-E-G (path length = 8), B-F-I (path length = 6), and B-F-H (path length = 8). The longest possible path in the network is B-D-G. Path B-D-G is the critical path of the network, and the length of this critical path is 11 days. It should be noted that in the case of forward pass calculations the earliest event time is calculated (and written on every event in bold numbers), as shown in Figure 11.1. In backward

DOI: 10.1201/9781003428992-11

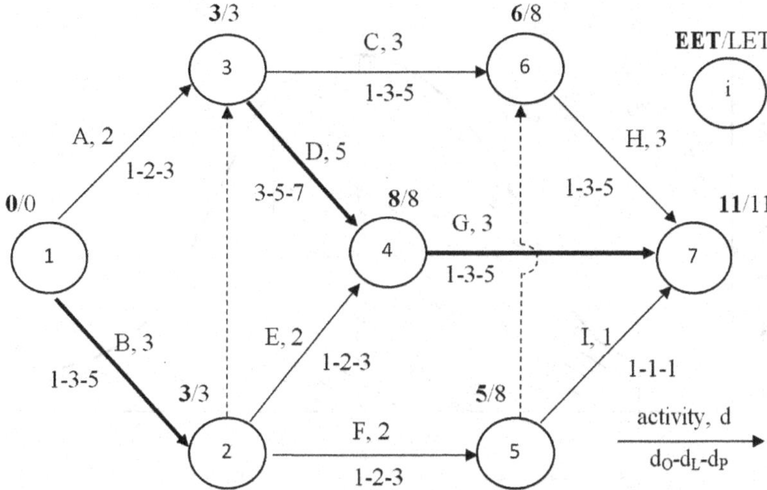

Figure 11.1 PERT network

pass calculations, the latest event time is calculated (and written on every event in normal numbers). For events which lie along the critical paths, the earliest event time is equal to the latest event time. For the network under consideration, the critical path is B-D-G and the critical events which lie on the critical path are 1, 2, 3, 4, and 7. For these events the earliest event time is equal to the latest event time.

The identification of a critical path is useful for effective project planning and control. It helps a planner allocate the required resources to their critical activities to ensure their timely completion. The difference between the latest event time and the earliest event time is called slack.

$$\text{Slack} = \text{LET} - \text{EET} \qquad 11.1$$

For the network under consideration, the slack values of events 1, 2, 3, 4, and 7, which lie along the critical path, are zero.

11.4 Critical Activities

All the activities along the critical path of a network are called critical activities. All the events which lie on the critical path are called critical events. The reason behind this criticality is that any delay to the time durations of critical activities will result in an equal delay to the scheduled completion time duration of a project. The identification of critical activities helps a planner allocate the required resources to their critical activities to ensure their timely completion. In the network shown in Figure 11.1, the critical path is B-D-G and the activities on the critical path are B, D, and G. Thus, activities B, D, and G are critical activities. The sum of the time durations of the critical activities (3 + 5 + 3) is equal to the critical path length. The length of the critical path is equal to the time duration for the completion of a project.

11.5 Probability Distribution

The three time duration estimates corresponding to a given activity are determined from its frequency distribution curve. The data used to draw a frequency distribution curve are collected from different sources. Frequency distribution curves have inherent uncertainties concerning the

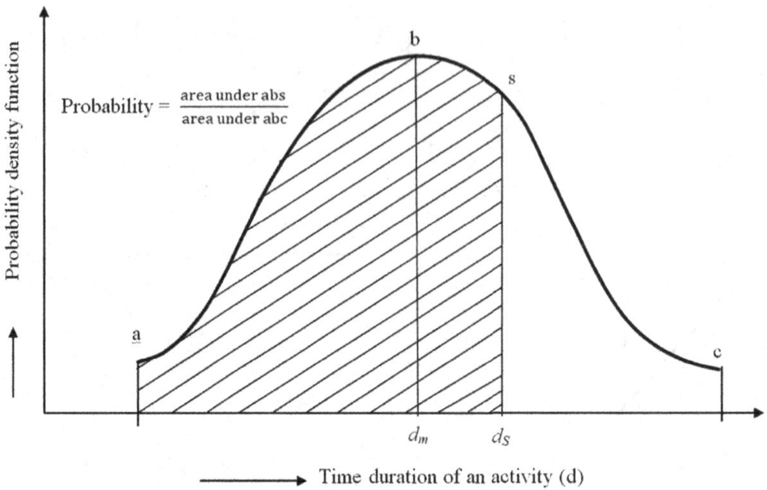

Figure 11.2 Probability distribution curve

time duration used to draw the curve. Probability is connected with these uncertainties. In probability theory, probability values are associated with all possible outcomes. Probability values have variations from 0 to 1. A probability value of 1 indicates that an event has a 100 percent chance of occurring while a probability value of 0 indicates that an event will not occur at all. A probability value nearer to 1 indicates a higher certainty of an event occurring.

A probability value is associated with the estimated time duration of an activity based on the data available. Generally, data collected from different sources are used to plot the *probability distribution curve*. The probability distribution curve is a frequency distribution curve; the values along the y-axis are standardized such that the area under the curve is equal to 1. The height or the ordinate of the curve at any point d_s is denoted by a function *f(d)*, usually called the *probability density function*. Figure 11.2 shows a probability distribution curve. The probability of completing an activity within the scheduled time duration (d_s) is equal to the ratio of the shaded area to the total area of the curve. Since the total area of the curve is equal to 1, the probability of completing an activity within the scheduled time duration is equal to the shaded area itself. The possibility of completing an activity within the optimistic time duration is minimum however maximum within the *pessimistic time duration*.

11.6 Central Limit Theorem

A project consists of several activities that have their own frequency distribution curves, as shown in Figure 11.3. Each frequency distribution curve has inherent uncertainties in its three time durations. The frequency distribution curve is used to construct the probability distribution curve. The probability of completing an activity within the scheduled time duration is calculated with the help of the probability distribution curve. However, in PERT, the determination of the probability of a project's completion is much more important than the probability of completion of an individual activity. To determine the probability of completing a project, the frequency or probability distribution curve of the project is required. PERT is used for projects that are not routine, therefore it is not possible to construct a frequency or probability distribution curve for the specific type of project. This necessitates the production of a

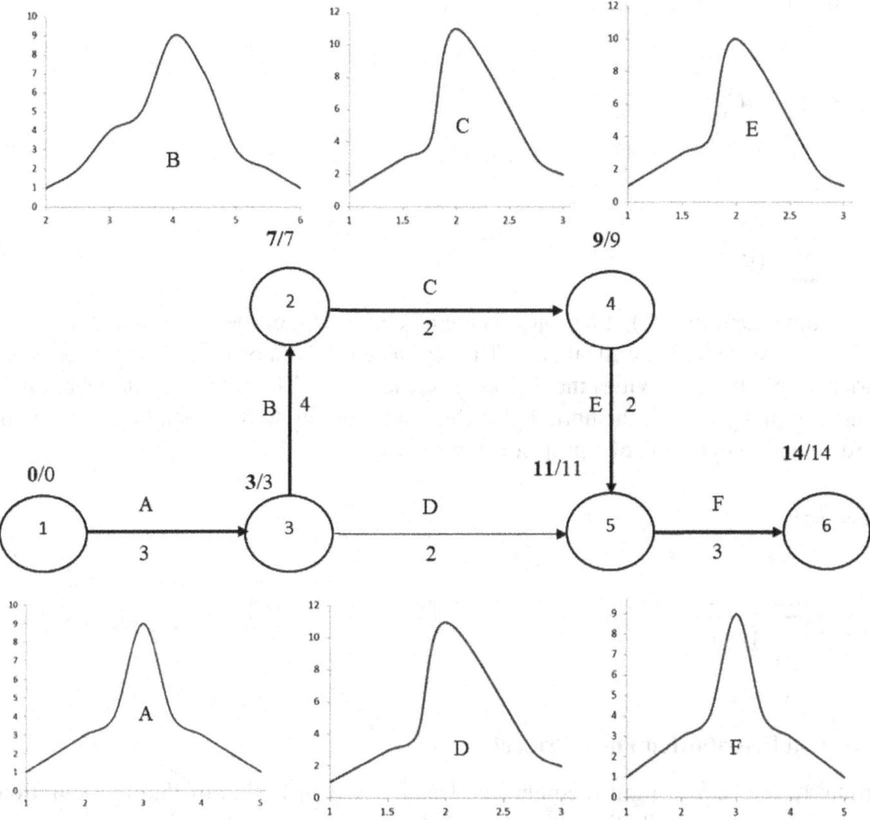

Figure 11.3 Different frequency distribution curves for different activities in the project

frequency or probability distribution curve for the project as a whole to determine the probability, in which the *central limit theorem* plays a major role. The theorem is stated as follows:

> The distribution of the mean of n independent observations from any distribution, or even from up to n different distributions, with finite mean and variance, approaches a normal distribution as the number of observations in the sample becomes large – that is, as n approaches infinity. The result holds irrespective of the distribution of each of the n elements making up the average.

The sum of the time durations of the *n* activities that lie along the critical path of a PERT network is equal to the project's time duration. Assume each activity in a project has a frequency distribution curve as shown in Figure 11.3. If there are 1, 2, 3, ..., *n* activities along the critical path of a PERT project, all the critical activities will have a mean time duration of d_1, d_2, d_3,.........................., d_n respectively, and will have a *standard deviation* of σ_1, σ_2, σ_3,.................., σ_n respectively. According to the central limit theorem, the frequency distribution of the time duration of a project as a whole is the *normal distribution,* also called *Gaussian distribution.* The mean time duration and *variance* values for the normal distribution of a project are given by:

$$d_m = d_1 + d_2 + d_3 + \dots\dots + d_n$$

$$d_m = \sum_{i=1}^{n} (d)_i \qquad 11.2$$

$$\sigma^2 = \sigma_1^2 + \sigma_2^2 + \sigma_3^2 + \dots\dots + \sigma_n^2$$

$$\sigma^2 = \sum_{i=1}^{n} (\sigma^2)_i \qquad 11.3$$

The mean time duration (d_m) of a project is the algebraic sum of the mean time durations of all the activities that lie on the critical path. The *variance* (σ^2) of a project is the algebraic sum of the variances of all the activities that lie on the critical path. The frequency distribution for the time duration of a project is the normal distribution according to the central limit theorem. The standard deviation (σ) value of a project is given by:

$$\sigma = \sqrt{\sigma^2}$$

$$\sigma = \sqrt{\sum_{i=1}^{n} (\sigma^2)_i} \qquad 11.4$$

11.7 Normal Distribution For a Project

According to the central limit theorem, the frequency distribution of the time duration of a project is the normal distribution. Figure 11.4 shows a normal distribution curve that is symmetrical to both sides of its peak. In the normal distribution, data are distributed symmetrically to both sides of the peak. The curve is not skewed toward the left or right side of the peak. The

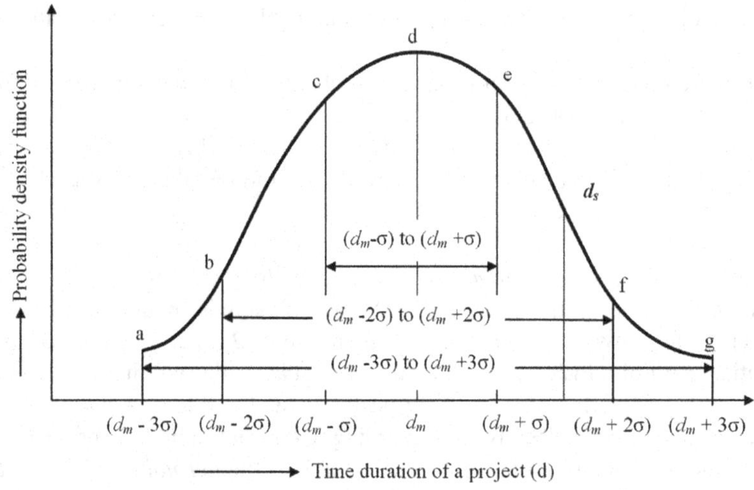

Figure 11.4 Normal distribution curve

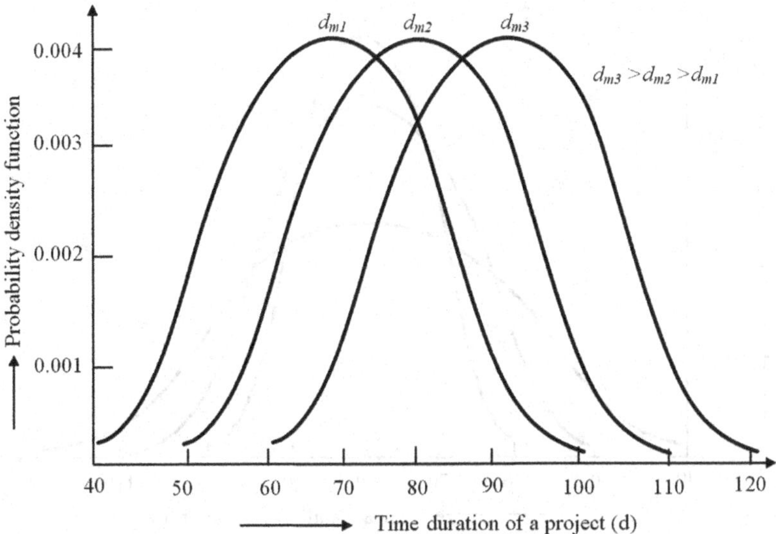

Figure 11.5 Normal distribution curves with different mean values

data distribution follows a bell shape, with most values clustered around the central region and tapering off away from the peak on both sides. Normal distributions are described by two values: mean and standard deviation. The value of the mean determines where the peak of the normal distribution curve is centred. The normal distribution curve is symmetrical around the mean value. Half the values fall below the mean value and half above the mean value. As shown in Figure 11.5, a curve shifts toward the right side with any increase in the value of the mean, while a curve shifts toward the left side with any decrease in the value of the mean. Thus, the value of the mean is the location parameter for a normal distribution curve, while the value of the standard deviation is its scale parameter. Standard deviation stretches or squeezes the normal distribution curve, as shown in Figure 11.6. A low standard deviation value results in a narrow curve while a high standard deviation value results in a wide curve.

The mean and standard deviation values of a project are used to develop its normal distribution curve. This is the curve between the probability density function and the variable (time duration in this case). For the given values of the mean (d_m), standard deviation (σ), and variance (σ^2), the probability density function (*pdf*) is given by:

$$f(d) = \frac{1}{\sigma\sqrt{2\pi}} e^{-\frac{(d-d_m)^2}{2\sigma^2}} \qquad 11.5$$

f (d) = *probability density function*
d = *time duration*
d_m = *mean time duration*
σ = *standard deviation*
σ^2 = *variance*

The normal distribution curve has a general shape such as is depicted in Figure 11.4. The curve gives the value of *pdf* for a given value of *d*. It is a probability distribution, thus the total

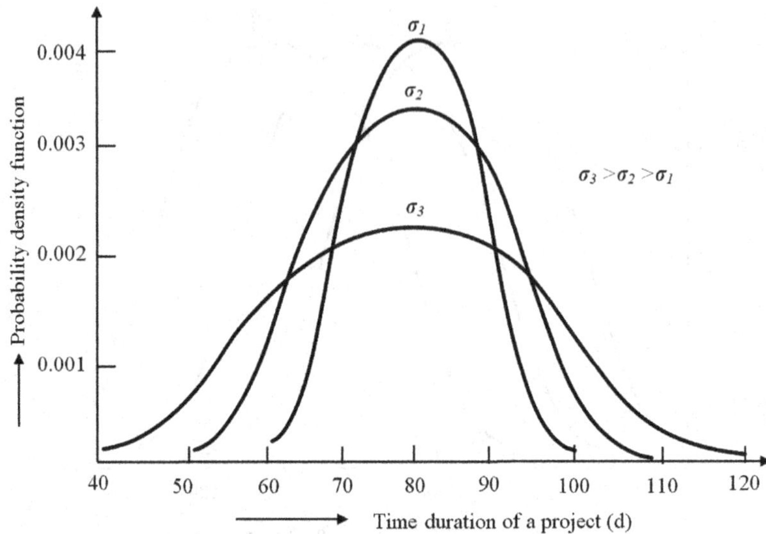

Figure 11.6 Normal distribution curves with different standard deviation values

area under the curve is equal to 1. For normal distribution curves, the following results are applicable:

Approximately 68 percent of the values of the normal distribution lie within $\pm\,\sigma$ from the mean time duration. This indicates that the area under the curve between $(d_m - \sigma)$ to $(d_m + \sigma)$ is 68 percent of the total area under the curve, where σ is the standard deviation and d_m is the mean time duration of a project.

Approximately 95 percent of the values of the normal distribution lie within $\pm\,2\sigma$ from the mean time duration. Thus, the area under the curve between $(d_m - 2\sigma)$ to $(d_m + 2\sigma)$ is 95 percent of the total area under the curve.

Approximately 99.7 percent of the values of the normal distribution lie within $\pm\,3\sigma$ from the mean time duration. Thus, the area under the curve between $(d_m - 3\sigma)$ to $(d_m + 3\sigma)$ is 99.7 percent of the total area under the curve.

It has already been discussed that the minimum time duration in which an activity or a project can be completed is the optimistic time duration (d_O), and the maximum time duration in which an activity or a project can be completed is the pessimistic time duration (d_P). 99.7 percent of the values lie between these minimum and maximum time durations. As shown in Figure 11.7, the range of the minimum and maximum time duration limits must be equal to $(d_m + 3\sigma) - (d_m - 3\sigma)$. In general:

$$(d_m + 3\sigma) - (d_m - 3\sigma) = (d_P - d_O)$$

$$\sigma = \left(\frac{d_P - d_O}{6}\right) \qquad 11.6$$

11.7.1 The Concept of Probability Calculation

Figure 11.4 depicts a probability distribution curve in which the probability of completing an activity within the scheduled time duration (d_s) is equal to the ratio between the area under the

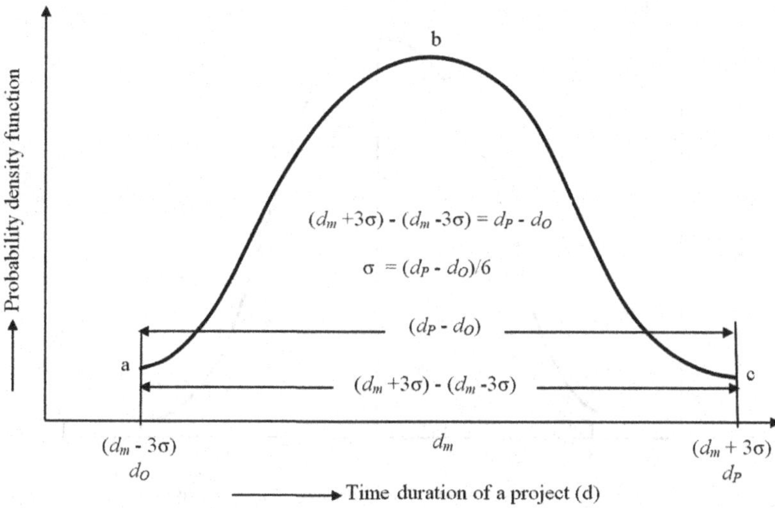

Figure 11.7 Normal distribution curve

curve between points a and d_s, and the total area beneath the curve. The total area beneath the curve is 1, therefore the probability of completing the activity within the scheduled time duration is equal to the area under the curve between points *a* and d_s. The calculation of the area underneath the curve can be simplified by using the *cumulative distribution function* (*cdf*). For the calculation of probability, the area under the curve is determined using the cumulative distribution function of the probability density function in the normal distribution as below:

$$\Phi(d) = \int_{-\infty}^{d} \frac{1}{\sigma\sqrt{2\pi}} e^{-\frac{(z-d_m)^2}{2\sigma^2}} dz \qquad 11.7$$

The cumulative distribution function provides the means for calculating the probability of completing a project. For any given value of the time duration (*d*), the probability of having a number less than or equal to *d* is obtained from the cumulative distribution function equation.

The cumulative distribution function requires mean time duration, standard deviation, and variance values. However, to make the calculation of probability more practical or simple, the values of $\Phi(d)$ are approximated and given in a table. The normal probability distribution is presented in the table with a mean time duration value equal to zero ($d_m = 0$), a standard deviation value equal to 1 ($\sigma = 1$), and a variance value equal to 1 ($\sigma^2 = 1$). If $d_m = 0$, $\sigma = 1$, and $\sigma^2 = 1$ the normal distribution is called the *standard normal distribution*. The equation of the *standard probability density functions* is:

$$f(d) = \frac{1}{\sqrt{2\pi}} e^{-\frac{d^2}{2}} \qquad 11.8$$

The curve of the standard normal distribution is shown in Figure 11.8. The equation for the *cumulative standard probability distribution function* of the standard normal distribution is:

$$\Phi(d) = \int_{-\infty}^{d} \frac{1}{\sqrt{2\pi}} e^{-\frac{z^2}{2}} dz \qquad 11.9$$

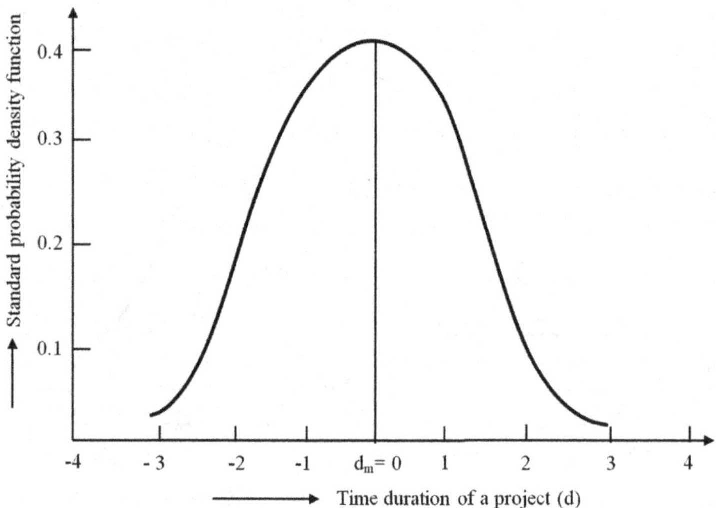

Figure 11.8 Standard normal distribution curve

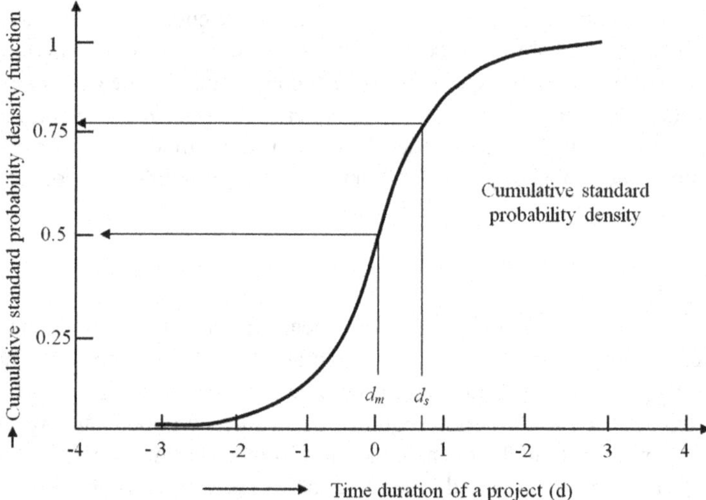

Figure 11.9 Cumulative standard normal distribution curve

The curve for the cumulative standard normal distribution is shown in Figure 11.9. The probability value is obtained on the y-axis by plotting a line from the x-axis to the cumulative standard probability density function as shown in the figure. To simplify the calculation of probability, the values of the cumulative standard normal distribution are given in Table 11.1. In order to use such tables, *Z transformation* is required. If *d* is normally distributed with a mean value equal to zero ($d_m = 0$), a standard deviation value equal to 1 ($\sigma = 1$), and a variance value equal to 1 ($\sigma^2 = 1$) then Z is referred to as a *standard normal deviate*, as given below:

$$Z = \frac{d_s - d_m}{\sigma} \qquad 11.10$$

Table 11.1 Normal deviate and the corresponding probability

Z (+)	*Probability (%)*	*Z (–)*	*Probability (%)*
0	50.0	0	50.0
+0.1	53.98	–0.1	46.02
+0.2	57.93	–0.2	42.07
+0.3	61.79	–0.3	38.21
+0.4	65.54	–0.4	34.46
+0.5	69.15	–0.5	30.85
+0.6	72.57	–0.6	27.43
+0.7	75.80	–0.7	24.20
+0.8	78.81	–0.8	21.19
+0.9	81.59	–0.9	18.41
+1.0	84.13	–1.0	15.87
+1.1	86.43	–1.1	13.57
+1.2	88.49	–1.2	11.51
+1.3	90.32	–1.3	9.68
+1.4	91.92	–1.4	8.08
+1.5	93.32	–1.5	6.68
+1.6	94.52	–1.6	5.48
+1.7	95.54	–1.7	4.46
+1.8	96.41	–1.8	3.59
+1.9	97.13	–1.9	2.87
+2.0	97.72	–2.0	2.28
+2.1	98.21	–2.1	1.79
+2.2	98.61	–2.2	1.39
+2.3	98.93	–2.3	1.07
+2.4	99.18	–2.4	0.82
+2.5	99.38	–2.5	0.62
+2.6	99.53	–2.6	0.47
+2.7	99.65	–2.7	0.35
+2.8	99.74	–2.8	0.26
+2.9	99.81	–2.9	0.19
+3.0	99.87	–3.0	0.13

The standard normal deviate in the standard normal distribution is also called the *probability factor*. Z represents the number of standard deviations away from the project's mean time duration. To find the probability of completing a project within the scheduled time duration (d_s) or less, the probability value is read as corresponding to the value of Z from Table 11.1.

11.8 Probability of Completing a Project

Step 1:

A project is divided into various activities, ensuring that information about each activity can be collected from different sources. The frequency distribution curve for each activity is plotted to determine the optimistic, pessimistic, and most likely time durations of an activity. Out of the three time durations, the optimistic and pessimistic time durations constitute the time duration range of each activity. The three time durations are used to calculate the mean time duration and variance of each activity. The mean time duration is calculated by using the weighted average method. In the weighted average method, weights are assigned to the three time durations; the denominator must be equal to the sum of the assigned weights. Here the mean time duration (d) of an activity is calculated by giving one weight to the optimistic time duration, four weights to the most likely time duration, and one weight to the pessimistic time duration, as shown below (Equation 10.1):

$$d = \frac{d_O + 4d_L + d_P}{6}$$

The variance of each activity is calculated from its time duration range using the following formula (from Equation 11.6).

$$\sigma^2 = \left(\frac{d_P - d_O}{6}\right)^2$$

Step 2:

The logical network of a project is developed using the PERT method previously discussed. The earliest and latest event times of all the events in the network in question are calculated using forward and backward pass calculations. The project mean time duration is calculated by adding together the mean time durations of all the activities which lie on the critical path or longest path of a PERT network (Equation 11.2).

$$\text{Project mean time duration}\left(d_m\right) = \sum_{i=1}^{n} (d)_i$$

Step 3:

The *variance* value of the critical path of a project network is calculated using the central limit theorem (Equation 11.3).

$$\text{variance}\,(\sigma^2) = \sum_{i=1}^{n} (\sigma^2)_i$$

Standard deviation is the square root of the variance. After the variance calculations, the standard deviation of the critical path of a project network is calculated using the equations given below:

$$\sigma = \sqrt{\sum_{i=1}^{n} (\sigma^2)_i}$$

$$\sigma = \sqrt{\sum_{i=1}^{n} \left(\frac{d_P - d_O}{6}\right)_i^2} \qquad 11.11$$

Step 4:

The scheduled completion time duration of a project (d_S), project mean time duration of a project (d_m), and standard deviation (σ) are used to calculate the probability factor (Z), expressed as below (Equation 11.10):

$$Z = \frac{d_s - d_m}{\sigma}$$

The value of Z is positive, zero, or negative. A positive Z value indicates that the schedule completion time duration is greater than the project mean time duration, as shown in Figure 11.9. The probability of completing a project in that time duration is higher than 50 percent. A Z value of zero indicates that the schedule completion time duration is equal to the project mean time duration. The probability of completing a project in that time duration is 50 percent. A negative Z value indicates that the schedule completion time duration is lower than the project mean time duration. The probability of completing a project in that time duration is less than 50 percent.

Step 5:

The probability corresponding to the Z value is determined from Table 11.1.

11.9 CPM vs PERT

CPM and PERT are two widely used network-based techniques for planning and scheduling projects. In both techniques, the execution sequence of a project is shown with the help of a network and the critical path is identified to allow one to calculate the time duration of a project. However, the most important differences between CPM and PERT are given below.

1. CPM is used for planning and scheduling projects which are routine. CPM is used for projects in which a planner has prior experience. On the other hand, PERT is used for projects which are not routine, generally being used for research and development projects.
2. CPM is used for projects in which a planner has prior experience, thus time durations estimates for a project's various activities have a fair degree of accuracy. Only one time duration is made for each activity in CPM. On the other hand, in PERT, three time duration estimates are made for each activity: an optimistic time duration, a most likely time duration, and a pessimistic time duration. Therefore, CPM uses a deterministic approach and PERT uses a probabilistic approach for the estimation of time durations.
3. All analyses in CPM are activity-based. Start times, finish times, and the various types of float corresponding to each activity in a project are calculated. On the other hand, PERT is event-based; event times and slack/float values corresponding to each event in a project are calculated.
4. CPM focuses on the trade-off between the total project cost and the time duration of a project with a primary emphasis is on minimizing the total project cost. On the other hand, PERT focuses on minimizing the time duration of a project. It is assumed that time duration is proportional to total project cost, thus a decrease in the project's time duration automatically decreases the total project cost.
5. CPM focuses on minimizing the total project cost. Therefore, CPM is used for projects where the main focus is on the total project cost, rather than on time duration. On the other hand, PERT focuses on minimizing the time duration of the project, thus it is used where time is more important than the total project cost.

11.10 Conclusion

PERT is used in R&D, complex, or non-routine projects in which time duration estimates for the various activities of a project lack a fair degree of accuracy. Project time durations are calculated in PERT the same way as they are in CPM. The probability of completing a project within a time duration longer or shorter than its mean time duration is also calculated. PERT is used when the focus is on minimizing the time duration of a project or when time is more important than the project cost.

Examples

Example 11.1: For the network given in Figure 11.10, three time estimates (in days) are mentioned for each activity. Determine the probability of completing the project by the end of days 18, 22, and 25. Also determine the time duration corresponding to a probability value of 75 percent.

Solution: The mean time duration and variance values of each activity are calculated and tabulated in Table 11.2.

Determination of the critical path and project mean time duration:

The critical path of the network is B-E-G-H.

$$\textit{The project mean time duration} = d_m = \sum\nolimits_{i=1}^{n} (d)_i = 22 \textit{ days.}$$

Standard deviation using central limit theorem:

$$\text{Variance of an activity} = \sigma^2 = \left(\frac{d_P - d_O}{6}\right)^2$$

$$\text{Variance of project} = \sigma^2 = \sum\nolimits_{i=1}^{n} (\sigma^2)_i = \sigma_B^{\ 2} + \sigma_E^{\ 2} + \sigma_G^{\ 2} + \sigma_H^{\ 2} = 1 + 0.44 + 0.11 + 0.44 = 1.9$$

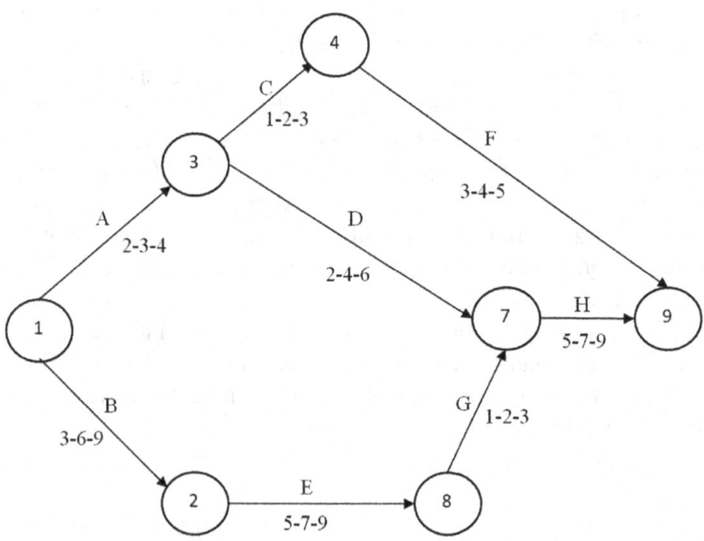

Figure 11.10 PERT project network

Table 11.2 Mean time duration and variance for each activity

Activities	d_O	d_L	d_P	$d=(d_O + 4d_L + d_P)6$	$\sigma^2 = [(d_P - d_O)/6]^2$
A	2	3	4	3	0.11
B	3	6	9	6	1
C	1	2	3	2	0.11
D	2	4	6	4	0.44
E	5	7	9	7	0.44
F	3	4	5	4	0.11
G	1	2	3	2	0.11
H	5	7	9	7	0.44

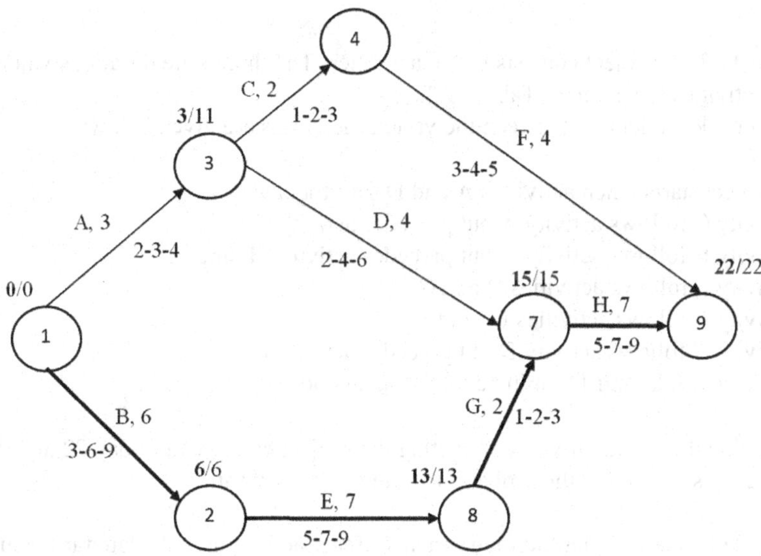

Figure 11.11 Determination of critical path and total project time duration

$$\textit{Standard deviation of the project} = \sqrt{\sum_{i=1}^{n} (\sigma^2)_i} = \sqrt{1.9} = 1.411$$

The probability of completing the project in ≤ 18 days:

$Z = (d_s - d_m) / \sigma = (18 - 22) / 1.411 = -2.82$

From Table 11.1, for Z = –2.82, the probability value is 0.26 percent.
The probability of completing the project in ≤ 18 days is 0.26 percent.
The probability of completing the project in ≤ 22 days:

$Z = (d_s - d_m) / \sigma = (22 - 22) / 1.411 = 0$

From Table 11.1, for Z = 0, the probability value is 50 percent.
The probability of completing the project in ≤ 22 days is 50 percent.
The probability of completing the project in ≤ 25 days:

$Z = (d_s - d_m) / \sigma = (25 - 22) / 1.411 = 2.12$

From Table 11.1, for Z = 2.12, probability value is 98.21 percent.
The probability of completing the project in ≤ 25 days is 98.21 percent.
The project time duration corresponding to a probability value of 75 percent:
From Table 11.1, the Z value corresponding to a probability value of 75 percent is 0.7.

$$Z = (d_s - d_m) / \sigma$$

$$0.7 = (d_s - 22) / 1.44$$

$$d_s = 23 \text{ (approximately)}$$

The project time duration corresponding to the probability value of 75 percent is ≤ 23 days approximately.

Example 11.2: A project consists of ten activities. The three time durations values of its various activities are given in Table 11.3.

The inter-dependencies between the various activities are given below:

- A project starts when activities A and D start together.
- Activity C follows activity A but precedes activity G.
- Activity B follows activity A but precedes activities E and F.
- Activity E follows activities B and C.
- Activity J follows activities F and L.
- Activity H follows activity D but precedes activity L.
- Activities G, E, and J terminate on the same node.

Determine the probability of completing the project by the end of day 22 and the time duration corresponding to the probability value of 75 percent.

Solution: The mean time duration and variance of each activity are calculated and tabulated in Table 11.4.

Table 11.3 The three time durations of the various activities in the project

Activities	*A*	*B*	*C*	*D*	*E*	*F*	*G*	*H*	*J*	*L*
d_O	2	1	3	2	4	1	1	2	5	3
d_L	5	2	4	4	7	4	2	5	6	4
d_P	8	3	5	6	10	7	3	8	7	5

Table 11.4 Mean time duration and variance of each activity

Activities	*A*	*B*	*C*	*D*	*E*	*F*	*G*	*H*	*J*	*L*
d_O	2	1	3	2	4	1	1	2	5	3
d_L	5	2	4	4	7	4	2	5	6	4
d_P	8	3	5	6	10	7	3	8	7	5
d	5	2	4	4	7	4	2	5	6	4
σ^2	1	0.11	0.11	0.44	1	1	0.11	1	0.11	0.11

Determination of the critical path and total project mean time duration:
The critical path of the network is D-H-L-J.

$$\textit{The project mean time duration} = d_m = \sum_{i=1}^{n} (d)_i = 4+5+4+6 = 19 \textit{ days.}$$

The standard deviation of the project using the central limit theorem:

$$\text{Variance of an activity} = \sigma^2 = \left(\frac{d_P - d_O}{6}\right)^2$$

$$\sigma^2 = \sum_{i=1}^{n} (\sigma^2)_i = \sigma_D^{\,2} + \sigma_H^{\,2} + \sigma_L^{\,2} + \sigma_J^{\,2} = 0.44 + 1 + 0.11 + 0.11 = 1.66$$

$$\textit{Standard deviationof the project} = \sqrt{\sum_{i=1}^{n} (\sigma^2)_i} = \sqrt{1.66} = 1.29$$

The probability of completing the project in ≤ 22 days:

$Z = (d_s - d_m) / \sigma = (22 - 19) / 1.29 = 2.32$

From the Table 11.1, for Z = 2.32, the probability value is 98.98 percent.
The probability of completing the project in ≤ 22 days is 98.98 percent.
The project mean time duration corresponding to the probability value of 75 percent:
From Table 11.1, the Z value corresponding to the probability value of 75 percent is 0.7.

$Z = (d_s - d_m) / \sigma$

$0.7 = (d_s - 19) / 1.29$

$d_s = 20$ (approximately)

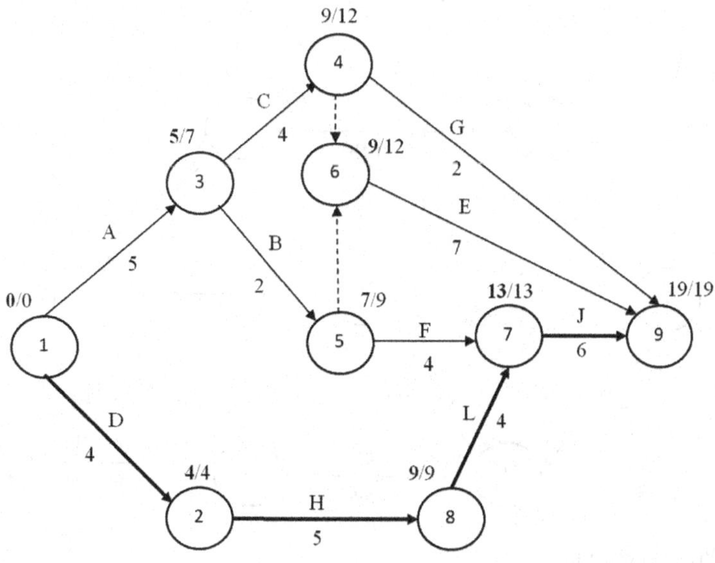

Figure 11.12 Determination of critical path and total project time duration

The project mean time duration corresponding to the probability value of 75 percent is ≤ 23 days approximately.

Exercises

Question 11.1: The fastest driver never takes less than 25 hours to complete the journey from station A to station B. However, a new learner on the same road will take a maximum of 48 hours. In general, this distance is covered within a time duration of 35 hours. What is the expected time duration?

Answer: 35.5 Hours.

Question 11.2: For the network given in Figure E11.1, three time estimates are given (in days) for each activity. Determine the probability of completing the project by the end of day 25. Also determine the time duration corresponding to the probability value of 75 percent.

Answer:

The critical path of the network is A-D-G, the project time duration is 23 days, and the standard deviation along the critical path is equal to 3.88.

The probability of completing the project by the end of day 25 is approximately equal to 69.75 percent.

The time duration corresponding to the probability value of 75 percent is 25.715 days, approximately equal to 26 days.

Question 11.3: For the network given in Figure E11.2, three time estimates are given (in days) for each activity. Determine the probability of completing the project by the end of day 25. Also determine the time duration corresponding to the probability value of 75 percent.

Answer:

The critical path of the network is A-I-E-G-H, the project time duration is 26 days, and the standard deviation along the critical path is equal to 1.234.

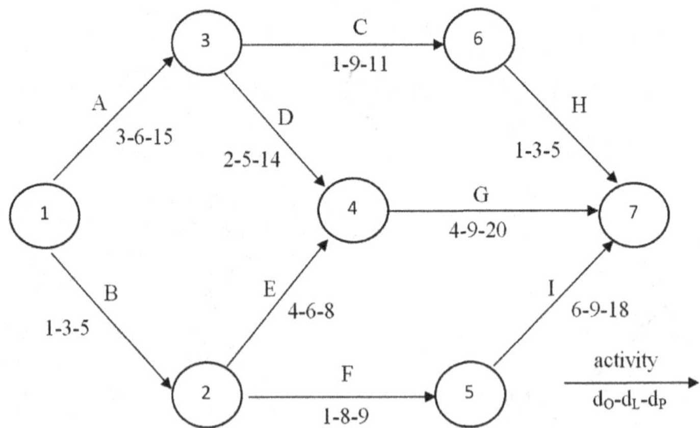

Figure E11.1 Project network

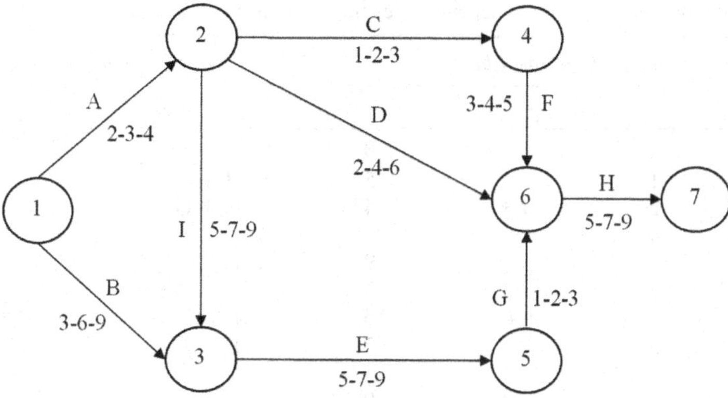

Figure E11.2 Project network

Table E11.1 Time durations and inter-dependencies between various activities

Activities	*Time durations (days)*			*Depends upon*
	d_o	d_L	d_p	
A	1	2	3	-
B	1	3	5	-
C	3	5	7	-
D	4	9	20	A, B
E	1	9	11	B, C
F	1	8	9	B
G	3	6	15	D, F
H	1	9	11	E, F
I	1	2	3	G, H

The probability of completing the project by the end of day 25 is approximately equal to 21.19 percent.

The time duration corresponding to the probability value of 75 percent is 26.86 days, approximately equal to 27 days.

Question 11.4: For the inter-dependencies and time estimates given in Table E11.1, determine the probability of completing the project by the end of days 20, 23, and 30.

Answer:

The critical path of the network is C-E-H-I, the project time duration is 23 days, and the standard deviation along the critical path is 2.46.

The probability of completing the project by the end of day 20 is approximately equal to 11.32 percent.

The probability of completing the project by the end of day 23 is 50 percent.

The probability of completing the project by the end of day 30 is approximately equal to 99.76 percent.

Question 11.5: For the inter-dependencies and time estimates given in Table E11.2, determine the probability of completing the project by the end of day 40.

Table E11.2 Time durations and inter-dependencies between various activities

Activities	*Time durations (days)*			*Depends upon*
	d_o	d_l	d_p	
B	5	7	9	-
C	1	3	5	B, F
D	2	3	4	C, I, G
F	4	9	20	-
G	5	7	9	F
H	1	2	3	G, J
I	3	6	15	L
J	3	5	7	L
K	1	2	3	-
L	1	8	9	F, K
E	5	7	9	D, H

Table E11.3 Time durations and inter-dependencies between various activities

Activities	*Depends upon*	*Time durations (days)*		
		d_o	d_L	d_p
A	-	1	8	9
B	-	1	3	5
C	-	2	3	4
D	A, B	3	5	7
E	B	3	4	5
F	B, C	1	2	3
G	D, F	3	6	15
H	D, F	5	7	9
I	F	1	2	3
J	E, H	5	7	9
K	G, J, I	4	9	20

Answer:

The critical path of the network is F-L-I-D-E, the project time duration is 34 days, and the standard deviation along the critical path is 3.66.

The probability of completing the project by the end of day 40 is approximately equal to 94.83 percent.

Question 11.6: For the inter-dependencies and time estimates given in Table E11.3, determine the probability of completing the project by the end of day 30.

Answer:

The critical path of the network is A-D-H-J-K, the project time duration is 36 days, and the standard deviation along the critical path is 3.197.

The probability of completing the project by the end of day 30 days is approximately equal to 2.71 percent.

12 Precedence Diagramming Method

12.1 Learning Objectives

After the completion of this chapter, readers will be able to:

- Represent inter-dependencies and time constraints between the different activities in a project,
- Do forward and backward pass calculations to determine project time durations and activity times in AON networks, and
- Do calendar date scheduling.

12.2 Introduction

Professor John W. Fondahl of Stanford University developed the precedence diagramming method in 1961, in which activity-on-node conventions were used to represent activities. The arrows connecting the activities (shown on the nodes of a network) define the inter-dependencies among them. The term *precedence diagramming* first appeared around 1964 in the User's Manual for an IBM 1440 computer program. One of the principal authors of the manual was J. David Craig of the IBM Corporation who was responsible for naming the technique as the *precedence diagramming method* (PDM).

The networks developed in CPM are based on a relationship in which an activity starts only after the completion of its preceding activities. In other words, the activities in a CPM network are connected to each other according to a finish-to-start relationship. In the case of the precedence diagramming method, three additional relationships (finish-to-finish, start-to-start, and start-to-finish), which will be discussed in subsequent sections, are also used in the development of a network. Further, a certain lag or lead may exist between preceding and subsequent activities. For example, the project of making a set of five chairs has three activities: purchasing materials, manufacturing the chairs, and polishing the chairs. CPM planners may show these three activities in a series. In practice, however, once a chair is manufactured, its polishing may start. The activity of polishing the chairs need not wait for the manufacturing of all five chairs. Such inter-dependencies are difficult to represent in CPM unless a network has a work breakdown structure going down to the finest degree of detail. In this case, both activities – manufacturing the chairs and polishing the chairs – need to be divided into five sub-activities. With such a small degree of detail, CPM networks become too large in size and thus become difficult to manage.

PDM represents practical situations by incorporating lags and leads between preceding and subsequent activities. Lag is a delay to the start of an activity after the end of the preceding activity. Lead is when an activity starts before the end of the preceding activity. It is an overlap between subsequent and preceding activities. Lead is also called *negative lag*. In this chapter,

DOI: 10.1201/9781003428992-12

the term *negative lag* has been used in place of *lead* to facilitate the calculations in PDM. PDM represents practical situations between various activities in a better way than CPM, and so PDM is best suited for planning and scheduling complex projects. PDM was originally developed by Professor Johan Fondhal (1924-2008) of Stanford University in the early 1960s.

12.3 Development of Networks

The development of a network in PDM is divided into the following three sub-sections. Each sub-section provides a basic idea about the topic, but each sub-section has been further elaborated with the help of suitable examples in the same chapter.

12.3.1 Activity-on-Node Representation

In PDM, activity-on-node (AON) representation is used to depict the various activities in a network. In AON representation, activities are shown on nodes. The inter-dependencies between a project's various activities are represented by arrows that connect the activities/nodes of a network. The lengths of the arrows have no significance. Generally, connecting arrows are drawn from left to right, but this is not a rule. Arrowheads are used to clearly depict the inter-dependencies between the various activities. Crossovers between the arrows representing the dependencies are always minimized. In case of unavoidable arrow crossovers, arrows may be depicted clearly using suitable symbols. Events and dummy activities are not generally used in PDM. However, if a network starts and/or ends with more than one activity, the start and/or end of a network are represented by a single start event and/or end event. AON representation is more flexible in the modeling of inter-dependencies between various activities than activity-on-arrow (AOA) representation.

12.3.2 Representation of Inter-dependencies

The inter-dependencies between the various activities in a project are used to develop an execution sequence. The development of a network to represent the execution sequence is similar to that used in CPM. However, in the case of CPM, only the finish-to-start (FS) relationship is used – that is, an activity starts only when the preceding activities are completed. FS is the only relationship used in AOA representation. However, in PDM, four types of relationships are used, as listed below.

- Finish-to-start (FS),
- Start-to-start (SS),
- Finish-to-finish (FF), and
- Start-to-finish (SF).

Each type of relationship has been elaborated in more detail with the help of suitable examples in subsequent sections.

12.3.3 Representation of Time Constraints

Time constraints are imposed due to the requirements of time before or after the start or finish of an activity. These are taken care of by positive and negative lag. Positive lag is the delay in the time duration between two activities; on the other hand, negative lag is an overlap of the time durations of two activities. No resources are used during positive and negative lag time durations. Time constraints arise from constraints of time at the starts and ends of dependent activities. Positive and negative lags are elaborated in more detail, in relation to the four types of relationship, in subsequent sections.

12.4 Development Of Networks: AOA to AON Representation

The development of networks has been discussed in previous chapters. In this section, the development of the same networks using AON and AOA representations will be discussed. In AOA representation, activities are represented by arrows and events by nodes. On the other hand, in AON representation, activities are represented by nodes and relationships by arrows. An activity is represented in AOA and AON representations as shown in Figure 12.1. The following are a few illustrations showing how logics represented using AOA representations are represented in AON representations.

Activity Y is controlled by activity X. Activity Y cannot begin until activity X is completed. Activities X and Y occur in a series. The network diagrams, in AOA and AON representations, are shown in Figure 12.2.

Activity Z is controlled by activities X and Y. Activity Z cannot begin until both activities X and Y are completed; the network diagrams are shown in Figure 12.3.

Activities Y and Z are controlled by activity X. Activities Y and Z cannot start unless activity X is completed; the network diagrams are shown in Figure 12.4.

Activities C and D are controlled by activity B. Activities C and D cannot start unless activity B is completed. Activity A precedes activity B. The network diagrams are shown in Figure 12.5.

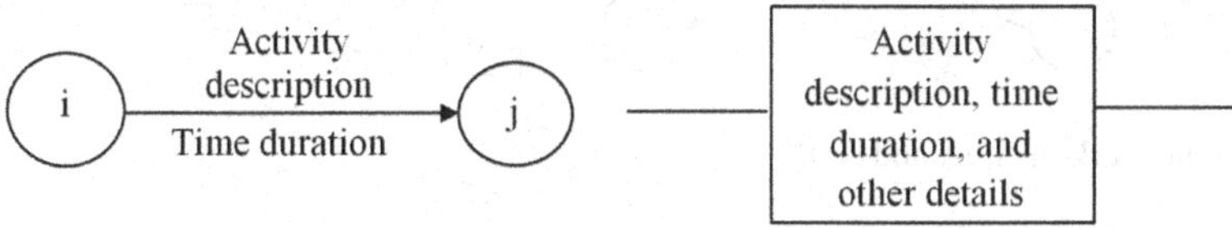

Figure 12.1 Representation of an activity

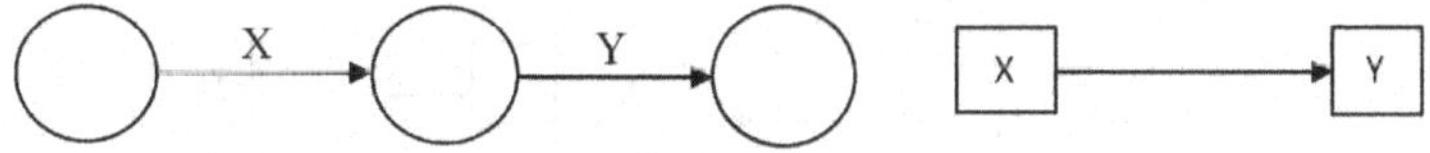

Figure 12.2 A network in AOA and AON

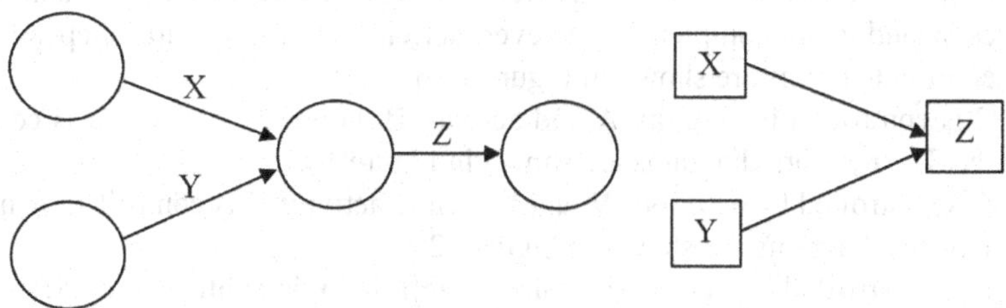

Figure 12.3 A network in AOA and AON

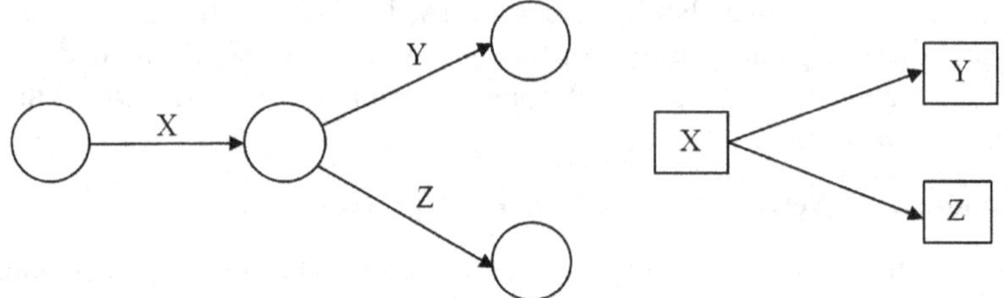

Figure 12.4 A network in AOA and AON

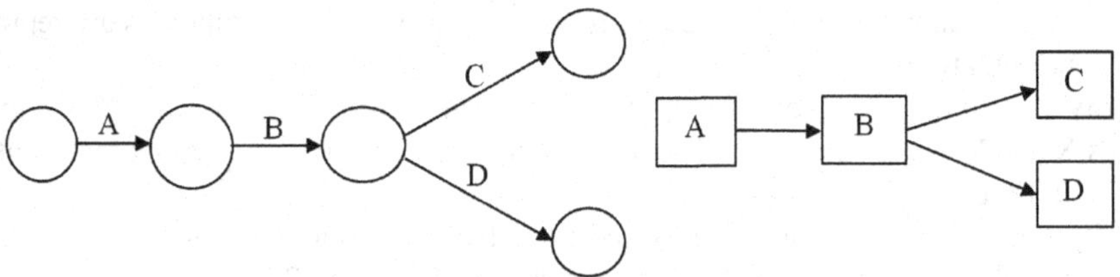

Figure 12.5 A network in AOA and AON

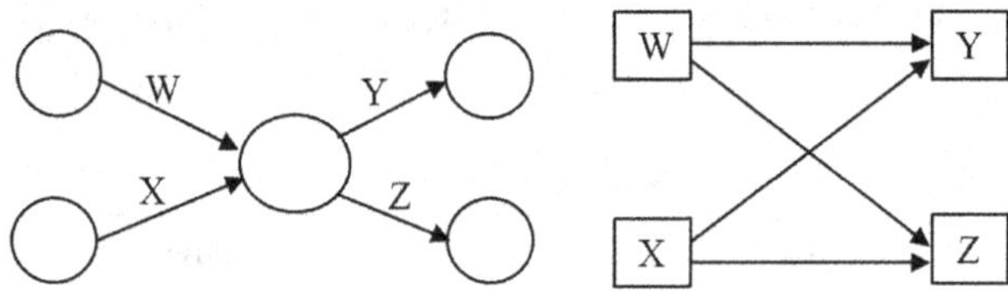

Figure 12.6 A network in AOA and AON

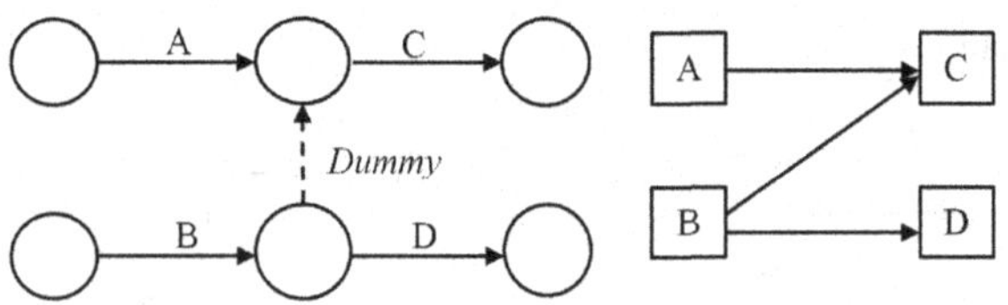

Figure 12.7 A network in AOA and AON

Activities Y and Z are controlled by activities W and X. Activities Y and Z cannot start until both activities W and X are completed. However, activities Y and Z start independent of each other. The network diagrams are shown in Figure 12.6.

Activity C is controlled by activity A and activity B, however, activity D is controlled by activity B only. The network diagrams are shown in Figure 12.7.

Activity Z is controlled by activities V and W, while activity Y is controlled by activities U and V. The network diagrams are shown in Figure 12.8.

Activity X is controlled by activities D and A; activity Y is controlled by activities A, B, and C, while activity Z is controlled by activity D only. The network diagrams are shown in Figure 12.9.

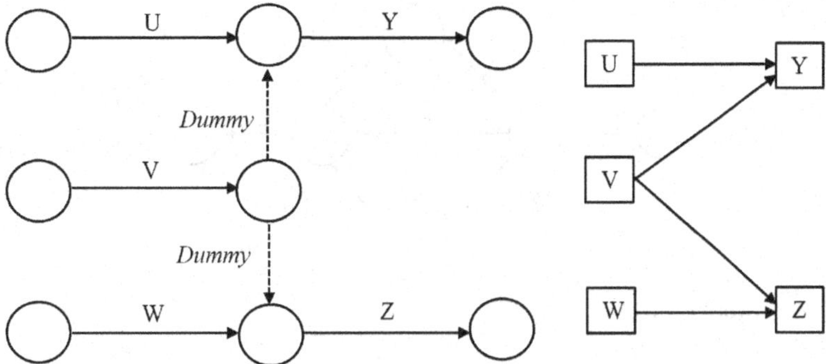

Figure 12.8 A network in AOA and AON

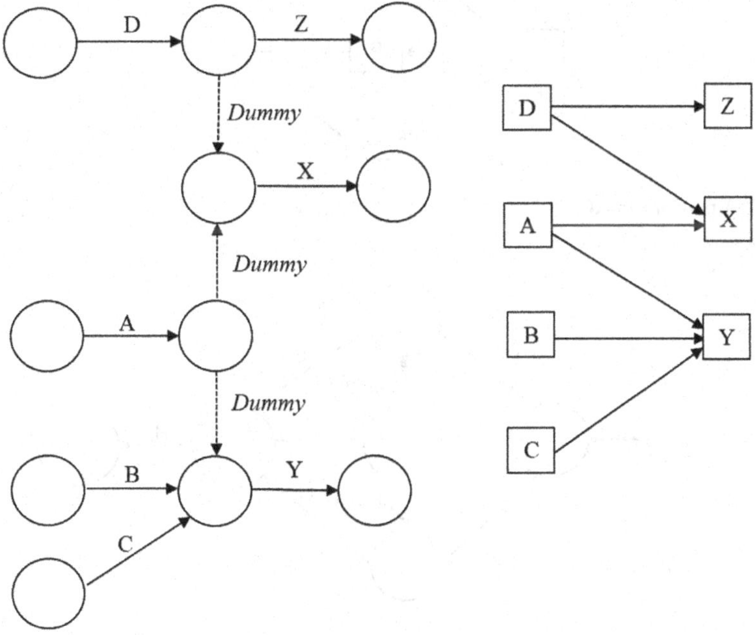

Figure 12.9 A network in AOA and AON

Example 12.1: Convert the AOA-based network representation shown in Figure 12.10 into an AON-based representation.

Solution: As shown in Figure 12.11.

Example 12.2: Convert the AOA-based network representation shown in Figure 12.12 into an AON-based representation.

Solution: As shown in Figure 12.13.

Example 12.3: Table 12.1 gives the list of inter-relationships among different activities of a project, draw AOA and AON-based network diagrams.

Solution: As shown in Figures 12.14 and 12.15.

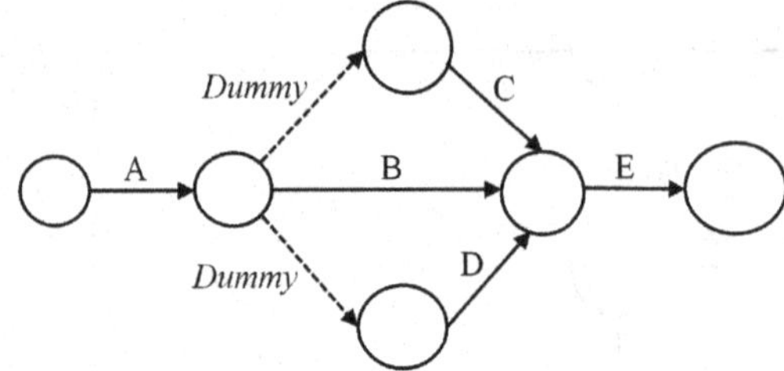

Figure 12.10 A network in AOA

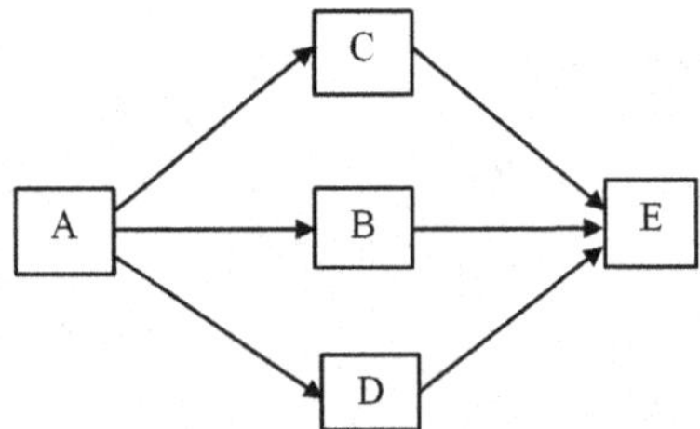

Figure 12.11 A network in AON

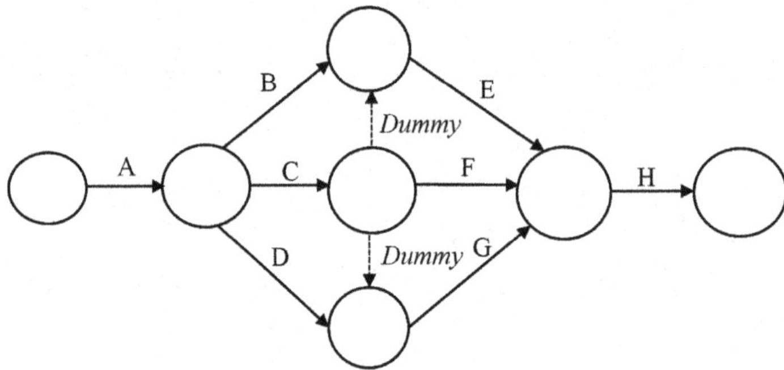

Figure 12.12 A network in AOA

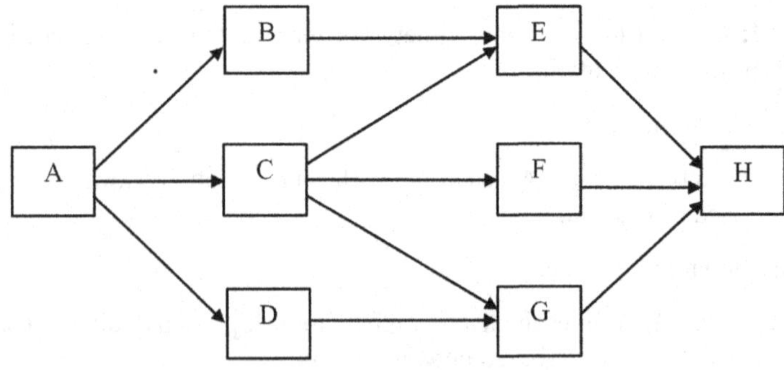

Figure 12.13 A network in AON

Table 12.1 The inter-relationships between different activities

Activities	*Immediately preceding activities*	*Immediately subsequent activities*
A	-	B, C
B	A	D, F
C	A	E, G
D	B	E, G
E	C, D	H, I
F	B	I
G	C,D	I
H	E	-
I	E, F, G	-

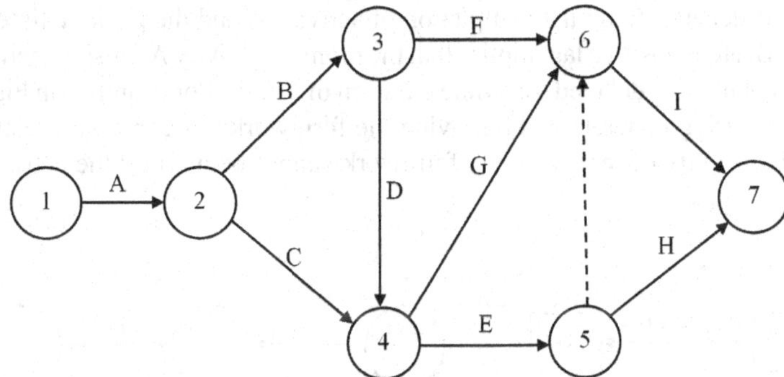

Figure 12.14 A network in AOA

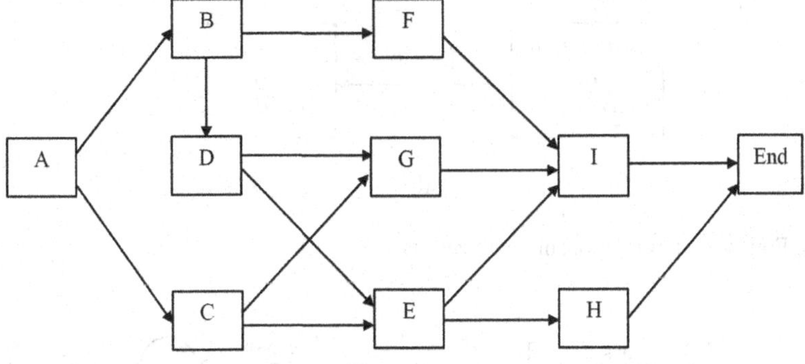

Figure 12.15 A network in AON

12.5 Development Of Networks: Inter-Dependencies and Time Constraints

As discussed earlier, four types of relationships between the various activities in a project are used in PDM. Further, positive and negative lags are elaborated (with reference to the four types of relationship) with the help of suitable examples in the following sections.

12.5.1 Finish-to-Start

An FS relationship with zero lag is the same as the relationship used in an activity-on-arrow relationship. An FS relationship with zero lag shows that a given activity cannot start unless the preceding activities are finished. In Figure 12.16(a), activity B cannot begin until activity A is completed. The start of activity B depends upon the completion of activity A. The zero lag value implies that the moment activity A finishes, activity B starts immediately within a time duration or lag of zero. For example, in Figure 12.16(b), two activities – installation of formwork and pouring concrete – are in an FS relationship with a time lag of zero. The activity of pouring concrete cannot begin until the activity of installing formwork is complete. The moment the activity of installing formwork is complete, the activity of pouring concrete will start immediately, without any delay.

An FS relationship with positive lag shows that the start of the subsequent activity has to be delayed, for a time duration equal to the value of the positive lag, after the end of the preceding activity. In Figure 12.17(a), the start of activity B has to be delayed for a time duration of 5 days after the end of activity A. Activity B cannot begin immediately after the end of activity A. The start of activity B depends upon the completion of activity A and the positive time lag between activities A and B. Here positive lag implies that the moment activity A finishes, activity B cannot start immediately but must delayed for a time duration of 5 days. For example, in Figure 12.17(b) two activities – pouring concrete and removing the formwork – are in an FS relationship with positive lag. The activity of removing the formwork cannot begin until the activity of pouring

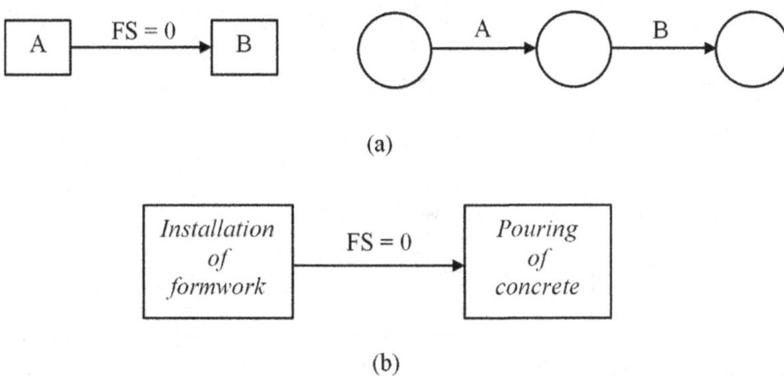

Figure 12.16 Finish-to-start relationship with zero lag

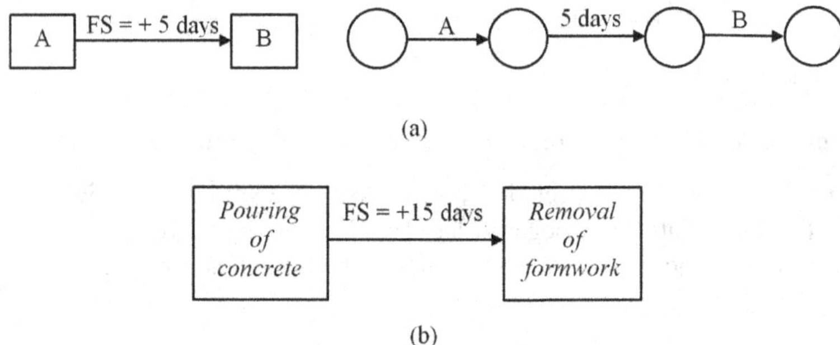

Figure 12.17 Finish-to-start relationship with positive lag

the concrete is complete; further, an FS lag of 15 days indicates that activity of removing the formwork must occur 15 days after the end of the activity of pouring the concrete. Positive lag has been used here to model a practical situation, the fact that when curing concrete, it requires 15 days to achieve full concrete strength. 15 days after the activity of pouring the concrete is complete, the activity of removing the formwork can start.

An FS relationship with negative lag shows that the subsequent activity has to start, by a time duration equals to the negative lag value, before the end of the preceding activity. This is the reason that negative lag is also sometimes called *lead*. In Figure 12.18(a), the start of activity B is preponed for a time duration of 5 days before the end of activity A. Activity B begins before activity A is completed. The start of activity B depends upon the negative lag. The negative lag of 5 days implies that when activity A finishes, activity B will have been started for 5 days beforehand. For example, in Figure 12.18(b) two activities can be selected, such as plastering the rooms and flooring the rooms. The start of the activity of flooring the rooms has been preponed for a time duration equal to 10 days before the end of the activity of plastering the rooms. The activity of flooring he rooms starts before the activity of plastering the rooms has been completed. The negative FS lag value of 10 days indicates that activity of flooring the rooms started 10 days before the end of the activity of plastering the rooms. This negative lag is used to model the fact that it is not necessary to wait for the completion of the plastering to start flooring the rooms, because once a few rooms are plastered, their flooring may be started.

An FS relationship with negative lag is depicted in AOA by breaking activity A into two parts with time durations equal to 'd_1-lag' and 'lag'. It is also depicted by breaking activity B into two parts with time durations equal to 'lag' and 'd_2-lag' as shown in Figure 12.19. FS relationships with zero, positive, and negative lags are depicted in the form of bar charts in Table 12.2.

12.5.2 Start-to-Start

An SS relationship with zero lag is the relationship between two activities that start simultaneously. An SS relationship with zero lag is the same as two activities starting simultaneously in AOA representation. In Figure 12.20(a), activity A and activity B start simultaneously. The lag

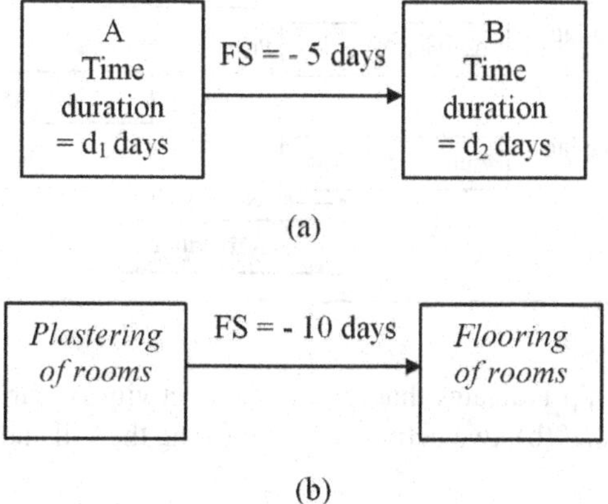

Figure 12.18 Finish-to-start relationship with negative lag

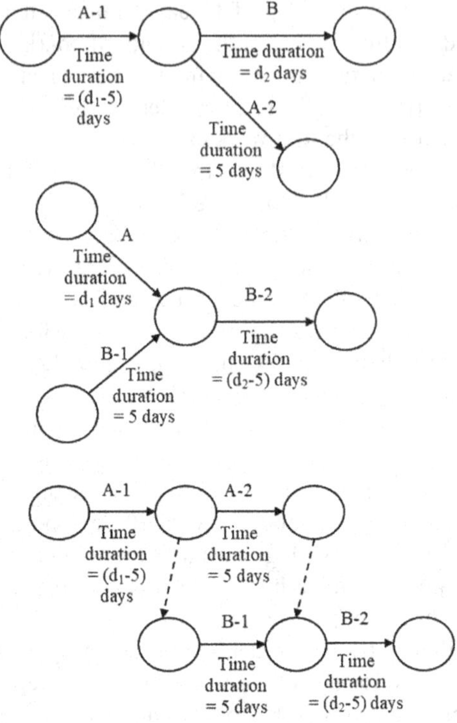

Figure 12.19 AOA representation of a finish-to-start relationship with negative lag

Table 12.2 FS relationships with zero, positive, and negative lags

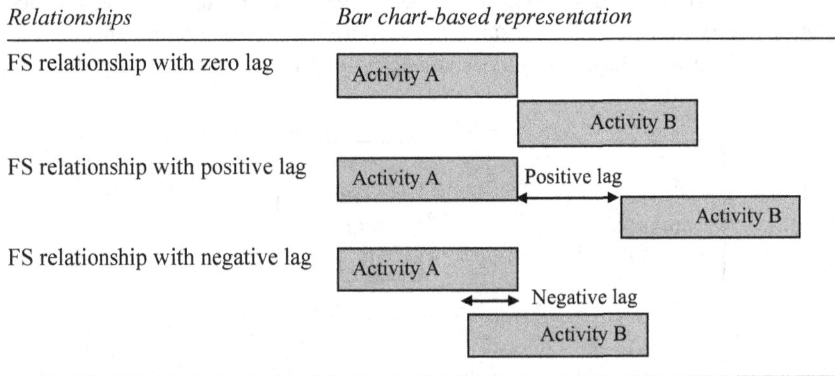

Relationships	*Bar chart-based representation*
FS relationship with zero lag	Activity A; Activity B
FS relationship with positive lag	Activity A; Positive lag; Activity B
FS relationship with negative lag	Activity A; Negative lag; Activity B

value in the SS relationship indicates that the moment activity A starts, activity B also starts. For example, in Figure 12.20(b), two activities – excavating the soil and dumping the soil – start simultaneously.

An SS relationship with positive lag shows that the start of the subsequent activity is delayed, by a time duration equal to the positive lag value, after the start of the preceding activity. In Figure 12.21(a and b), the start of activity B is delayed by a time duration equal to 5 days after

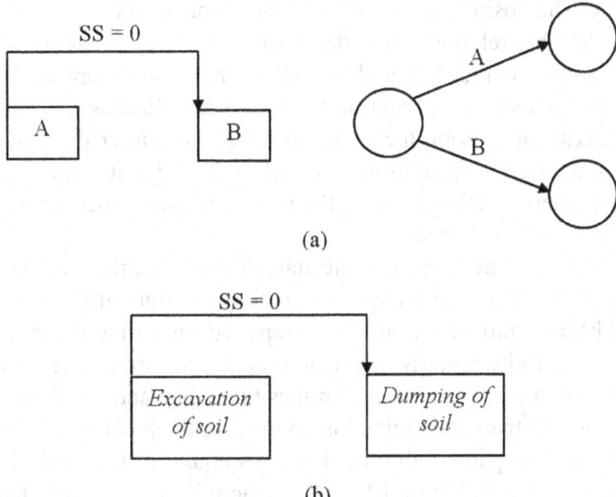

Figure 12.20 Start-to-start relationship with zero lag

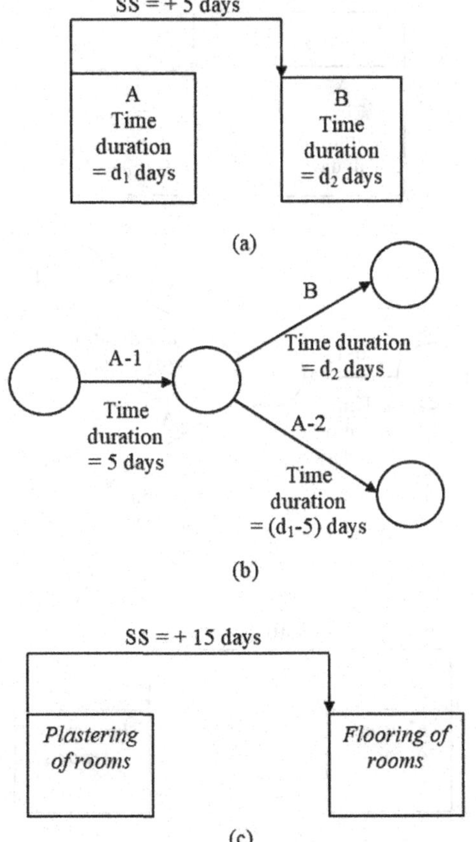

Figure 12.21 Start-to-start relationship with positive lag

the start of activity A. The positive value of the SS lag implies that activity B starts 5 days after the start of activity A. This relationship is depicted in AOA by breaking activity A into two parts with time durations equal to 'lag' and 'd_1-lag' as shown in Figure 12.21(b). For example, in Figure 12.21(c), two activities – plastering the rooms and flooring the rooms – are in an SS relationship. The activity of flooring the rooms starts 15 days after the start of the activity of plastering the rooms. SS lag has been used to model the fact that it is not necessary to wait for the all rooms to be plastered before flooring the rooms, because once a few rooms have been plastered their flooring may be started.

In an SS relationship with negative lag, the start of the subsequent activity is preponed, by a time duration equal to the negative lag value, before the start of the preceding activity. In Figure 12.22(a and b), the start of activity B is preponed for a time duration of 5 days before the start of activity A. Activity B starts before activity A. The start of activity B depends upon the negative lag. The negative lag of 5 days implies that when activity A starts, activity B will have started 5 days before that. This relationship may be depicted in AOA representation by breaking activity B into two parts with time durations equal to 'lag' and 'd_2-lag' as shown in Figure 12.22(b). For example, in Figure 12.22(c) two activities can be selected, such as flooring

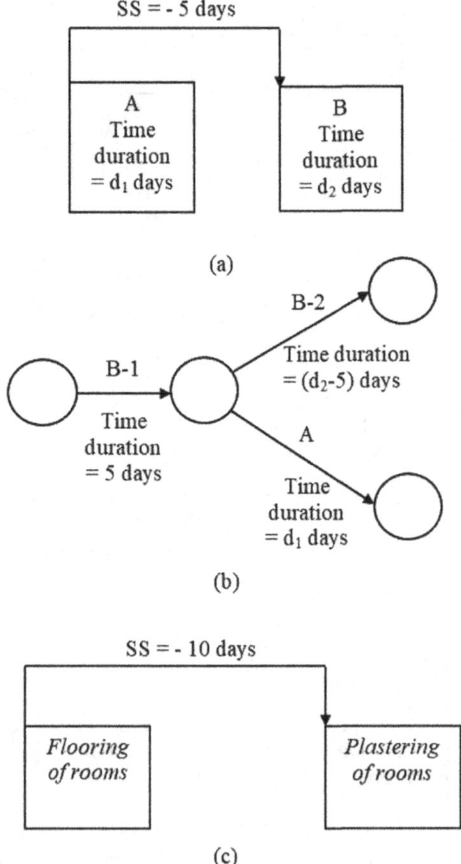

Figure 12.22 Start-to-start relationship with negative lag

Table 12.3 SS relationships with zero, positive, and negative lags

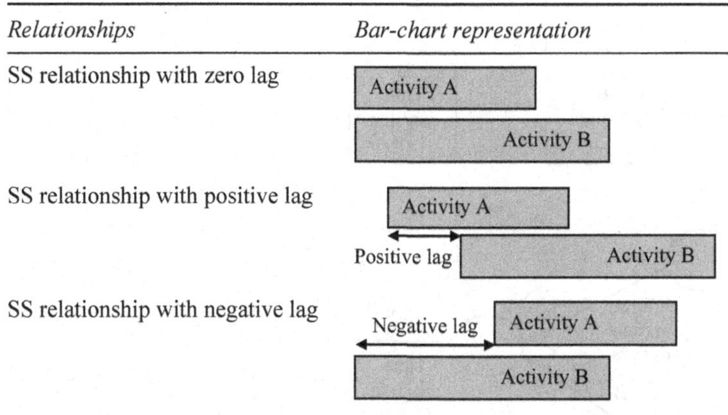

Relationships	*Bar-chart representation*
SS relationship with zero lag	Activity A; Activity B
SS relationship with positive lag	Activity A; Positive lag; Activity B
SS relationship with negative lag	Negative lag; Activity A; Activity B

the rooms and plastering the rooms. The start of the activity of plastering the rooms has been preponed by a time duration of 10 days before the start of the activity of flooring the rooms. The activity of plastering the rooms starts before the activity of flooring the rooms. The negative f SS lag value of 10 days indicates that the plastering started 10 days before the start of the flooring. SS relationships with zero, positive, and negative lags are shown in the form of bar charts in Table 12.3.

12.5.3 Finish-to-Finish

An FF relationship with zero lag is the same as two activities finishing together in AOA representation. An FF relationship with zero lag is used to depict the relationship between two activities that finish simultaneously. In Figure 12.23, activity A and activity B finish simultaneously. The zero lag in this FF relationship implies that the moment activity A finishes, activity B also finishes.

An FF relationship with positive lag shows that the end of the subsequent activity is delayed, by the time duration equals to the positive lag value, after the end of the preceding activity. In Figure 12.24(a and b), the end of activity B has been delayed by a time duration of 5 days after the end of activity A. This positive FF lag implies that activity B ends 5 days after the end of activity A. This relationship is depicted in AOA by breaking activity B into two parts with time durations equal to 'd_2-lag' and 'lag' as shown in Figure 12.24(b). For example, in Figure 12.24(c), the activity of flooring the rooms ends 15 days after the end of the activity of plastering the rooms.

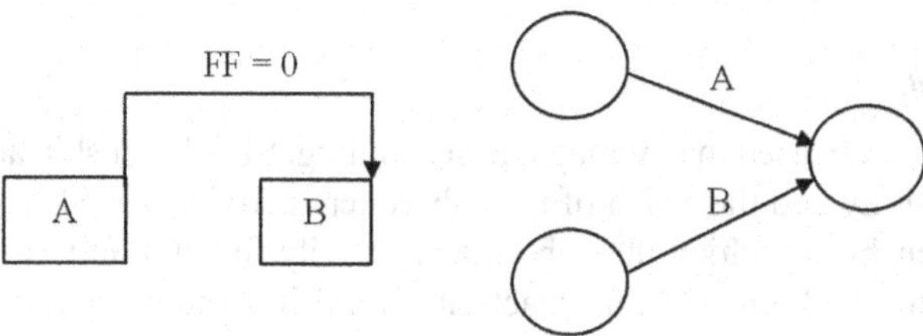

Figure 12.23 Finish-to-finish relationship with zero lag

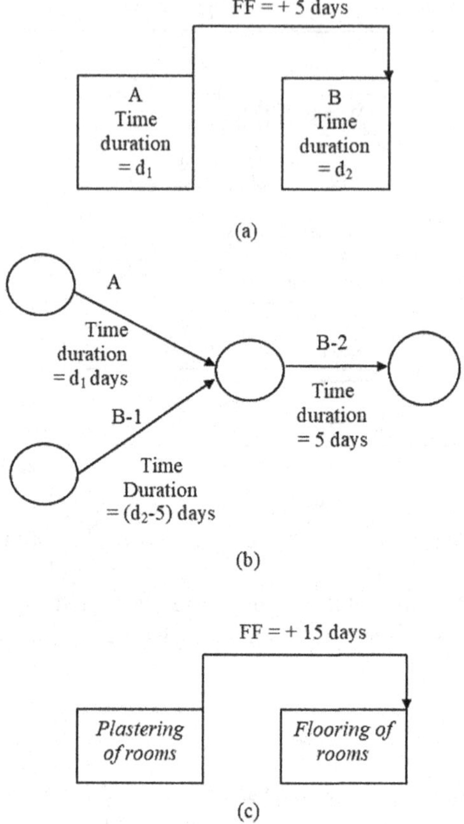

Figure 12.24 Finish-to-finish relationship with positive lag.

In an FF relationship with negative lag, the end of the subsequent activity is preponed, by a time duration equal to the negative lag value, before the end of the preceding activity. In Figure 12.25(a and b), the end of activity B has been preponed by a time duration of 5 days before the end of activity A. This negative FF lag implies that activity B ends 5 days before the end of activity A. This relationship is depicted in AOA by breaking activity A into two parts with time durations equal to 'd_1-lag' and 'lag' as shown in Figure 12.25(b). For example, in Figure 12.25(c), the activity of laying a pipeline ends 10 days before the end of the activity of filling a trench. The activity of filling a trench ends 10 days after the end of the activity of laying a pipeline. FF relationships are shown in Table 12.4.

12.5.4 Start-to-Finish

SF relationships are rarely used in planning and scheduling. SF relationship are between the start of the preceding activity and the finish of the subsequent activity. An SF with zero lag is used to depict the relationship in which the subsequent activity finishes and the preceding activity starts at the same time. In Figure 12.26(a), activity B finishes and preceding activity A starts at the same time. In Figure 12.26(b), a project has two activities: assembling the roof and the truss supply. The activity of assembling the roof cannot begin unless the truss supply is completed.

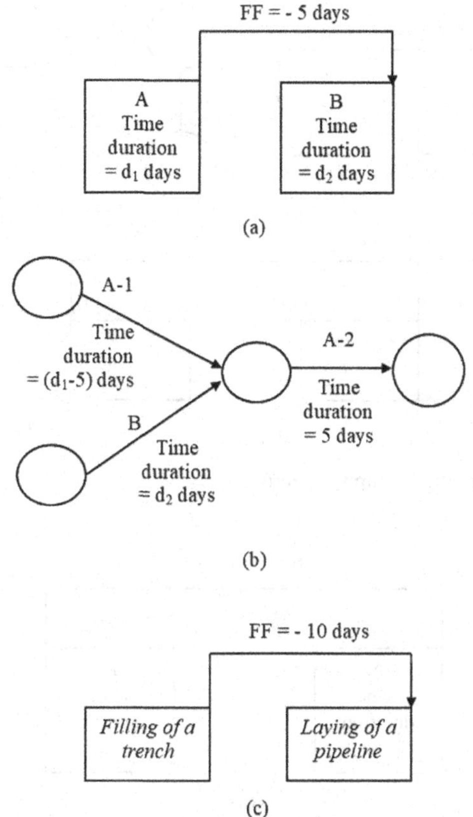

Figure 12.25 Finish-to-finish relationship with negative lag

Table 12.4 FF relationships with zero, positive, and negative lags

Relationships	*Bar-chart representation*
FF relationship with zero lag	Activity A Activity B
FF relationship with positive lag	Activity A — Positive Activity B
FF relationship with negative lag	Activity A Activity B — Negative FF lag

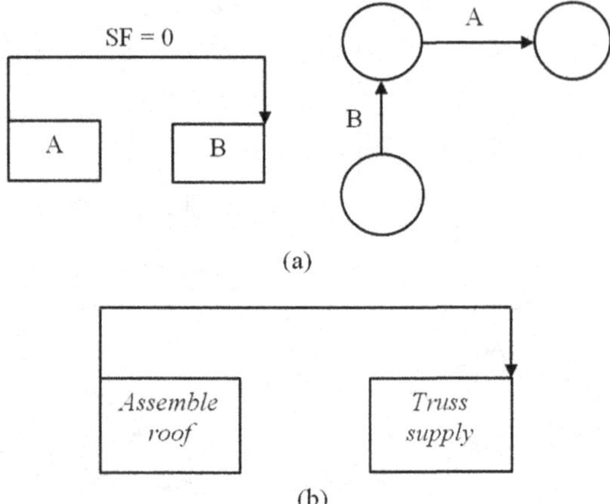

Figure 12.26 Start-to-finish relationship with zero lag

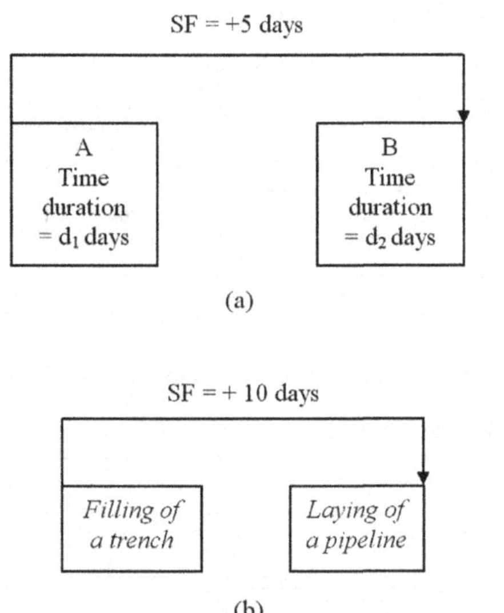

Figure 12.27 Start-to-finish relationship with positive lag

In an SF relationship with positive lag, the end of the subsequent activity is delayed, by a time duration equal to the positive lag value, after the start of the preceding activity. In Figure 12.27(a), the end of activity B has been delayed by a time duration of 5 days after the start of activity A. This positive SF lag implies that activity B ends 5 days after the start of activity A. For example, in Figure 12.27(b), the activity of laying a pipeline ends 10 days after the start of the activity of filling a trench. This relationship is depicted in AOA representation by breaking activity A into two parts with time durations equal to 'lag' and 'd_1-lag'. It is also depicted by breaking activity B into two parts with time durations equal to 'd_2-lag' and 'lag' as shown in Figure 12.28.

In an SF relationship with negative lag, the end of the subsequent activity is preponed, by a time duration equals to the negative lag value, before the start of the preceding activity. In

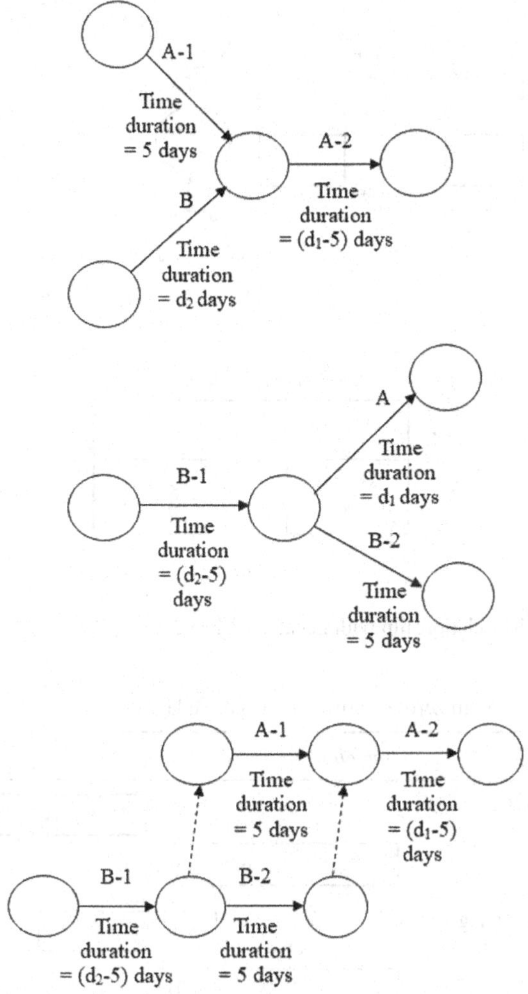

Figure 12.28 AOA representation of a start-to-finish relationship with positive lag

Figure 12.29(a), the end of activity B has been preponed for a time duration of 5 days before the start of activity A. This negative SF lag implies that activity B ends 5 days before the start of activity A. For example, in Figure 12.29(b), the truss supply ends 10 days before the start of the roof assembly. SF relationships are shown in Table 12.5.

The symbol used to represent an activity on a node in this chapter is shown in Figure 12.30.

12.6 Activity Times

The four types of activity time – earliest start time, earliest finish time, latest start time, and latest finish time – discussed in the critical path method chapters are also calculated in PDM. The earliest start time is the earliest possible time at which an activity can start. It is the earliest time before which all preceding activities can be completed to allow for the activity under consideration to start at its earliest. It is calculated through forward pass calculations by moving from the

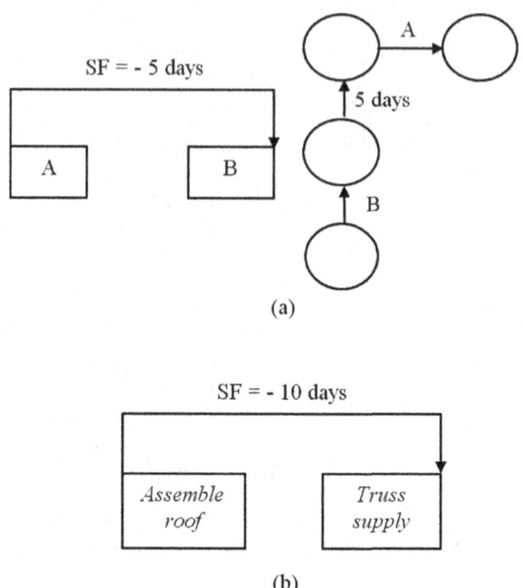

Figure 12.29 Start-to-finish relationship with negative lag

Table 12.5 SF relationships with zero, positive, and negative lags

Relationships	*Bar-chart representation*
SF relationship with zero lag	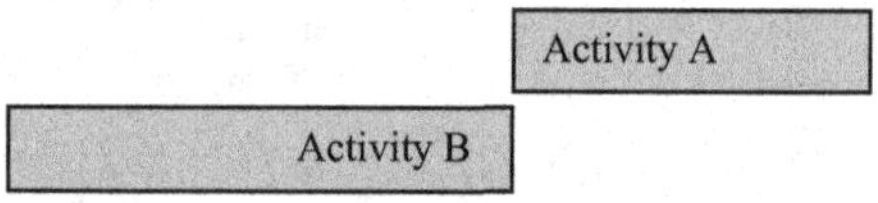 Activity A Activity B
SF relationship with positive lag	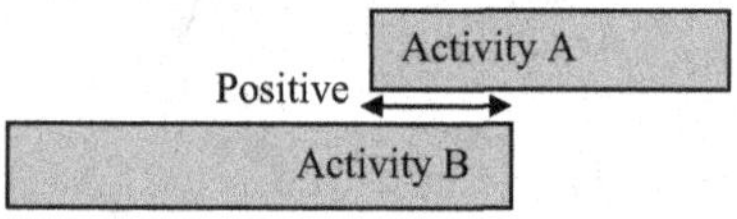 Activity A Positive Activity B
SF relationship with negative lag	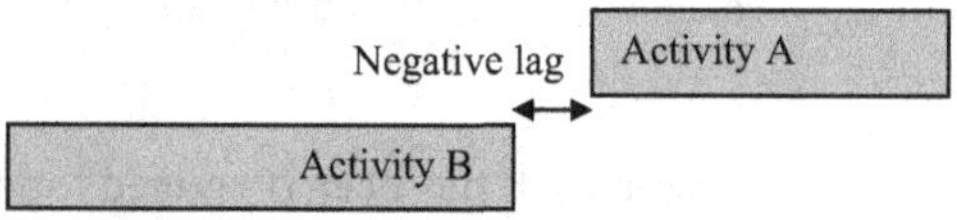 Negative lag Activity A Activity B

Earliest start time	Time duration	Earliest finish time
Activity description		
Latest start time	Total float	Latest finish time

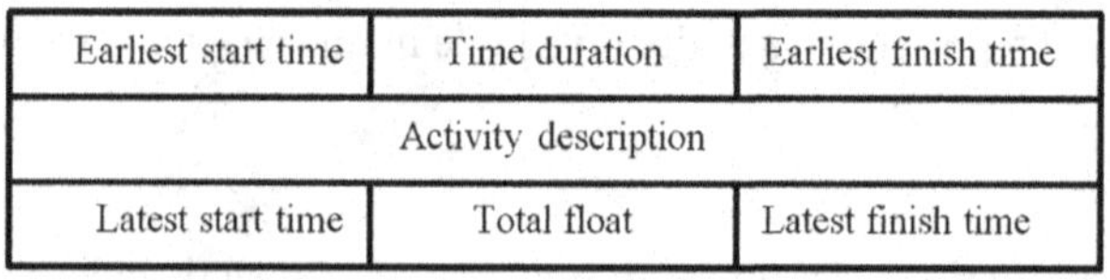

Figure 12.30 The symbol used to represent an activity in AON representation

starting activity to the last activity in a network. The earliest finish time is the earliest possible time at which an activity can finish. If an activity starts at its earliest start time and takes its time duration (d) then it will finish at its earliest finish time. Mathematically it is calculated by adding the time duration to the earliest start time.

$$EFT = EST + d \qquad 12.1$$

The latest start time is the latest possible time by which an activity can start without any delay to the completion of a project. Mathematically, the latest start time is calculated by subtracting the time duration from the latest finish time.

$$LST = LFT - d \qquad 12.2$$

If an activity starts as late as possible and it ends as late as possible, this is called its latest finish time. The latest finish time of an activity is the latest time at which it can finish without any delay to the project's completion. It is calculated through backward pass calculations by moving from the last activity to the starting activity of a network.

12.7 Forward Pass Calculations

Activity time calculations involve first a forward pass and then a backward pass through a network. Forward pass calculations proceed sequentially from the starting to the terminal activities in a network by moving along the direction of the connecting arrows. Forward pass calculations provide the earliest start times and earliest finish times of each activity. Forward pass calculations are initiated by assigning an arbitrary earliest start time to the starting activities of the network. A zero value is usually assigned to the starting activity as its earliest start time. Forward pass calculations proceed sequentially by following the steps given below.

Step 1: Assume a project starts at time zero. The earliest start time of the starting activity of the project is thus taken as zero.

Earliest start time of the starting activity = 0

Step 2: The earliest start time (EST) of an activity under consideration is given by:

EST = Maximum EST value obtained through the immediately preceding paths of the activity under consideration.

Step 3: The earliest finish time (EFT) of an activity is the algebraic sum of its earliest start time and time duration (d). For a given activity it is given by (Equation 12.1):

$$EFT = EST + d$$

The computations are done sequentially from the starting to the last activity of a network by moving along the direction of the arrows.

12.7.1 Calculation of EST and EFT Using Forward Pass

Consider the relationships between the different activities given in Table 12.6 for the demonstration of the use of forward pass calculations to find EST and EFT of all the activities in the network. An AON representation of the given relationships is shown in Figure 12.31. The same representation, along with positive and negative lags, is shown in Figure 12.32. The starting activity begins at time zero. Thus, the start time of the project is zero (as discussed in step 1).

Activity A: The project starts with the start of activity A at time zero. The EST of activity A is 0. The EFT of an activity is calculated by adding its time duration to its EST. Therefore, the EFT of activity A is the end of day 2 (0 + 2).

Activity B: Activity A is connected to activity B through an FS relation with a positive lag of 2 days. Thus, the EST of activity B is the end of day 4 (2 + 2). The EFT of an activity is calculated by adding its time duration to its EST. The EFT of activity B is the end of day 8 (4 + 4).

Activity C: Activity A is connected to activity C through an SS relation with a positive lag of 3 days. Thus, the EST of activity C is the end of day 3. The EFT of activity C is the end of day 6 (3 + 3).

Activity E: Activity B is connected to activity E through 2 days of positive FF lag. Hence, the EFT of activity E through this path is the end of day 10 (8 + 2). The EST of activity E through this path is the end of day 8 (10 – 2). Activity C is also connected to activity E through 1 day of positive FS lag. Thus, the EST of activity E through this path is the end of day 7 (6 + 1). Activities B and C precede activity E, thus the maximum EST value obtained through the

Table 12.6 Activities, inter-relationships, time constraints, and their time durations

Activities	*Immediately preceding activities*	*Time constraints*	*Time durations (days)*
A	-	-	2
B	A	FS + 2 days	4
C	A	SS + 3 days	3
D	C	SS + 4 days	4
	E	FS + 2 days	
E	B	FF + 2 days	2
	C	FS + 1 day	
F	D	FS + 3 days	3
	E	FF + 1 day	

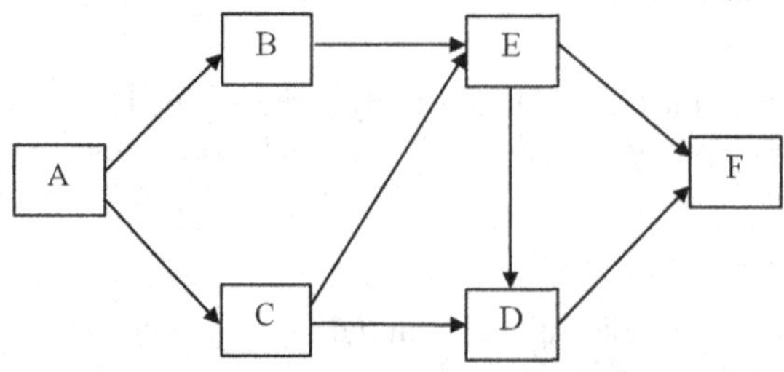

Figure 12.31 Activity-on-node representation

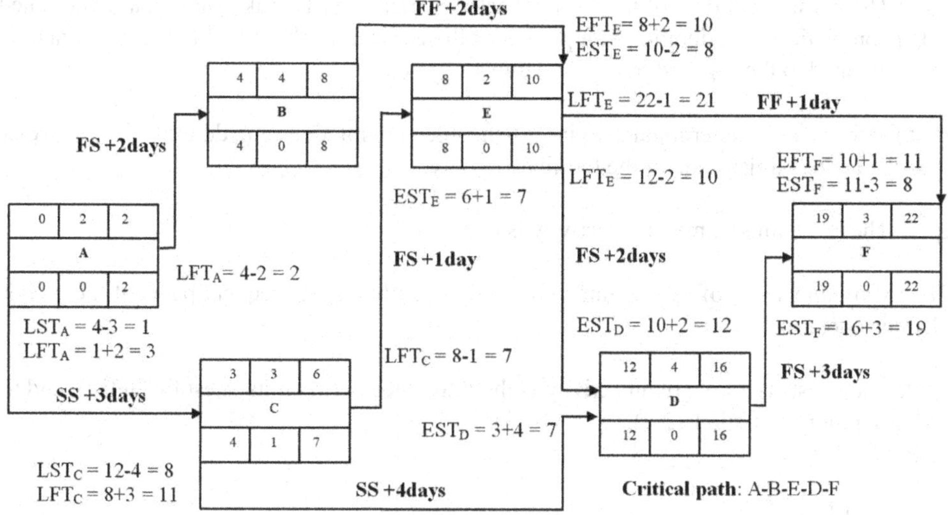

Figure 12.32 Activity-on-node representation with time constraints

immediately preceding paths is taken as the EST of activity E, that is, the end of day 8. The EFT of activity E is the end of day 10 (8 + 2).

Activity D: Activity E is connected to activity D through 2 days of positive FS lag. Thus, the EST of activity D through this path is the end of day 12 (10 + 2). Activity C is also connected to activity D through 4 days of positive SS lag. Thus, the EST of activity D through this path is the end of day 7 (3 + 4). Activities E and C precede activity D, thus the maximum EST value obtained through the immediately preceding paths is taken as the EST of activity D, that is, the end of day 12. The EFT of activity D is the end of day 16 (12 + 4).

Activity F: Activity E is connected to activity F through 1 day of positive FF lag. Thus, the EFT of activity F through this path is the end of day 11 (10 + 1) and the EST of activity F through this path is the end of day 8 (11 – 3). Activity D is also connected to activity F through 3 days of positive FS lag. Thus, the EST of activity F through this path is the end of day 19 (16 + 3). Activities E and D precede activity F, thus the maximum value of EST obtained through the immediately preceding paths is taken as the EST of activity F, that is, the end of day 19. The EFT of activity F is the end of day 22 (19 + 3). Therefore, the time duration of the project is 22 days.

12.8 Backward Pass Calculations

Backward pass calculations are done sequentially from the terminal activity to the starting activity of a project by moving against the direction of the arrows. Backward pass calculations provide the latest finish time and latest start time of each activity. Backward pass calculations start by deciding the value of the scheduled completion time duration for the completion of a project. The scheduled completion time duration (d_s) value is taken as the latest finish time of the last activity in a project. If the scheduled completion time duration of a project is not specified, the latest finish time of the terminal activity is taken as equal to its earliest finish time as calculated in the forward pass calculations. The calculations proceed sequentially from the terminal activity to the starting activity of the network by following the steps given below.

Step 1: The latest finish time of the terminal activity of a project is taken as equal to the scheduled completion time duration of a project or the earliest finish time of the terminal activity as calculated in the forward pass calculations.

Latest finish time of the terminal activity = Scheduled completion time duration for the project (d_s) or the earliest finish time of the terminal activity.

Step 2: The latest finish time of an activity is equal to:

LFT = Minimum value of LFT obtained through immediately subsequent paths of the activity under consideration.

Step 3: The latest start time of an activity is the difference between its latest finish time and the time duration (Equation 12.2).

$$LST = LFT - d$$

12.8.1 Calculation of LST and LFT Using Backward Pass

The steps discussed above are used on the network shown in Figure 12.32 for the demonstration of backward pass calculations. The backward pass calculations start from the last activity of the network, activity F in this case. The earliest finish time of activity F as computed in forward pass calculations is taken as its latest finish time, in this case the end of day 22 (step 1).

Activity F: The EFT of activity F is the end of day 22, thus the LFT of activity F is also the end of day 22. The LST of activity F is the end of day 19 (22 – 3).

Activity D: Activity D is connected to activity F through 3 days of positive FS lag. Thus, the LFT of activity D is the end of day 16 (19 – 3). The LST of activity D is the end of day 12 (16 – 4).

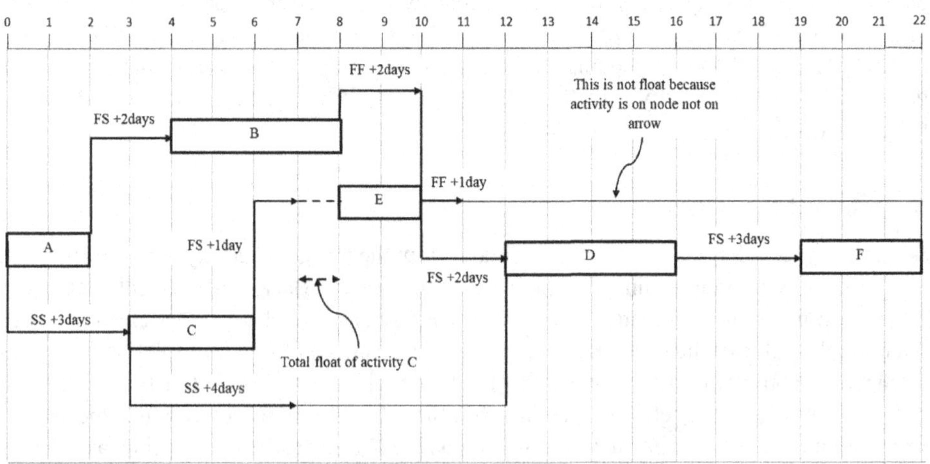

Figure 12.33 Time scaled version of a network based upon EST

Activity E: Activity F is connected to activity E through 1 day of positive FF lag. Thus, the LFT of activity E through this path is the end of day 21 (22 – 1). Activity D is also connected to activity E through 2 days of positive FS lag. The LFT of activity E through this path is the end of day 10 (12 – 2). Activities F and D succeed activity E, thus the minimum LFT value obtained through the immediately subsequent paths is taken as the LFT of activity E, that is, the end of day 10 in this case. The LST of activity E is the end of day 8 (10 – 2).

Activity C: Activity E is connected to activity C through 1 day of positive FS lag. Thus, the LFT of activity C through this path is the end of day 7 (8 – 1). Activity D is also connected to activity C through 4 days of positive SS lag. Thus, the LST of activity C through this path is the end of day 8 (12 – 4). The LFT of activity C through this path is the end of day 11 (8 + 3). The activities E and D succeed activity C, thus the minimum LFT value obtained through the immediately subsequent paths is taken as the LFT of activity C, that is, the end of day 7. The LST of activity C is the end of day 4 (7 – 3).

Activity B: Activity E is connected to activity B through 2 days of positive FF lag. Thus, the LFT of activity B is the end of day 8 (10 – 2). The LST of activity B is the end of day 4 (8 – 4).

Activity A: Activity B is connected to activity A through 2 days of positive FS lag. Thus, the LFT of activity A through this path is the end of day 2 (4 – 2). Activity C is also connected to activity A through 3 days of positive SS lag. Thus, the LST of activity A through this path is the end of day 1 (4 – 3). The LFT of activity A through this path is the end of day 3 (1 + 2). Activities B and C succeed activity A, thus the minimum LFT value obtained through the immediately subsequent paths is taken as the LFT of activity A, that is, the end of day 2. The LST of activity A is 0 (2 – 2). The time-scaled version of the same network is shown in Figure 12.33.

12.9 Critical Paths and Critical Activities

A network consists of many paths of different lengths between its starting and end points. The length of a path is the time duration required to complete it. Out of all the possible paths through a network, the critical path is the longest path that connects the start and the end of a network. Out of all possible paths through a network, at least one path is critical. A network may have more than one critical path, and in the most extreme circumstance, all the paths through a network may become critical when the lengths of all the critical paths are equal.

The activities along the critical path of a network are called critical activities. The reason behind this criticality is that any delay in the time duration of the critical activities results in a delay to the completion of the project under consideration. The identification of the critical path is useful for the effective planning and control of a project. It helps a planner allocate adequate resources to the critical activities to ensure their timely completion. In the network shown in Figure 12.32, the critical path is A-B-E-D-F and the critical activities are A, B, E, D, and F. The length of the critical path is 22 days. The length of the critical path is the project time duration. For the activities which lie on the critical paths, the earliest start time is equal to the latest start time, and the earliest finish time is equal to the latest finish time.

12.10 Total Float

The total float is the total amount of time by which the completion time of an activity lying on a non-critical path can exceed its EFT without affecting any time of any activity lying on the critical path. For an activity lying on a non-critical path, the total float is the difference between its latest start time and earliest start time, or its latest finish time and earliest finish time. Mathematically, the total float (TF) of an activity is calculated by:

$$TF = LST - EST \tag{12.3}$$

or

$$TF = LFT - EFT \tag{12.4}$$

or

$$TF_{\text{under-consideration}} = LST_{\text{under-consideration}} - EST_{\text{under-consideration}} \tag{12.5}$$

Example 12.4: For each activity in the network shown in Figure 12.34, calculate the EST, EFT, LST, LFT, and total float values. Also identify the critical path of the network.

Solution: The EST, EFT, LST, LFT, and total float values of the various activities in the network given in Figure 12.34 are given in Figure 12.35. The critical path of the network is A-C-F-H-I.

Example 12.5: For each activity in the network shown in Figure 12.36, calculate the EST, EFT, LST, LFT, and total float values. Also identify the critical path of the network.

Solution: The EST, EFT, LST, LFT, and total float values of the various activities in the network given in Figure 12.36 are given in Figure 12.37. The critical path of the network is A-B-F-H-I.

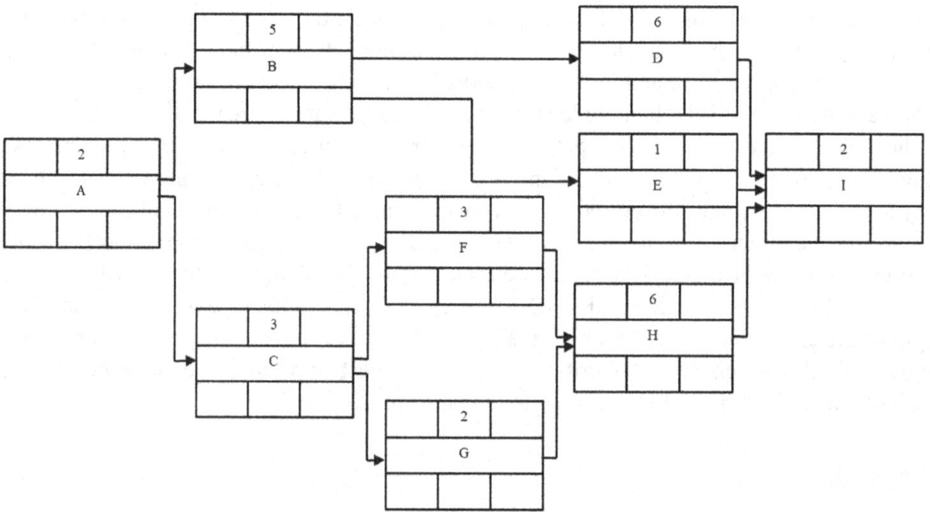

Figure 12.34 Activity-on-node representation

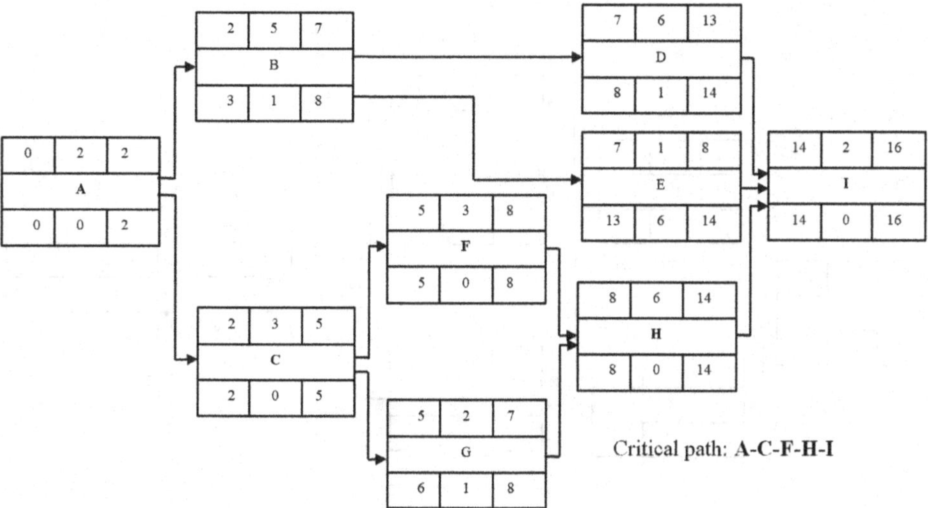

Figure 12.35 Activity-on-node representation

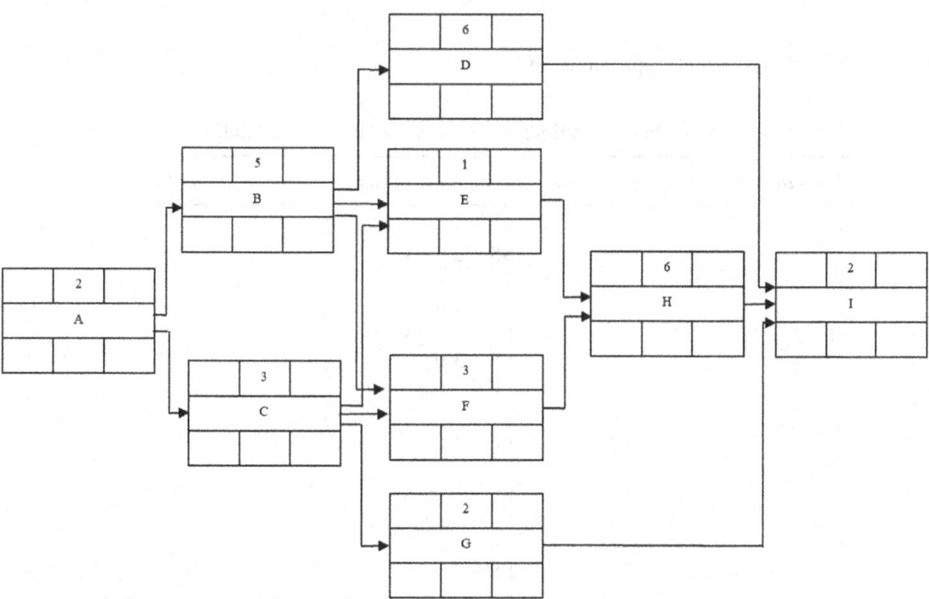

Figure 12.36 Activity-on-node representation

Example 12.7: For the relationships given in Table 12.7, draw the network diagram and calculate the EST, EFT, LST, LFT, and total float values of each activity. Also identify the critical path and represent the network on a time scale.

Solution: An AON representation of the project is shown in Figure 12.38. This helps in developing the network along with all the types of relationships shown in Figure 12.39. The EST, EFT, LST, LFT, and total float values of the various activities in the network are given in Figure 12.39. The critical path of the network is A-C-F-H-J. A time-scaled version of the same network is shown in Figure 12.40.

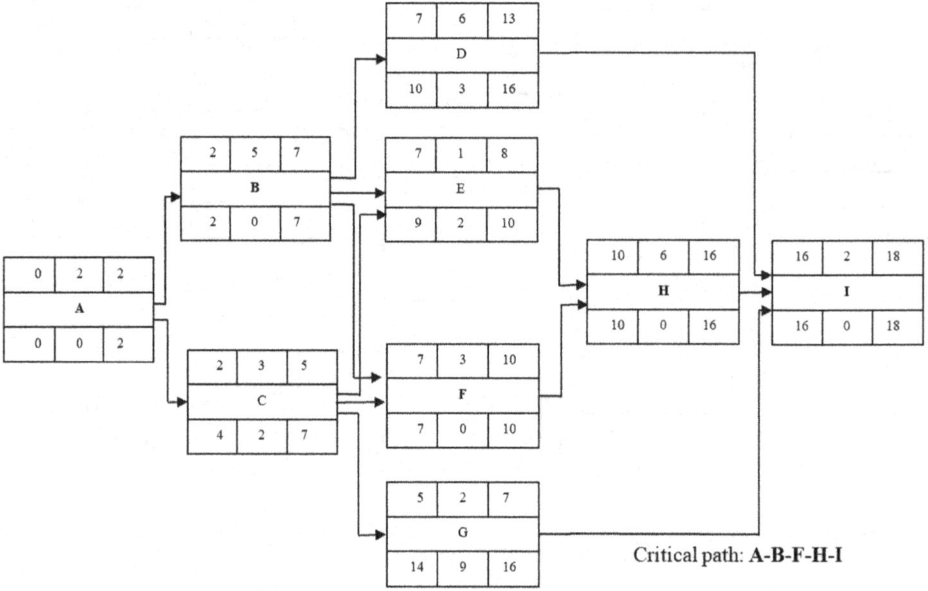

Figure 12.37 Activity-on-node representation

Table 12.7 Activities, inter-relationships, time constraints, and their time durations

Activities	*Immediately preceding activities*	*Time constraints*	*Time durations (days)*
A	-	-	2
B	A	SS + 2 days	2
C	A	FS	2
D	A	FF + 1 day	1
E	B	FF + 2 days	2
F	C	SS + 3 days	4
G	C	FF + 1 day	3
H	F	FS	2
	G	FF + 1 day	
I	D	FS	3
	G	FF + 3 days	
J	E	FF + 4days	2
	H	FS	
	I	FS + 2 days	

Example 12.8: For the relationships between the activities given in Table 12.8, draw the network diagram and calculate the EST, EFT, LST, LFT, and total float values of each activity. Also identify the critical path and represent the network on a time scale.

Solution: An AON representation of the project is shown in Figure 12.41. This helps in developing the network along with all the types of relationships shown in Figure 12.42. The EST, EFT, LST, LFT, and total float values of the various activities on the network are given in Figure 12.42. The critical path of the network is A-C-B-E-F-G-H. A time-scaled version of the same network is shown in Figure 12.43.

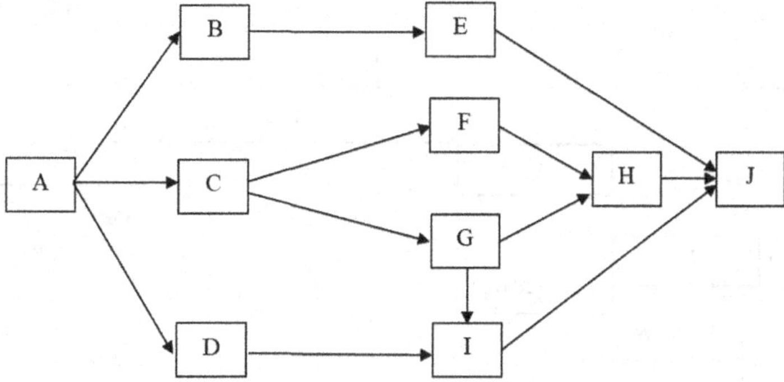

Figure 12.38 Activity-on-node representation

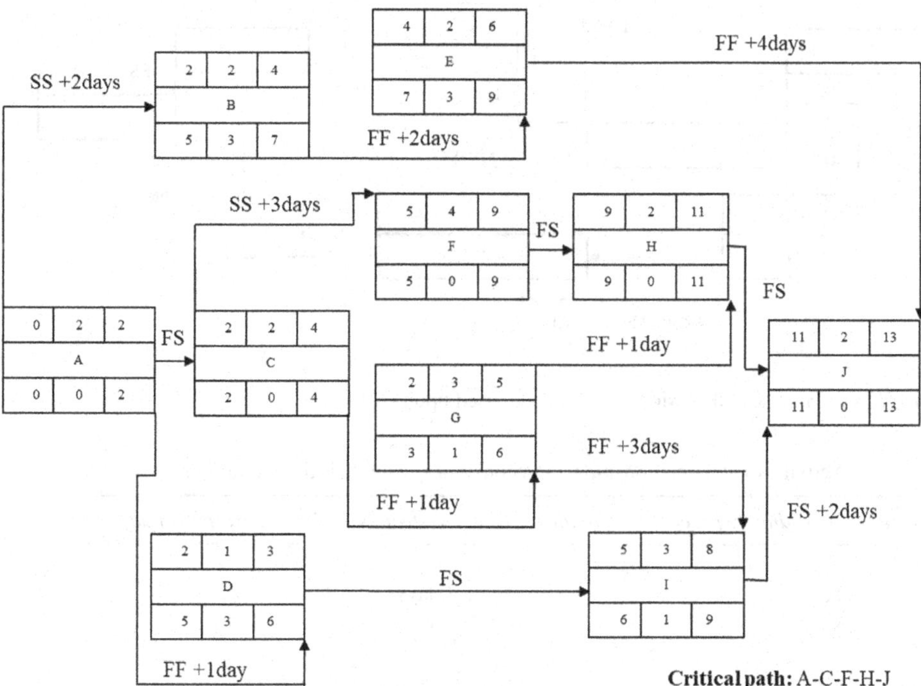

Figure 12.39 Activity-on-node representation with time constraints

12.11 Calendar Date Scheduling

In the following sections, calendar date scheduling will be discussed with reference to the four types of relationships between activities used in PDM. A month from a calendar year, used to schedule activities, is given in Table 12.9. Assume that 6 days a week are working days, Sundays are non-working days, and day 6 of the month is a holiday.

12.11.1 Finish-to-Start

The FS relationship with zero lag in Figure 12.44 shows that activity G cannot begin until activity D is completed. Activity D starts with the start of day 4 and takes 5 days to complete. It ends

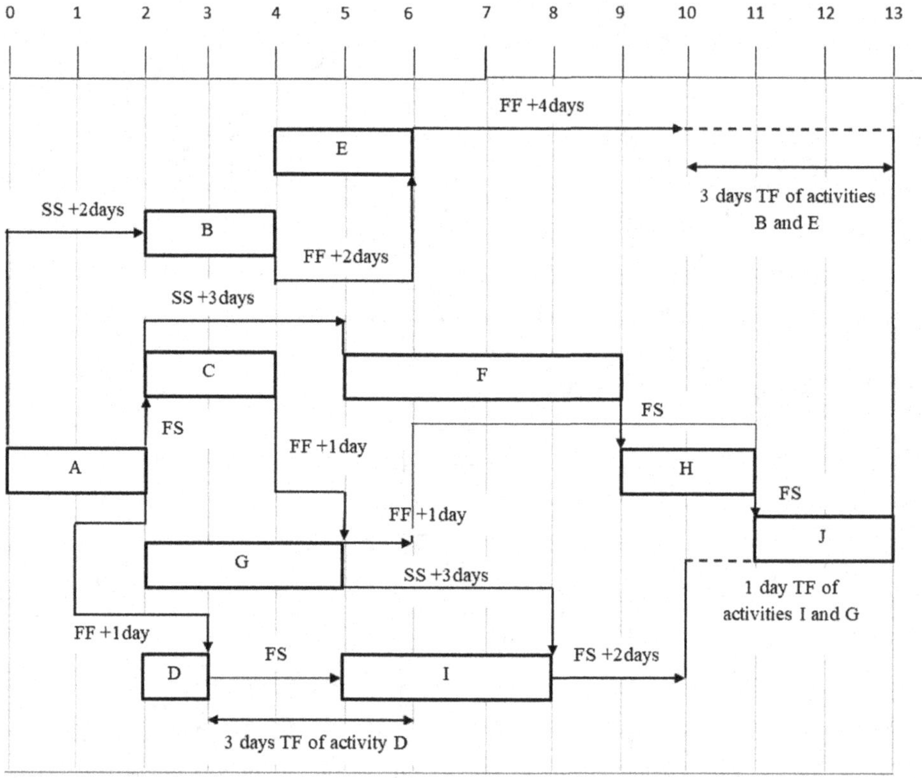

Figure 12.40 Time scaled version of a network based upon EST

Table 12.8 Activities, inter-relationships, time constraints, and their time durations

Activities	*Immediately preceding activities*	*Time constraints*	*Time durations in days*
A	-	-	2
B	A	SS + 2 days	3
	C	SS + 1 day	
C	A	FS + 2 days	2
D	A	FF + 1 day	1
E	B	FF + 2 days	2
F	C	SS + 3 days	4
	E	FS + 2 days	
G	D	FF + 3 days	2
	F	FF + 1 day	
H	E	FF + 4 days	2
	F	FS	
	G	SS + 3 days	

at the end of day 9; day 6 is assumed to be a holiday. Day 10 is a Sunday, therefore activity G starts at the start of day 11 and ends at the end of day 12.

If activity G ends at the end of day 16 and takes 2 days to complete, it has to start at the start of day 15. If activity G starts at the start of day 15, activity D has to end at the end of day 14. Day 10 is a Sunday, therefore the start of day 9 is the start of activity D, as shown in Figure 12.44.

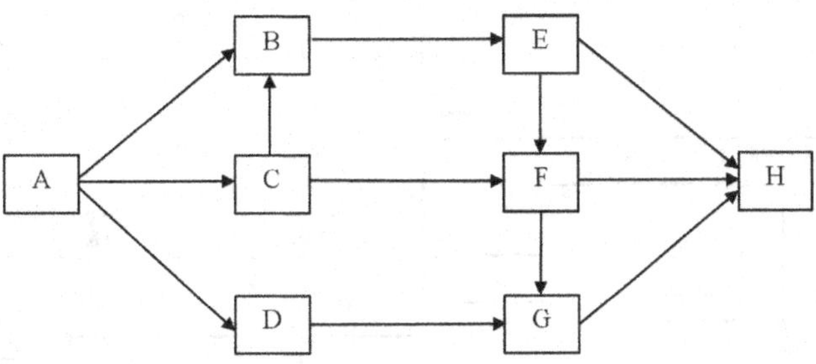

Figure 12.41 Activity-on-node representation

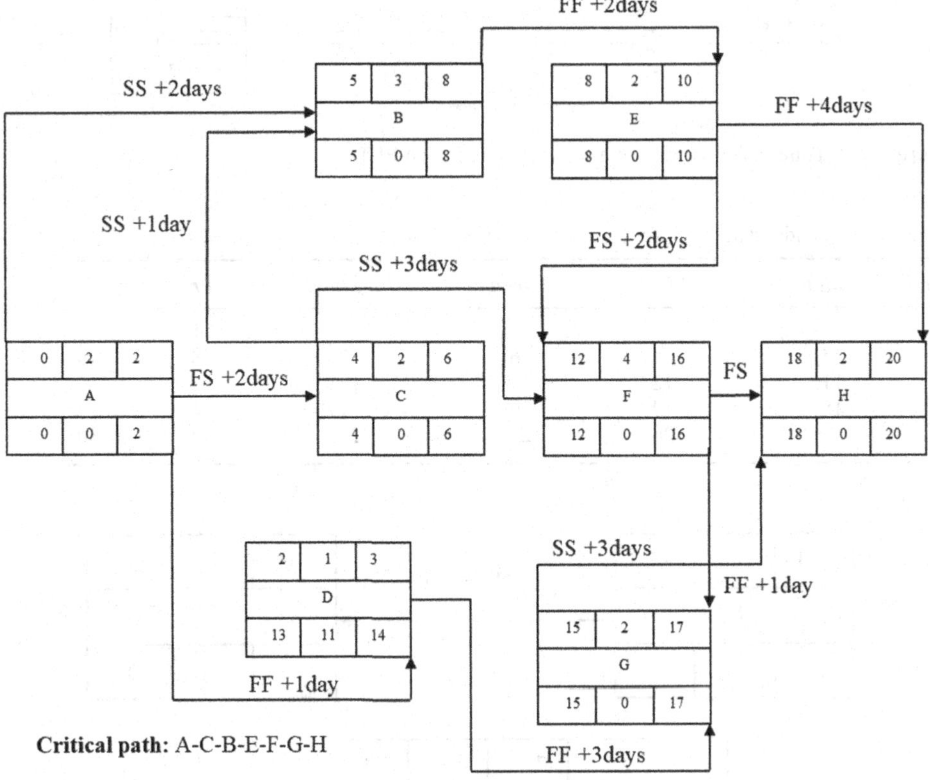

Figure 12.42 Activity-on-node representation with time constraints

An FS relationship with positive lag shows that the start of the subsequent activity has to be delayed, by a time duration equal to the positive lag value, after the end of the preceding activity. Activity D starts at the start of day 4 and ends at the end of day 9. This positive lag implies that the start of activity G has to be delayed by a time duration of 3 days after the end of activity D. Day 10 is a Sunday, therefore the three days of positive lag are days 11, 12, and 13. Activity G starts at the start of day 14 and ends at the end of day 15.

If activity G ends at the end of day 16, activity G has to start with the start of day 15. The three days of positive lag are 14, 13, and 12. In this case activity D has to end with the end of

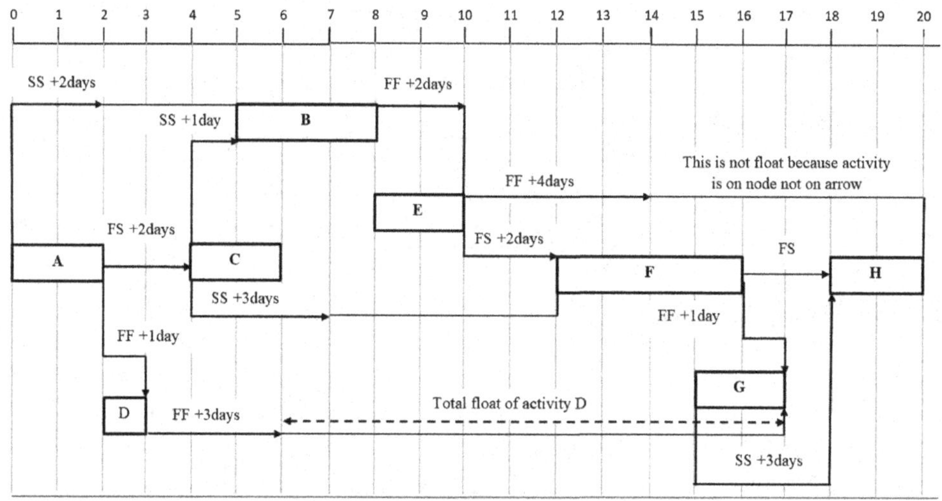

Figure 12.43 Time scaled version of a network based upon EST

Table 12.9 Calendar dates

Sun	*Mon*	*Tue*	*Wed*	*Thu*	*Fri*	*Sat*
					1	2
3	4	5	**6**	7	8	9
10	11	12	13	14	15	16
17	18	19	20	21	22	23
24	25	26	27	28	29	30

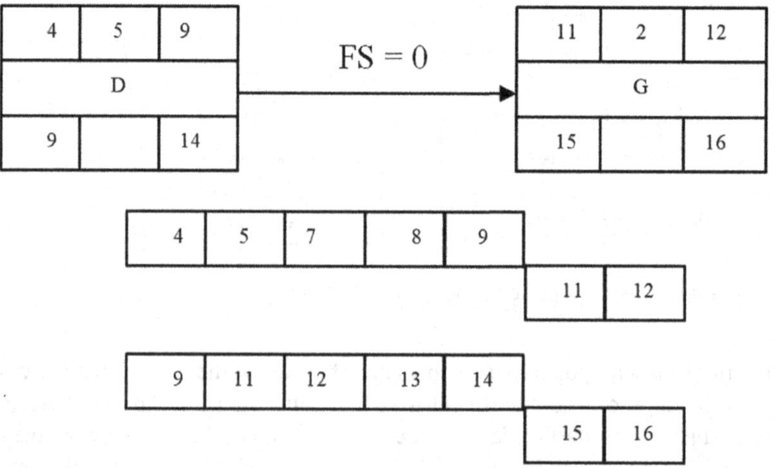

Figure 12.44 FS relationship with zero lag

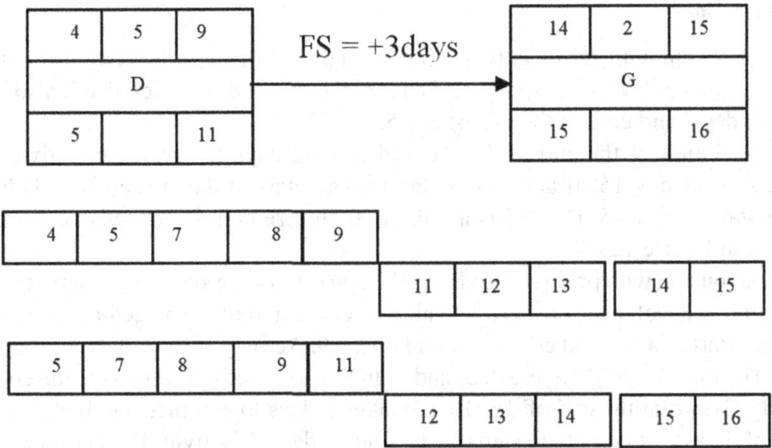

Figure 12.45 FS relationship with positive lag

day 11. Days 6 and 10 are holidays, thus the start of day 5 is the start of activity D, as shown in Figure 12.45.

An FS relationship with negative lag shows that the subsequent activity has to start, by a time duration equal to the negative lag value, before the end of the preceding activity. Activity D starts at the start of day 4 and ends at the end of day 9. The negative lag implies that the start of activity G has to be preponed by a time duration of 2 days before the end of activity D. The two days of negative lag are 9 and 8. Activity G has to start at the start of day 8 and end at the end of day 9.

If activity G ends at the end of day 16 and takes 2 days to complete, it has to start at the start of day 15. The two days of lag are 15 and 16. In this case activity D has to end at the end of day 16, therefore the start of day 12 is the start of activity D, as shown in Figure 12.46.

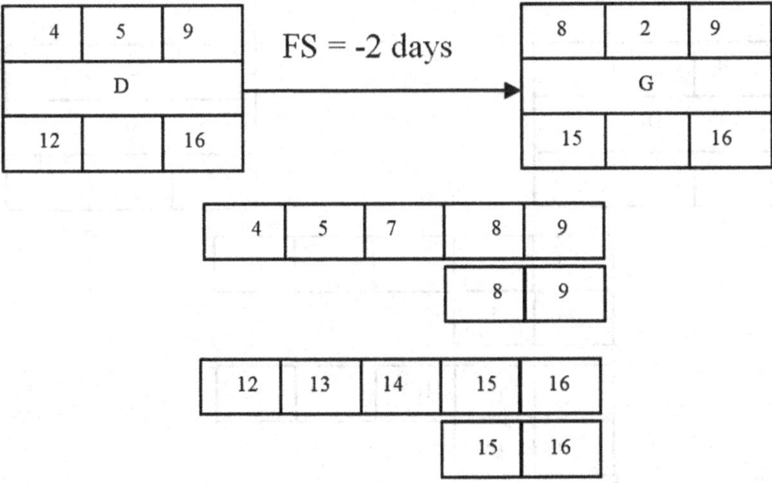

Figure 12.46 FS relationship with negative lag

12.11.2 Start-to-Start

In an SS relationship with zero lag, two activities start simultaneously. Activity D starts at the start of day 4, takes 5 days to complete, and ends at the end of day 9. Activity G also has to start at the start of day 4 and ends at the end of day 5.

If activity G ends at the end of day 16 and takes 2 days to complete, activity G has to start at the start of day 15. If activity G starts at the start of day 15, activity D has also to start at the start of day 15. Day 17 is a holiday therefore activity D ends at the end of day 20 as shown in Figure 12.47.

An SS relationship with positive lag shows that the start of the subsequent activity is delayed, by a time duration equal to the positive lag value, after the start of the preceding activity. Activity D starts at the start of day 4 and ends at the end of day 9. Activity G starts 3 days after the start of activity D. The three days of lag are 4, 5, and 7. In this case activity G ends at the end of day 9.

If activity G ends at the end of day 16 and takes 2 days to complete, activity G has to start at the start of day 15. If activity G starts at the start of day 15, activity D has to start three days before the start of activity G. The three days of positive lag are 14, 13, and 12. Thus, activity D starts at the start of day 12 and ends at the end of day 16, as shown in Figure 12.48.

In an SS relationship with negative lag, the start of the subsequent activity is preponed, by a time duration equal to the negative lag value, before the start of the preceding activity. Activity D starts at the start of day 4 and ends at the end of day 9. Activity G starts 2 days before the start of activity D. The two days of negative lag are 1 and 2. In this case activity G starts at the start of day 1 and ends at the end of day 2.

If activity G ends at the end of day 16 and takes 2 days to complete, it has to start at the start of day 15. If activity G starts at the start of day 15, activity D has to start two days after the start of activity G. The two days of negative lag are 15 and 16. Thus, activity D has to start at the start of day 18 and end at the end of day 22, as shown in Figure 12.49.

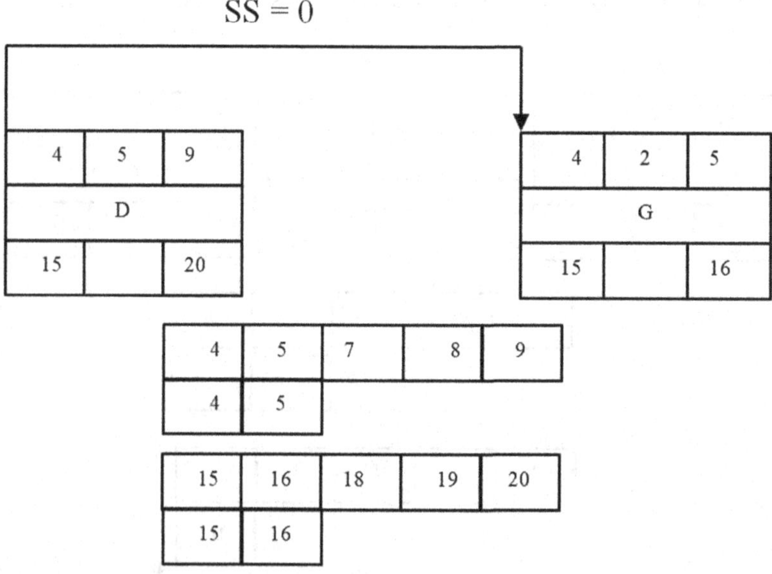

Figure 12.47 SS relationship with zero lag

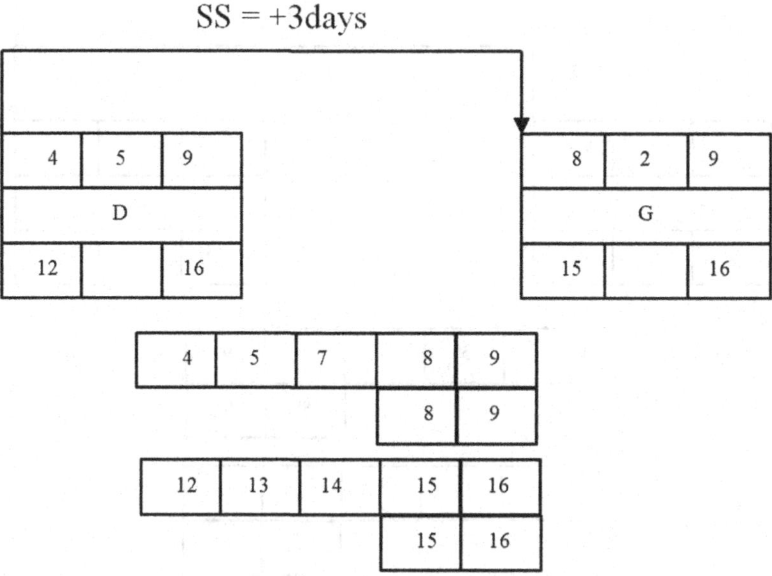

Figure 12.48 SS relationship with positive lag

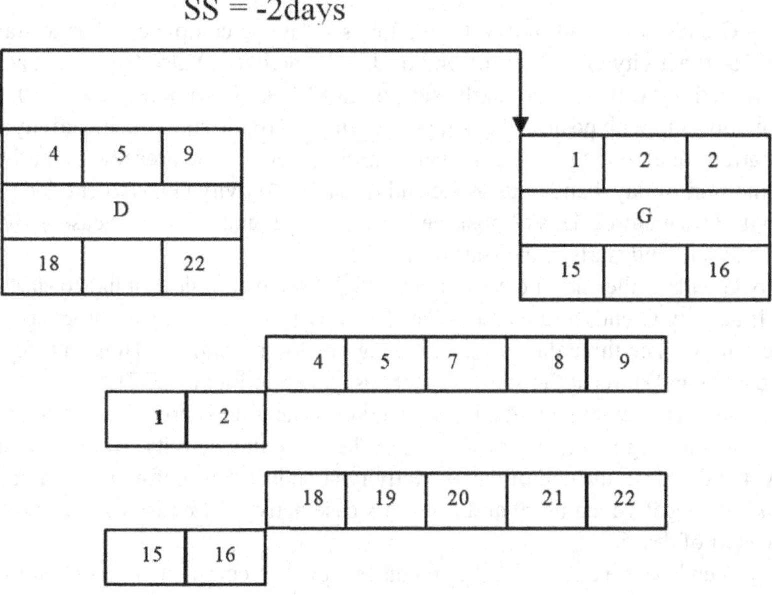

Figure 12.49 SS relationship with negative lag

12.11.3 Finish-to-Finish

An FF relationship with zero lag is used to depict the relationship between two activities that finish simultaneously. Activity D starts at the start of day 4, takes 5 days to complete, and ends at the end of day 9. Activity G also has to end at the end of day 9 and start at the start of day 8.

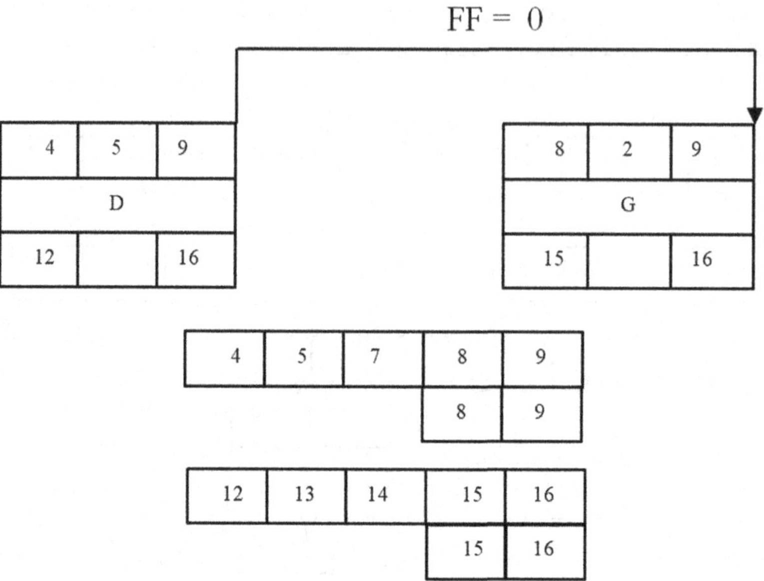

Figure 12.50 FF relationship with zero lag

If activity G ends at the end of day 16 and takes 2 days to complete, it has to start with the start of day 15. If activity G ends at the end of day 16, activity D also has to end at the end of day 16. Thus, activity D has to start at the start of day 12, as shown in Figure 12.50.

An FF relationship with positive lag shows that the end of the subsequent activity is delayed, by time duration equal to the positive lag value, after the end of the preceding activity. Activity D starts at the start of day 4 and ends at the end of day 9. Activity G has to end 3 days after the end of activity D. The three days of positive lag are 11, 12, and 13. In this case activity G ends at the end of day 13 and starts at the start of day 12.

If activity G ends at the end of day 16 and takes 2 days to complete, it has to start at the start of day 15. If activity G ends at the end of day 16, activity D has to finish three days before the finish of activity G. The three days of positive lag are 14, 15, and 16. Thus, activity D ends at the end of day 13 and starts at the start of day 8, as shown in Figure 12.51.

In an FF relationship with negative lag, the end of an activity is preponed, by a time duration equal to the negative lag value, before the end of the preceding activity. Activity D starts at the start of day 4 and ends at the end of day 9. Activity G ends 2 days before the end of activity D. The two days of negative lag are 9 and 8. In this case activity G ends at the end of day 7 and starts at the start of day 5.

If activity G ends at the end of day 16 and takes 2 days to complete, it has to start at the start of day 15. If activity G ends at the end of day 16, activity D has to end two days after the end of activity G. The two days of lag are 18 and 19. Thus, activity D ends at the end of day 19 and starts at the start of day 14, as shown in Figure 12.52.

12.11.4 Start-to-Finish

An SF relationship with zero lag is used to depict a relationship in which the subsequent activity finishes and the preceding activity starts at the same time. Activity D starts at the start of day 4

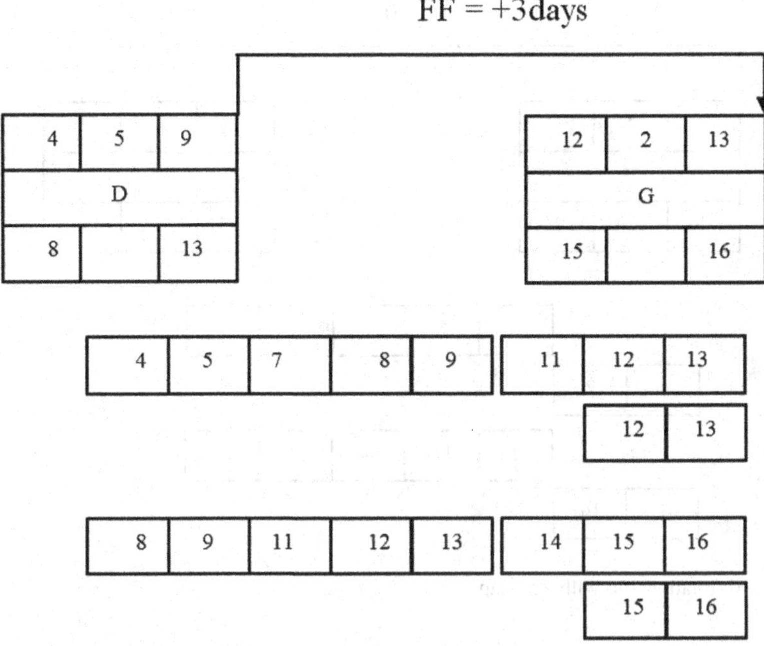

Figure 12.51 FF relationship with positive lag

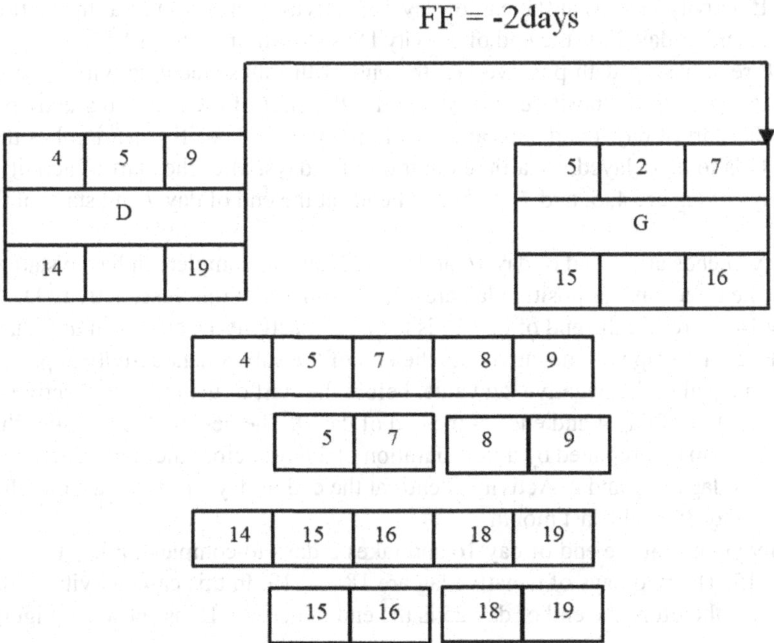

Figure 12.52 FF relationship with negative lag

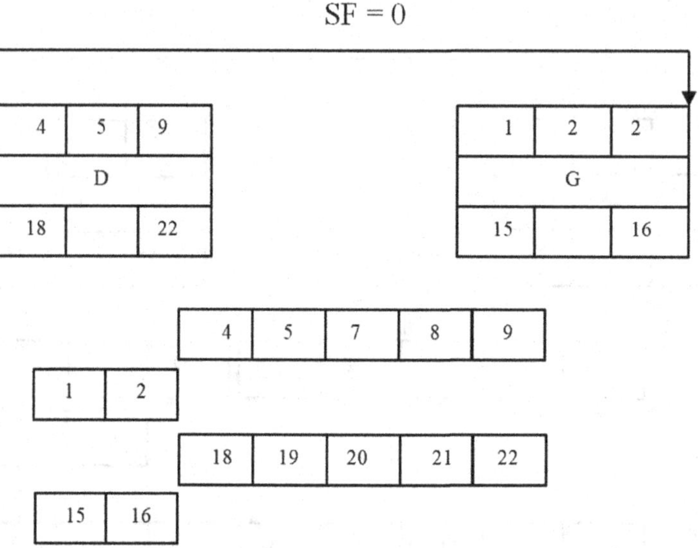

Figure 12.53 SF relationship with zero lag

and ends at the end of day 9. Activity D starts at the start of day 4, therefore activity G ends at the end of day 2 and starts at the start of day 1.

If activity G ends at the end of day 16 and takes 2 days to complete, it has to start at the start of day 15. If activity G ends at the end of day 16, activity D has to start at the start of day 18, therefore the end of day 22 is the end of activity D as shown in Figure 12.53.

In an SF relationship with positive lag, the end of the subsequent activity is delayed, by a time duration equal to the positive lag value, after the start of the preceding activity. Activity D starts at the start of day 4 and ends at the end of day 9. The positive lag implies that the end of activity G is to be delayed, by a time duration of 3 days, after the start of activity D. Thus, the three days of lag are 4, 5, and 7. Activity G ends at the end of day 7 and starts at the start of day 5.

If activity G ends at the end of day 16 and takes 2 days to complete, it has to start at the start of day 15. The three days of positive lag are 14, 15, and 16. In this case, activity D starts at the start of day 14, therefore the end of day 19 is the end of activity D, as shown in Figure 12.54.

In an SF relationship with negative lag, the end of the subsequent activity is preponed, by a time duration equal to the negative lag value, before the start of the preceding activity. Activity D starts at the start of day 4 and ends at the end of day 9. The negative lag implies that the end of activity G has to be preponed by a time duration of 2 days before the start of activity D. Thus, the two days of lag are 1 and 2. Activity G ends at the end of day 31 of the last month and starts at the start of day 30 of the last month.

If activity G ends at the end of day 16 and takes 2 days to complete, it has to start with the start of day 15. The two days of negative lag are 18 and 19. In this case activity D starts at the start of day 20, therefore the end of day 25 is the end of activity D, as shown in Figure 12.55.

Example 12.9: Draw a network diagram for the precedence relationships given in Table 12.10. Table 12.11 provides calendar dates for the scheduling of the project. The project starts at the start of the month, assuming 6 working days per week, and that all Sundays

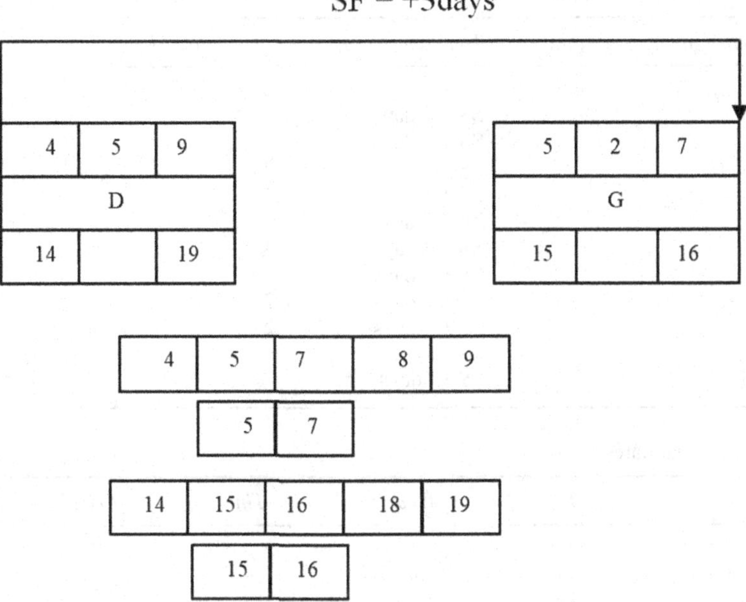

Figure 12.54 SF relationship with positive lag

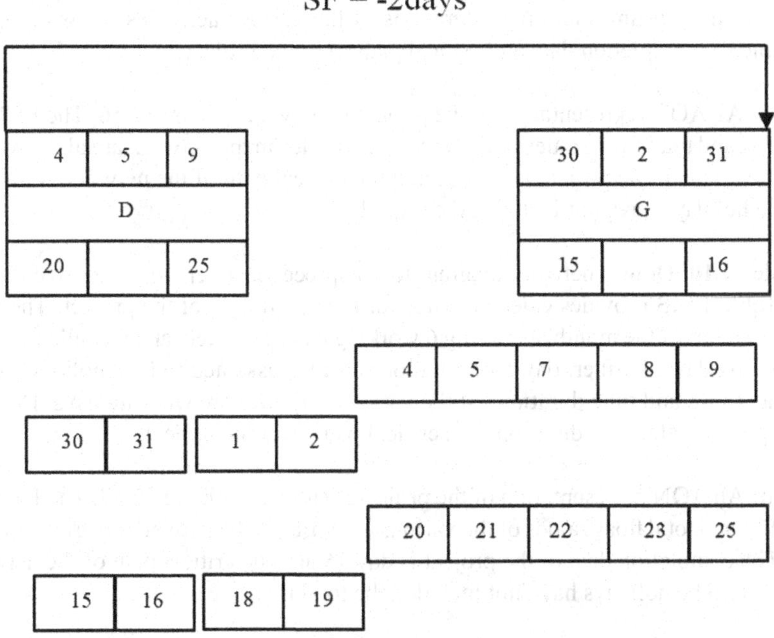

Figure 12.55 SF relationship with negative lag

Table 12.10 Activities, inter-relationships, time constraints, and their time durations

Activity	*Immediately preceding*	*Time constraints*	*Duration (days)*
A	-	-	2
B	A	SS + 2 days	3
	C	SS + 1 day	
C	A	FS + 2days	2
D	A	FF + 1 day	1
E	B	FF + 2 days	2
F	C	SS + 3 days	4
	E	FS + 2 days	
G	D	FF + 3 days	2
H	E	FF + 4 days	2
	F	FS	
	G	SS +3 days	

Table 12.11 Calendar dates

Sun	*Mon*	*Tue*	*Wed*	*Thu*	*Fri*	*Sat*
						1
2	3	4	5	6	7	8
9	10	11	12	13	**14**	15
16	17	18	19	20	21	22
23	24	25	26	27	28	29
30	31					

are non-working days. Further, day 14 of the month is also assumed to be a holiday. Assume all the lag times and time durations of the various activities are working days. Determine the completion date and critical path(s) of the project.

Solution: An AON representation of the project is shown in Figure 12.56. The EST, EFT, LST, LFT, and total float values of the various activities in the network are also given. The completion date of the project is day 22 and the critical path of the network is A-C-B-E-F-H. The holidays have not included the total float.

Example 12.10: Draw a network diagram for the precedence relationships given in Table 12.12. Table 12.13 provides calendar dates for the scheduling of the project. The project starts at the start of the month, assuming 6 working days per week, and that all Sundays are non-working days. Further, day 6 of the month is also assumed to be a holiday. Assume all the lag times and time durations of the various activities are working days. Determine the completion date, time duration, and critical path(s) of the project.

Solution: An AON representation of the project is shown in Figure 12.57. The EST, EFT, LST, LFT, and total float values of the various activities in the network are given in Figure 12.57. The completion date of the project is day 16 and the critical path of the network is A-B-D-F-H. The holidays have not included the total float.

12.12 PDM for Repetitive Projects

PDM is also sometimes used for the planning and scheduling of repetitive projects. Consider a project that involves the construction of three similar parking sheds. The project has three

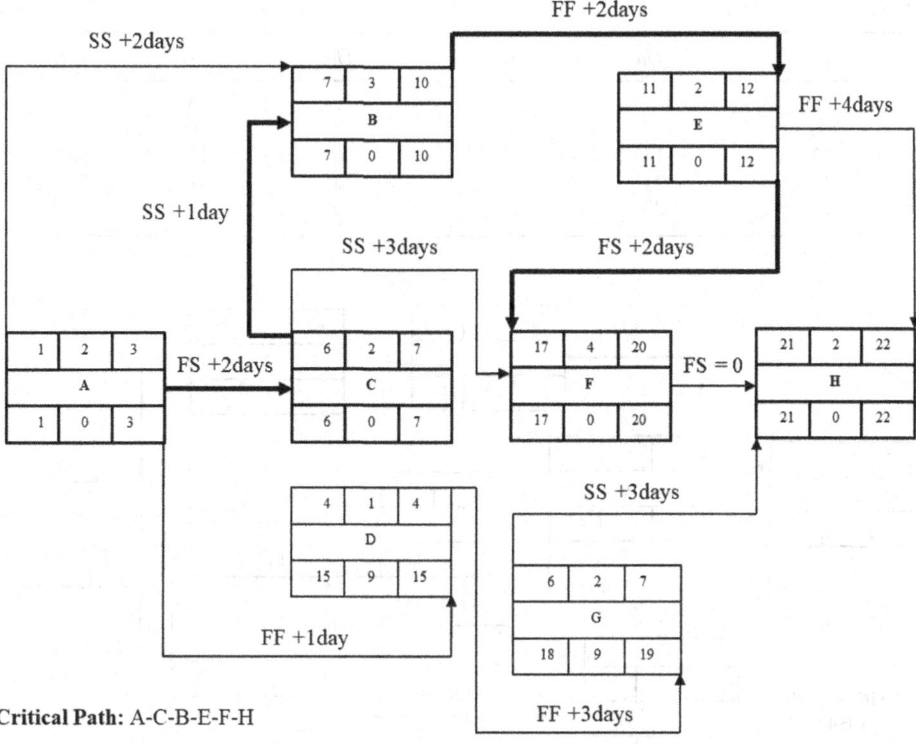

Figure 12.56 Activities, inter-relationships, time constraints, and EST, EFT, LST, LFT, and total float values

Table 12.12 Activities, time durations, and relationships among them

Activity	*Immediately preceding*	*Time constraints*	*Duration (days)*
A	-	-	2
B	A	FS	2
C	A	FF + 1 day	1
D	B	SS + 3 days	4
E	B	FF + 1 day	3
F	D	FS	2
	E	FF + 1 day	
G	C	FS	3
	E	FF + 3 days	
H	F	FS	2
	G	FS + 2 days	

similar parking sheds, therefore it is a repetitive project. The activities involved in the construction of a parking shed are laying the foundation, building the structure, roofing, and finishing. The construction of one parking shed involves four activities which are repeated three times in the construction of three similar parking sheds, making a total of twelve activities. In summary, the project has four repetitive activities which are repeated three times, making a total of twelve activities. Figure 12.58 shows the execution sequence, using PDM, for the construction of three

Table 12.13 Calendar dates

Sun	*Mon*	*Tue*	*Wed*	*Thu*	*Fri*	*Sat*
			1	2	3	4
5	**6**	7	8	9	10	11
12	13	14	15	16	17	18
19	20	21	22	23	24	25
26	27	28	29	30	31	

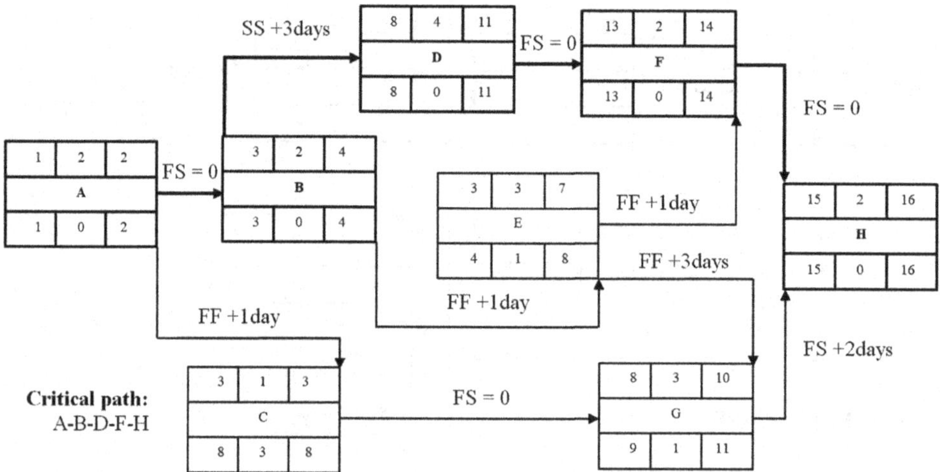

Figure 12.57 Activities, inter-relationships, time constraints, and EST, EFT, LST, LFT, and total float values

similar parking sheds. In other words, the figure represents the execution sequence of the construction of three repeated units of a repetitive project.

Initially, the execution sequence of the activities involved in the construction of the first shed is finalized. The order finalized for the construction of the first shed has been kept the same for the construction of the other two sheds. The first shed takes 35 days to complete. In repetitive projects, each repetitive activity is generally assigned the same resources to keep its time duration the same across each repeated unit. The resources assigned to a repetitive activity continue working on all the other repeated units. The second shed is completed at the end of day 44. The project ends with the completion of the third shed, which is completed at the end of day 53.

12.13 Network Development Procedure

The procedure for developing a network in PDM is almost the same as in CPM. However, in PDM, activities are represented on nodes, and four types of relationships between the various activities are used. The time constraints are additional to the precedence relationships in PDM. The network development procedure is discussed below, step-by-step.

Step 1: The activities involved in a project are identified and defined clearly. A brief description of each activity involved is also provided.

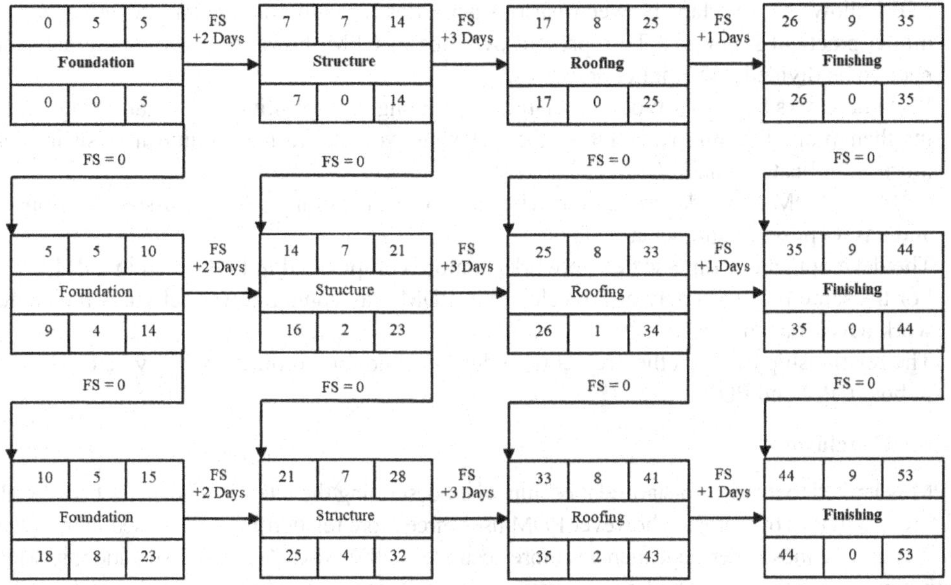

Figure 12.58 Scheduling of a repetitive project

Step 2: The time durations of all the activities in the network are estimated. Generally, a single estimate is made of the time duration of each activity.

Step 3: A network is developed directly by experienced planners to represent the execution sequence of the various activities defined earlier. The inter-relationships between the various activities are sometimes listed in a tabular form and the network is then developed from the table.

Step 4: Time constraints are decided to facilitate the availability of time durations before and/or after the starts or ends of the various activities. These are taken care of by positive and negative lags. These time constraints are represented on a network in the form of FS, SS, FF, and SF time lags. The units of the lags and time durations are kept the same.

Step 5: The developed network is analyzed to determine the activity times. The network analysis has two passes: a forward pass for the calculation of earliest times, and a backward pass for the calculations of latest times. The critical path of the network and the floats of the individual activities are also determined. Finally, the time-scaled version of the network may also be developed.

12.14 CPM and PDM: Similarities and Dissimilarities

The following are points of similarity and dissimilarity between the critical path method and the precedence diagraming method.

- Project planning and scheduling are done separately in both CPM and PDM; project planning is done before the project scheduling.
- PDM uses four types of relationships – finish-to-start, finish-to-finish, start-to-start, and start-to-finish – to represent the inter-relationships/logics between the various activities. CPM, however, uses only a finish-to-start relationship.
- Alterations of logic are easier in PDM than in CPM because only the connecting arrows, representing the inter-relationships, are changed.

- PDM allows for overlaps between various activities through positive and negative lags, to model practical situations. To represent overlaps in CPM, however, overlapping activities need to be divided into smaller activities.
- PDM networks have four types of relationships along with positive and negative lags, making them more difficult to understand than CPM networks. This makes time analysis in PDM more complicated than in CPM.
- CPM and PDM networks are similar when a network uses only a finish-to-start relationship and has no positive and negative lags.
- The development of time-scaled networks is more complicated in PDM than in CPM.
- For the same project, a network developed in PDM may contain fewer activities than a network developed in CPM.
- The relationship between the project time duration and total project cost may be established in both CPM and PDM.

12.15 Conclusion

CPM is limited to a finish-to-start relationship when describing the inter-dependencies between the different activities of a project, however PDM uses three more relationships, discussed in this chapter, to describe those inter-dependencies more accurately. PDM uses AON representation, in which activities are represented on nodes and arrows connect those nodes to represent the inter-dependencies between them. Time constraints are also modeled through positive and negative lags in PDM, due to the time requirements before or after the start or end of an activity. PDM also allows for overlaps between project activities through positive and negative lags, to model practical situations. The calculation of the time duration of a project and the EST, EFT, LST, LFT, and total float of the various activities in a project, have also been discussed. The chapter also covers the calendar date scheduling of a project. CPM and PDM networks are similar when a network uses only a finish-to-start relationship and the network has no positive and negative lags.

Exercises

Question 12.1: For the relationships given in Table E12.1, draw the network diagram, calculate the EST, EFT, LST, LFT, and total float values of each activity, and identify the critical path.

Answer: The EST, EFT, LST, LFT, and total float values of the various activities in the network are given in Figure E12.1. The critical path of the network is A-C-B-E-F-G-H.

Table E12.1 Activities, inter-relationships, time constraints, and their time durations

Activities	*Immediately preceding activities*	*Time constraints*	*Time durations in days*
A	-	-	4
B	A	SS + 2 days	2
	C	SS + 1 day	
C	A	FF + 2 days	2
D	A	SF + 1 day	1
E	B	SF + 2 days	3
F	C	SS + 3 days	4
	E	FS + 2 days	
G	D	FF + 3 days	5
	F	FS + 3 days	
H	E	FF + 4 days	3
	F	FS	
	G	FF + 1 day	

Question 12.2: For the relationships given in Table E12.2, draw the network diagram and calculate the EST, EFT, LST, LFT, and four floats values of each activity, and identify the critical path.

Answer: The EST, EFT, LST, LFT, and total float values of the various activities in the network are given in Figure E12.2 and Table E12.3. The critical path of the network is A-C-F-H.

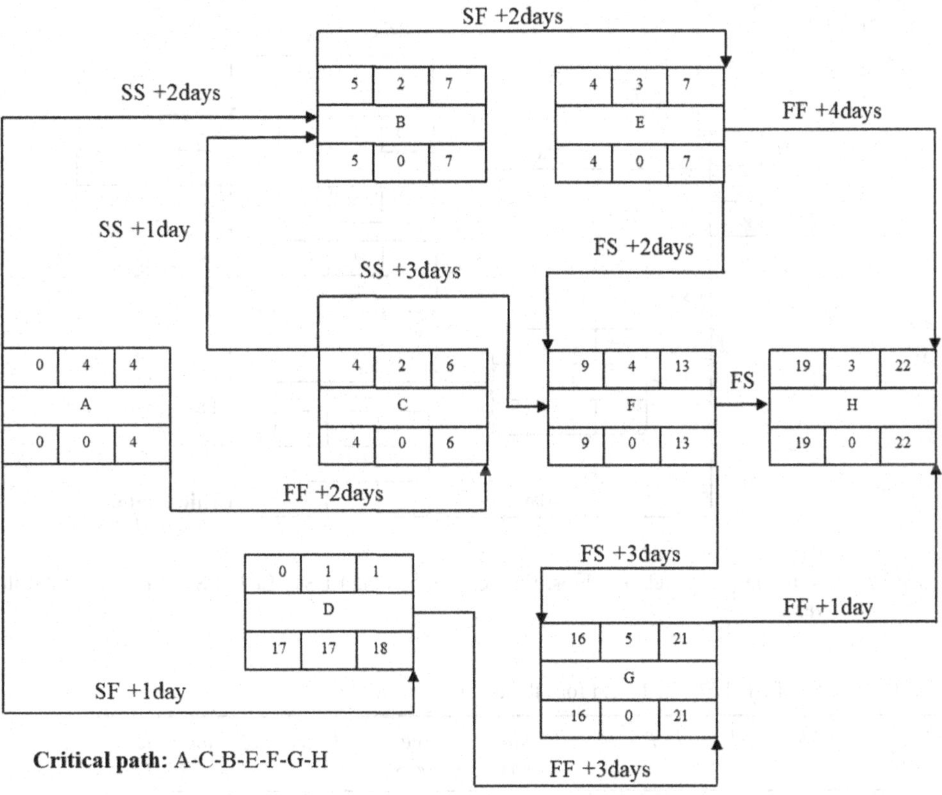

Figure E12.1 Activities, inter-relationships, time constraints, and EST, EFT, LST, LFT, and total float values

Table E12.2 Activities, inter-relationships, time constraints, and their time durations

Activity	*Immediately preceding*	*Time constraints*	*Duration (days)*
A	-	-	3
B	A	SS + 2 days	2
C	A	FS + 3 days	4
D	B	FS + 1 day	3
E	B	SS + 2 days	2
F	C	FF + 3 days	4
G	A	FS + 3 days	2
H	D	SF + 4 days	2
	E	SS + 3 days	
	F	FS + 2 days	
	G	FS + 3 days	

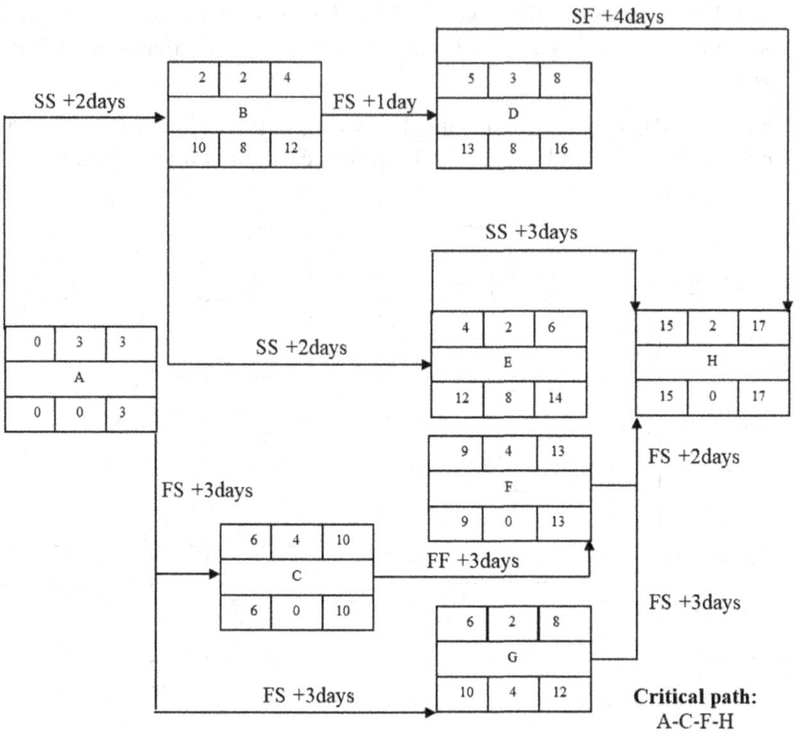

Figure E12.2 Activities, inter-relationships, time constraints, and EST, EFT, LST, LFT, and total float values

Table E12.3 EST, EFT, LST, LFT, and four floats for each activity

Activities	*EST*	*EFT*	*LST*	*LFT*	*Total floats*	*Free floats*	*Interfering*	*Independent*
A	0	3	0	3	0	0	0	0
B	2	4	10	12	8	0	8	0
C	6	10	6	10	0	0	0	0
D	5	8	13	16	8	8	0	0
E	4	6	12	14	8	8	0	0
F	9	13	9	13	0	0	0	0
G	6	8	10	12	4	4	0	4
H	15	17	15	17	0	0	0	0

Question 12.3: Draw a network diagram for the precedence relationships given in Table E12.4. Table E12.5 provides calendar dates for scheduling the project. The project starts on day 3 of the month, assuming six working days per week, and that all Sundays are non-working days. Assume that all lag times and time durations of the various activities are working days. Determine the completion date, time duration, and critical path(s) of the project.

Answer: The EST, EFT, LST, LFT, and total float values of the various activities in the network are given in Figure E12.3. The completion date of the project is day 27. The critical path of the network is A-B-E-F-H-I. The holidays have not been included in the total float.

Table E12.4 Activities, inter-relationships, time constraints, and their time durations

Activity	*Immediately preceding*	*Time constraints*	*Duration (days)*
A	-	-	3
B	A	SS + 4 days	4
C	A	FS	2
D	A	FF + 1 day	1
E	B	FS + 3 days	5
F	C	FS + 1 day	3
	E	FF + 3 days	
G	C	FF + 2 days	2
	D	FS + 2 days	
H	E	FF +2days	4
	F	FS	
I	G	FS + 2 days	3
	H	FF + 2 days	

Table E12.5 Calendar dates

Sun	*Mon*	*Tue*	*Wed*	*Thu*	*Fri*	*Sat*
3	4	5	6	7	8	9
10	11	12	13	14	15	16
17	18	19	20	21	22	23
24	25	26	27	28	29	30

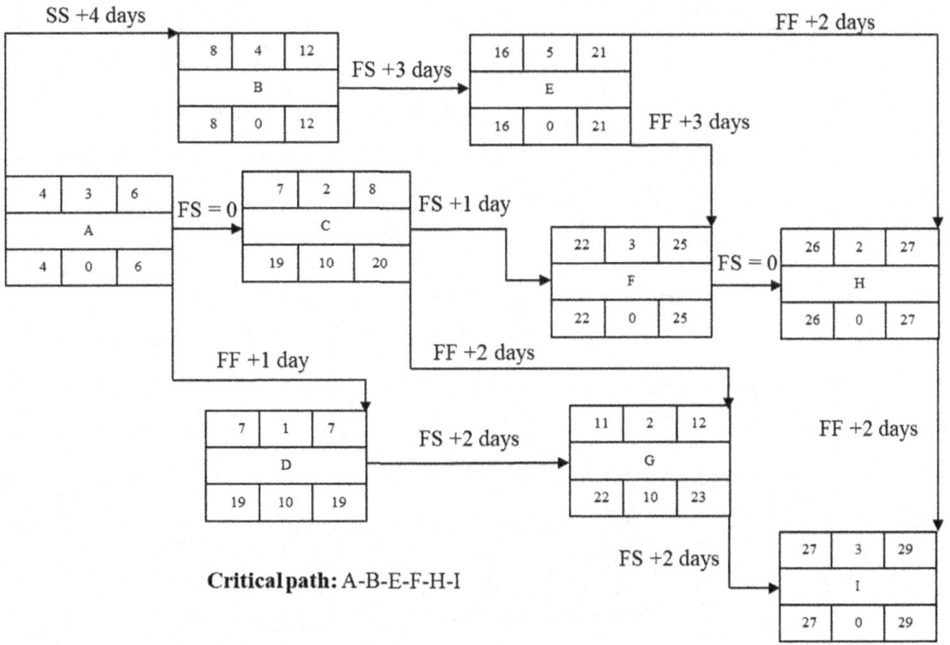

Figure E12.3 Activities, inter-relationships, time constraints, and EST, EFT, LST, LFT, and total float values

Question 12.4: What are the main differences between activity-on-node representation and the representation used in PDM?

Question 12.5: List five practical examples for each type of relationship given below.

i. Start-to-start relationship,
ii. Finish-to-finish relationship, and
iii. Start-to-finish relationship.

13 Line of Balance for Repetitive Projects

13.1 Learning Objectives

After the completion of this chapter, readers will be able to:

- Understand repetitive projects and their scheduling,
- Use the line of balance technique for scheduling of repetitive projects, and
- Understand the benefits of the line of balance technique for scheduling of repetitive projects.

13.2 Introduction

The simplest tool for project planning and scheduling is the bar chart. However, bar charts have many limitations, as discussed in previous chapters. Network-based techniques are commonly used for planning and scheduling large and complex projects but have limitations when applied to repetitive projects. The main limitations are their non-continuous crew engagement, lack of information about work location, lack of information about progress direction, and lack of information about task production rates at any given point during the project's execution. For planning and scheduling repetitive projects, line of balance (LOB) is the most widely used technique. It has distinct advantages over network-based techniques when applied to repetitive projects. This chapter focuses on the planning and scheduling of repetitive projects.

13.3 Repetitive Projects

Repetitive projects are projects where various activities are repeated. A repetitive project may consist of both non-repetitive and repetitive parts. For example, in the construction of a multi-story building, the foundation is constructed once only, and is a non-repetitive part. The construction of similar floors across the multi-story building would be a repetitive part. Depending upon the nature of a repetitive project, activities may be large or small in terms of the volume of work involved. Repetitive projects with a small number of repetitive activities usually involve large volumes of work. Projects containing repetitive activities include the construction of roads, laying of pipelines, tunneling, multi-story buildings, etc.

Consider a project that involves the construction of three similar parking sheds. The project involves three similar parking sheds; hence, it is a repetitive project. The activities involved in the construction of a parking shed are the construction of the foundation and the structure, roofing, and finishing. The construction of a parking shed involves four activities which are repeated three times in the construction of three similar parking sheds, making a total of twelve activities. The project has four repetitive activities repeated three times, making a total of twelve activities.

DOI: 10.1201/9781003428992-13

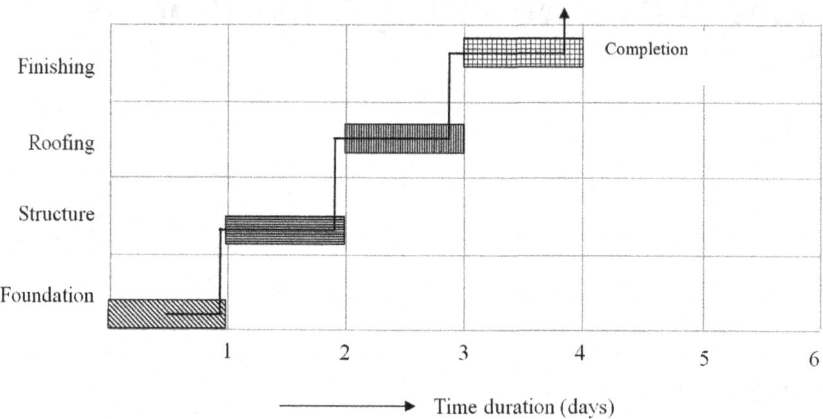

Figure 13.1 Execution sequence for the construction of a parking shed

Figure 13.1 shows the execution sequence for the construction of a parking shed in the form of a bar chart, in which all four activities are in a sequence. The figure represents the execution sequence for the construction of a repetitive unit within the repetitive project. Figure 13.2 shows the execution sequence for the construction of three parking sheds in the form of a bar chart in which all twelve activities are shown. In other words, it represents the execution sequence for the construction of three repetitive units within the repetitive project.

13.4 Scheduling of Repetitive Projects: Line of Balance

LOB is a widely used technique for planning and scheduling repetitive projects. An activity to be carried out multiple times at multiple locations is treated as a single activity in LOB. This reduces the total number of activities represented on an LOB schedule. The LOB-based schedule of a repetitive project is represented in the form of a graph. In an LOB graph, a single line represents a set of similar repetitive activities. In other words, an entire repetitive task is represented as a single activity using a single line. For example, the construction of one parking shed involves four activities, as shown in Figure 13.1. The four activities repeat three times in

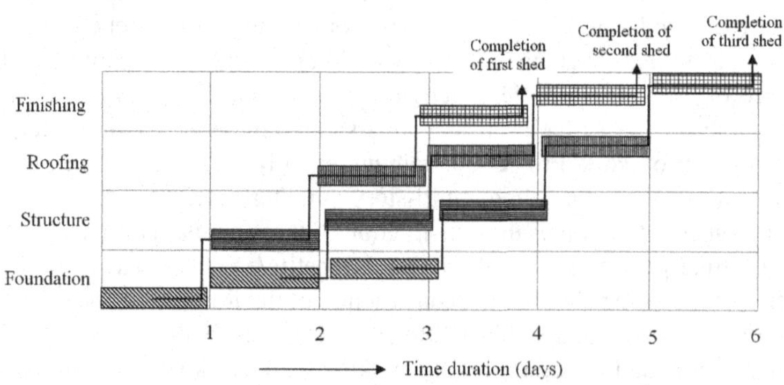

Figure 13.2 Execution sequence for the construction of three similar parking sheds

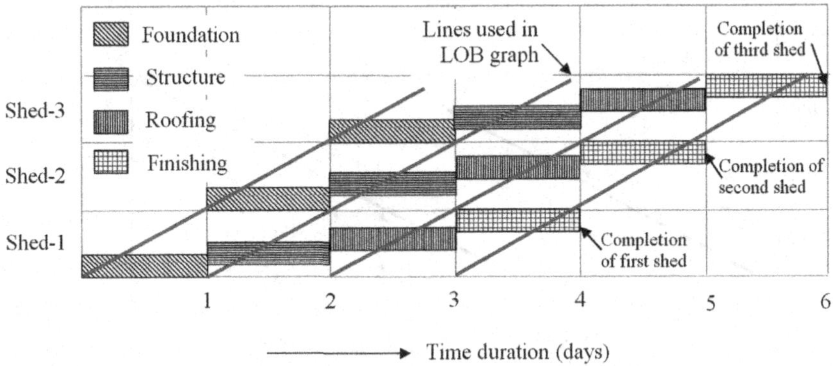

Figure 13.3 Execution sequence for the construction of three similar parking sheds. The inclined lines shown are used in LOB graphs for representing the execution sequence

the construction of three similar parking sheds, making a total of twelve activities. In an LOB graph, three repetitions of an activity across the three repetitive units are represented as a single activity. Thus, the twelve activities are represented as four activities in the LOB graph.

Figure 13.2 shows the execution sequence for the construction of three similar parking sheds, in which all twelve activities are shown. Figure 13.3 represents the execution sequence for the construction of three similar parking sheds, in which inclined lines are used to represent the four repetitive activities in the LOB graph. Figure 13.3 is similar to Figure 13.2 in terms of the execution sequence. In Figure 13.3, repetitive units are shown along the y-axis. Figure 13.4 shows the execution sequence for the construction of three similar parking sheds in the form of an LOB graph in which four lines represent the four activities that repeat three times in the construction of three similar parking sheds.

In an LOB graph, an execution schedule is represented in the form of a 2D graph. The graph has two axes: time and distance, or *units*. In general, the time duration of a project is represented along the x-axis. Distance is plotted along the y-axis. Distance is used for the scheduling of linear projects like highways, pipelines, or tunneling, while units are used for multi-unit projects such as multi-story buildings. For a linear project like a highway, the LOB graph is also plotted

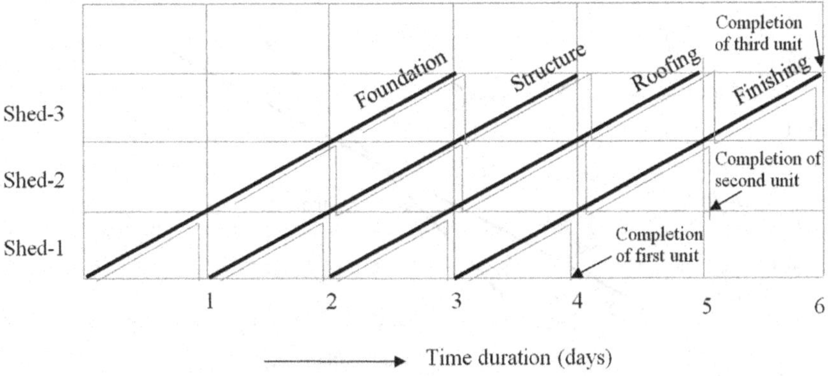

Figure 13.4 LOB graph representing an execution sequence for the construction of three parking sheds

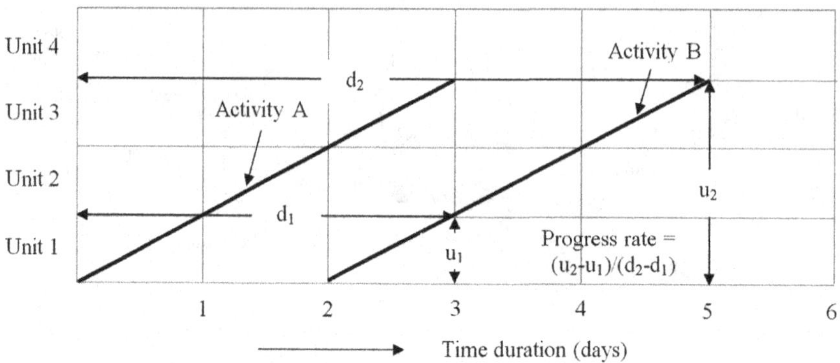

Figure 13.5 Progress rate of activity B as represented on an LOB graph

between distance and time, with distance represented along the x-axis and time along the y-axis. Activities in an LOB graph are represented by lines with constant or sometimes changing slopes.

The slope of a line in an LOB graph represents the progress rate, execution rate, production rate, or productivity of an activity. In Figure 13.5, the progress rate of an activity B is computed as $(u_2\text{-}u_1)/(d_2\text{-}d_1)$. The slope of a line in an LOB graph represents the rate at which an activity is carried out so as to stay on schedule. In Figure 13.5, two repetitive activities, A and B, are shown. Activity A has the same production rate as that of activity B: one repetitive unit is produced per day. Activity A runs parallel to activity B because they have the same production rate. The progress of an activity at any point of time is controlled by comparing the position of the activity on the graph and its progress onsite. Activity B, as shown in Figure 13.5, has progressed by u_2 units within the time duration d_2.

13.4.1 Activities with Different Progress Rates

Consider a project that consists of four repetitive units. Each repetitive unit has two activities, making a total of eight activities. In Figure 13.6, two activities, A and B, are shown. The relationship between the two activities is such that activity B starts when activity A is completed. In Figure 13.6, the first unit of activity B starts when the first unit of activity A is completed. The production rate of activity A is one unit per day and the production rate of activity B is two

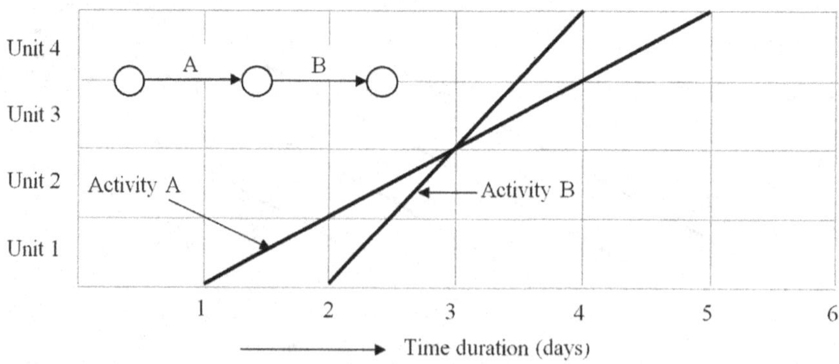

Figure 13.6 Activities with different progress rates on an LOB graph

units per day. Activity A has half the production rate of activity B. With these production rates, activity B crosses activity A after the second repetitive unit, which is not possible in practice, as shown in Figure 13.6. To maintain the inter-dependency between activities A and B, a time lag of half a day is set between each repetitive unit of activity B, as shown in Figure 13.7. Due to the time lag between each repetitive unit of activity B, resource continuity is not maintained, which results in an inefficient use of resources in activity B. Activity B is not performed continuously in this case; rather, breaks are given so as to maintain the inter-dependency between activities A and B.

13.5 Work/Resource Continuity

In Figure 13.7, activity A and activity B of a repetitive project are shown. The first unit of activity B starts when the first unit of activity A is completed. To maintain the inter-dependency between activity A and activity B, a time lag of half a day is provided between each repetitive unit of activity B. Due to the lag between each repetitive unit of activity B, resource continuity is not maintained, which results in an inefficient use of resources in activity B. Activity B is not being performed continuously; rather, breaks have been provided to maintain the logic. However, in the case of activity A, resources are continuously employed until the four units are completed. In other words, resource continuity is maintained in the case of activity A but not in the case of activity B.

To maintain resource continuity in the case of activity B, an initial time delay/lag of one and half days at the start of its first repetitive unit is set, as shown in Figure 13.8. This delay is equal to the sum of the three time delays set between each repetitive unit of activity B – that is, 3 times the half-day delay/lag set between each unit in activity B in Figure 13.7. In this way, activity B can be completed uninterrupted, maintaining its resource continuity and its inter-dependency with activity A.

LOB is a resource-driven planning and scheduling technique for projects with repetitive activities. The main objective of the LOB technique is to estimate the required resources for each activity in such a way that resource continuity is maintained. A project with repetitive activities is scheduled in such a way that a continuous use of resources is maintained for all the repetitive activities in a project. It enables the continuous use of resources throughout repetitive activities and improves work efficiency. Unlike network-based techniques, the LOB technique

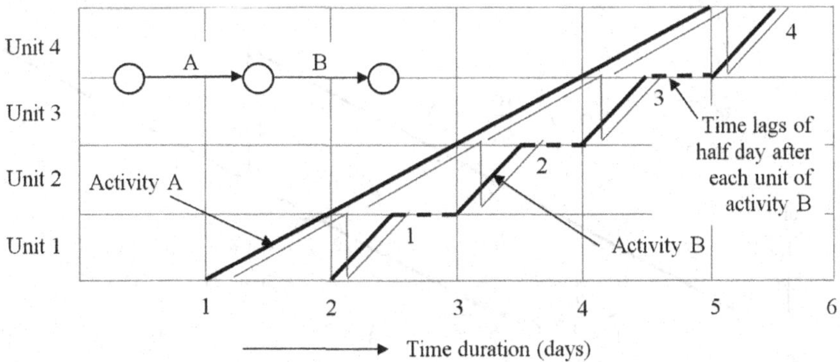

Figure 13.7 Time lags given to activity B to maintain the required inter-dependency between activity A and activity B

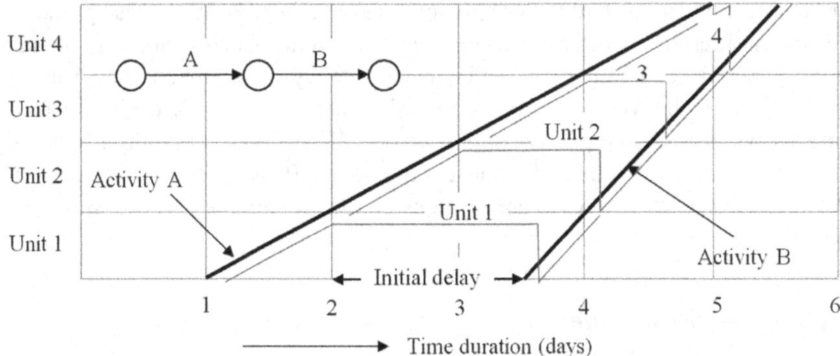

Figure 13.8 An initial delay of one and a half days at the start of the first repetitive unit of activity B to maintain resource continuity in activity B

does not emphasize the minimization of a project's time duration, but focuses instead on productivity enhancement, ensuring resource continuity through balanced activity production rates.

13.6 Time and Space Buffers

In Figure 13.9, two repetitive activities, A and B, are shown. The first unit of activity B can start when the first unit of activity A is completed. In Figure 13.9, the first unit of activity B is started one day after the completion of the first unit of activity A. This time delay of one day at the start of activity B is called *lag time*, *buffer time*, or simply a *buffer*. Buffer time is the time interval between two successive activities. In Figure 13.9, a buffer time of one day is provided between activity A and activity B. *Buffer space* is a gap in distance or units between successive activities. Figure 13.9 shows a readily visible buffer space between two activities, activity A and activity B, on an LOB graph.

13.7 Types of Repetitive Projects

Repetitive projects are classified, based on their productivity, into *typical* and *non-typical repetitive projects*. Typical repetitive projects consist of activities which have the same quantity of work

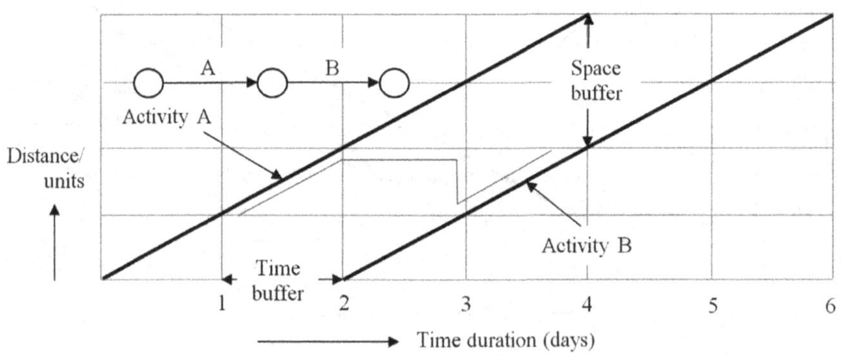

Figure 13.9 Buffer times and spaces

in each of their repetitive units and use resources with the same productivity for each repetitive unit. In a typical repetitive project, straight lines with constant slopes represent the activities in a repetitive project. Non-typical repetitive projects have different quantities of work in their different repetitive units and use resources with different productivities in different repetitive units. In a non-typical repetitive project, lines with changing slopes represent the activities in a repetitive project.

Repetitive projects are classified, based on the nature of the work involved, into two categories: *discrete* and *continuous*. *Discrete repetitive projects* have discrete repetitive units, like similar houses on different locations, or similar stories of a multi-story building. Thus, discrete repetitive projects are also referred to as vertical projects. *Continuous repetitive projects* involve continuous work like the construction of highways, pipelines, tunnels, or canals, in which progress is expressed in terms of meters, kilometers, or any other suitable measure of accomplishment. Continuous repetitive projects are also often called *horizontal* or *linear projects*.

13.8 Activities With Different Productivities

Consider a project that involves the construction of eleven similar operator rooms. The activities involved in the construction of an operator room are those of site preparation, laying the foundation, producing the structure, electrical work, and finishing. The construction of one operator room involves five activities; each activity is repeated eleven times in the construction of eleven similar operator rooms, making a total of fifty-five activities. Figure 13.10 shows an execution sequence for the construction of an operator room in which all activities are in a sequence. Figure 13.10 also shows an LOB graph representing an execution sequence for the construction of 11 operator rooms. The repetitive activity of site preparation is represented by a single line that starts at time zero and ends at the end of day 11. The slope of a line is one unit per day. The repetitive activity of laying the foundation is also represented by a single line that starts at the end of day 1 and ends at the end of day 12; the slope of the line is one unit per day. The repetitive activity of producing the structure starts at the end of day 2 and ends at the end of day 24; the slope of the line is half a unit per day.

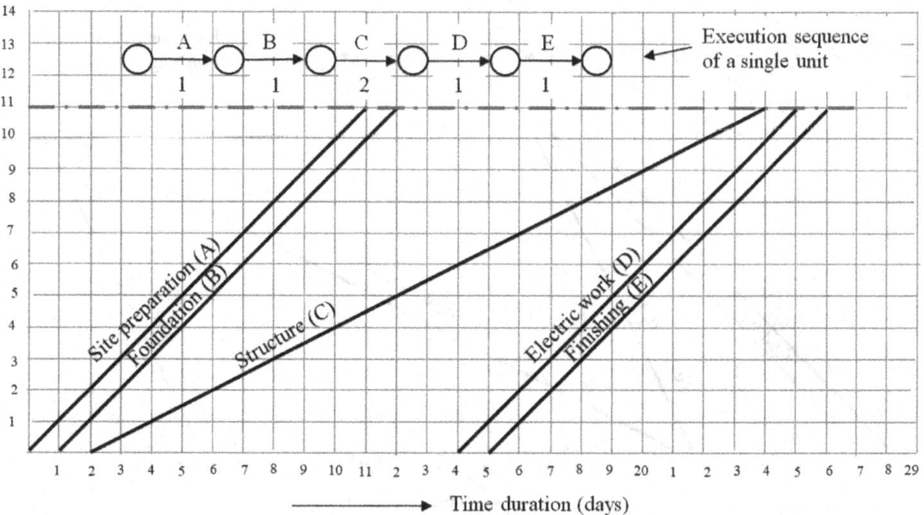

Figure 13.10 LOB graph representing the construction schedule of eleven similar operator rooms

If the first unit of electrical work is started when the first unit of the structure is completed, with the given production rate, the activity of electrical work will cross over the activity of producing the structure in a way that is not possible in practice. In this case, the last unit of electrical work is started when the last unit of the *structure* is completed. With the given production rate, the repetitive activity of the electrical work starts at the end of day 14 and ends at the end of day 25; the slope of the line is one unit per day. The electrical work is performed continuously in this case. The last repetitive activity, the finishing, starts at the end of day 15 and ends at the end of day 26; the slope of the line is one unit per day. The time duration of the project is 26 days. The LOB-based schedule for the construction of eleven similar operator rooms is provided in Table 13.1.

Consider another project that involves the development of 10 similar coding units. The activities involved in the development of one unit are designing, coding, testing, implementation, and testing. Each unit involves five activities; each activity is repeated ten times, making a total of fifty activities. Figure 13.11 shows the execution sequence for the development of one unit in which all activities are in a sequence. Figure 13.11 also shows an LOB graph

Table 13.1 LOB-based schedule for the construction of eleven similar operator rooms

Activities	*Time durations of a single unit (in days)*	*Number of units involved*	*Time durations of eleven units (in days)*	*Rate of execution (units per day)*	*Start time of first unit (in days)*	*Finish time of last unit (in days)*
Site preparation(A)	1	11	11	1	0	11
Foundation (B)	1	11	11	1	1	12
Structure (C)	2	11	22	0.5	2	24
Electric works (D)	1	11	11	1	14	25
Finishing (E)	1	11	11	1	15	26

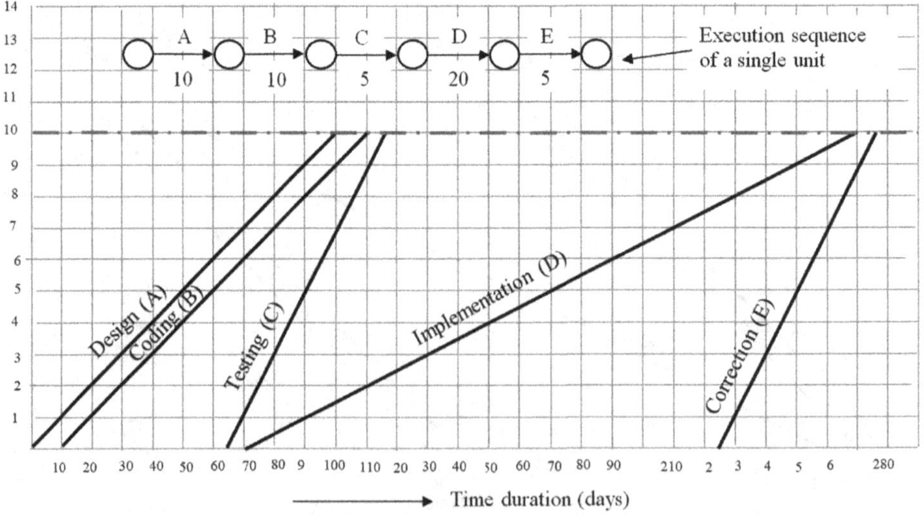

Figure 13.11 LOB graph showing the schedule for the development of ten similar coding units

representing the execution sequence for the development of 10 coding units. The repetitive activity of designing is represented by a line that starts at time zero and ends at the end of day 100. The slope of the line is 0.1 units per day. The repetitive activity of coding is also represented by a line that starts at the end of day 10 and ends at the end of day 110; the slope of the line is 0.1 units per day.

If the first unit of the activity of testing is started when the first unit of the activity of coding is completed, the activity of testing will cross over the activity of coding which, with the given production rate, is not possible. Thus, the last unit of testing is started when the last unit of coding is completed. With the given production rate, the repetitive activity of testing starts at the end of day 65 and ends at the end of day 115; the slope of the line is 0.2 units per day. Testing is performed continuously in this case. The first repetitive unit implementation is started when the first unit testing is completed. Thus, the implementation starts at the end of day 70 and ends at the end of day 270; the slope of the line is 0.05 units per day.

If the first unit of correction is started when the first unit of activity implementation is completed, with the given production rate, correction will cross over with implementation. Thus, the last unit of correction is started when the last unit of implementation is completed. With the given production rate, the repetitive activity of correction starts at the end of day 225 and ends at the end of day 275; the slope of the line is 0.2 units per day. In this case, the project duration is 275 days. The LOB-based schedule for the development of ten similar coding units is provided in Table 13.2.

13.9 Linear Interpretation of Precedence Relationships

Four types of inter-dependencies/logics between the various activities in a project are used in PDM. The zero, positive, and negative lags in the four types of relationships are elaborated in a linear interpretation in the following sub-sections.

13.9.1 Finish-to-Start

A finish-to-start relationship with zero lag shows that an activity cannot start unless the preceding activity is finished. In Figure 13.12(a), activity B cannot begin until activity A is completed. The start of activity B depends upon the completion of activity A. The zero lag value implies that the moment activity A finishes, activity B starts immediately, with no lag.

An FS relationship with positive lag shows that the start of an activity has to be delayed, by a time duration equal to the positive lag value, after the end of the preceding activity.

Table 13.2 LOB-based schedule for the development of ten similar coding units

Activities	*Time durations of a single unit (in days)*	*Number of units involved*	*Time durations of 10 units (in days)*	*Rate of execution (unit per day)*	*Start time of first unit (in days)*	*Finish time of last unit (in days)*
Design (A)	10	10	100	0.1	0	100
Coding (B)	10	10	100	0.1	10	110
Testing (C)	05	10	50	0.2	65	115
Implement-ation (D)	20	10	200	0.05	70	270
Correction (E)	05	10	50	0.2	225	275

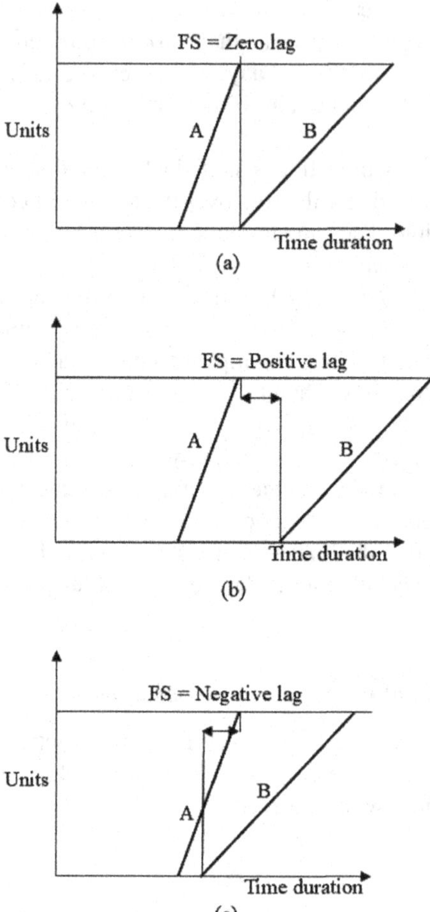

Figure 13.12 A finish-to-start relationship with zero, positive, and negative lags

In Figure 13.12(b), the start of activity B has to be delayed, by a time duration equal to the positive lag value, after the end of activity A. The start of activity B depends upon the completion of activity A and the lag between them.

An FS relationship with negative lag shows that an activity has to start, by a time duration equal to the negative lag value, before the end of the preceding activity. This is the reason that negative lag is also sometimes called lead. In Figure 13.12(c), the start of activity B is preponed, by a time duration equal to the negative lag value, before the end of activity A. Activity B begins before activity A is completed.

13.9.2 Start-to-Start

A start-to-start relationship with zero lag is a relationship between two activities in which they start simultaneously. In Figure 13.13(a), activity A and activity B start simultaneously. The zero lag value in this SS relationship indicates that the moment activity A starts, activity B also starts.

An SS relationship with positive lag shows that the start of an activity is delayed, by a time duration equal to the positive lag value, after the start of the preceding activity. In

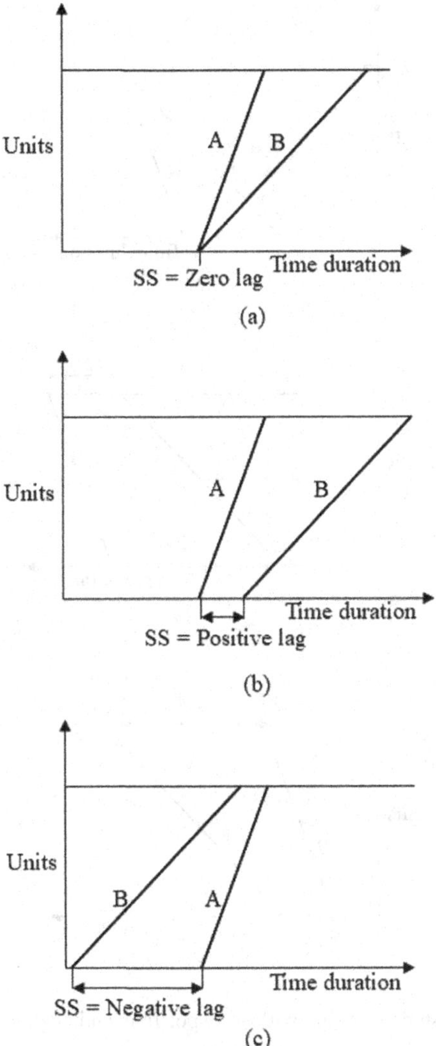

Figure 13.13 A start-to-start relationship with zero, positive, and negative lags

Figure 13.13(b), the start of activity B is delayed, by a time duration equal to the positive lag value, after the start of activity A.

In an SS relationship with negative lag, the start of the subsequent activity is preponed, by a time duration equal to the negative lag value, before the start of the preceding activity. In Figure 13.13(c), the start of activity B is preponed, by a time duration equal to the negative lag value, before the start of activity A. The start of activity B depends upon the negative lag.

13.9.3 Finish-to-Finish

A finish-to-finish relationship with zero lag is used to depict a relationship between two activities that finish simultaneously. In Figure 13.14(a), activity A and activity B finish simultaneously.

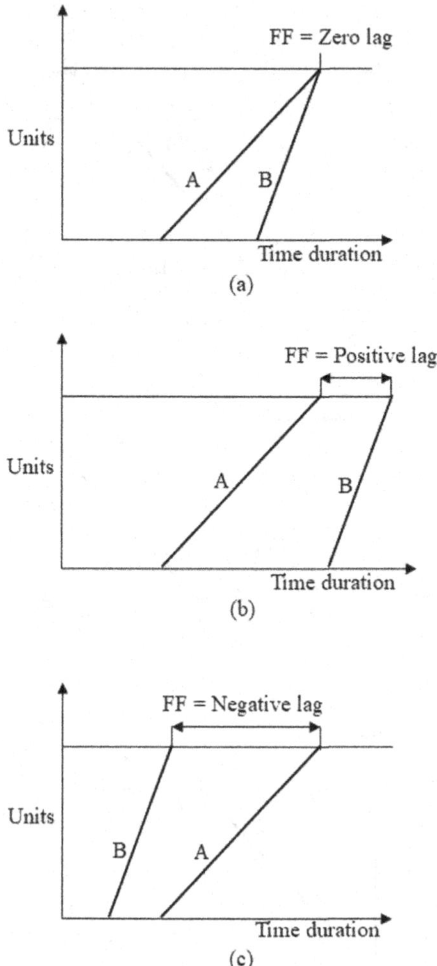

Figure 13.14 A finish-to-finish relationship with zero, positive, and negative lags

The zero lag value in this FF relationship implies that the moment activity A finishes, activity B also finishes.

An FF relationship with positive lag shows that the end of the subsequent activity is delayed, by a time duration equal to the positive lag value, after the end of the preceding activity. In Figure 13.14(b), the end of activity B is delayed, by a time duration equal to the positive lag value, after the end of activity A. This positive FF lag implies that activity B will have been ended by a time duration equal to the positive lag value after the end of activity A.

In an FF relationship with negative lag, the end of an activity is preponed, by a time duration equal to the negative lag value, before the end of the preceding activity. In Figure 13.14(c), the end of activity B is preponed, by a time duration equal to the negative lag value, before the end of activity A.

13.9.4 Start-to-Finish

A start-to-finish relationship with zero lag shows that an activity should already be finished when the preceding activity starts. SF with zero lag is used to depict a relationship in which an

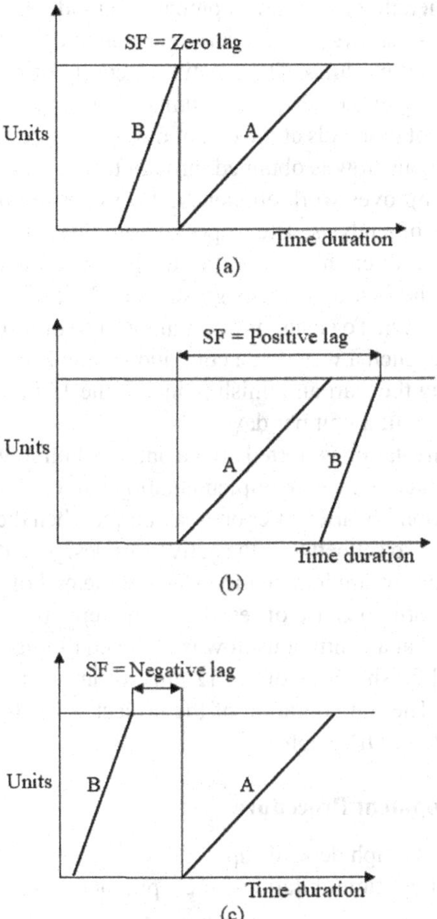

Figure 13.15 A start-to-finish relationship with zero, positive, and negative lags

activity finishes and its preceding activity starts at the same time. In Figure 13.15(a), activity B finishes and preceding activity A starts at the same time.

In an SF relationship with positive lag, the end of an activity is delayed, by a time duration equal to the positive lag value, after the start of the preceding activity. In Figure 13.15(b), the end of activity B is delayed, by a time duration equal to the positive lag value, after the start of activity A. Positive SF lag implies that activity B ends after the start of activity A by a time duration equal to the positive lag value.

In an SF relationship with negative lag, the end of an activity is preponed, by a time duration equal to the negative lag value, before the start of the preceding activity. In Figure 13.15(c), the end of activity B is preponed, by a time duration equal to the negative lag value, before the start of activity A. This negative SF lag implies that activity B ends before the start of activity A by a time duration equal to the negative lag.

13.10 Alternate Representation of an LOB Graph

Consider a project that involves the development of twelve similar modules. The activities involved in the development of a module are the analysis, the design, and the implementation.

Figure 13.16 shows the schedule for the development of a module, in which all activities are in a sequence. Figure 13.16 also shows the bar chart representation of the execution sequence for the development of 12 similar modules. The repetitive activity of analysis is represented by 12 horizontal bars corresponding to the 12 different units of the activity. The first unit of analysis starts at time zero and the last unit ends at the end of day 12. The repetitive activity is scheduled in such a way that a continuous flow is obtained through the 12 units. This enables the continuous use of resources and improves work efficiency. Two lines are drawn joining the start and finish points of the 12 units of analysis. The slopes of both lines are one unit per day.

A one-day buffer is set between the first unit of analysis and the first unit of the subsequent activity, the design. Thus, the first unit of design starts at the end of day 2 and the last unit of design ends at the end of day 26. To ensure the continuous use of resources, the repetitive activity of design is scheduled in such a way that a continuous flow is obtained through its 12 units. Two lines are drawn joining the start and finish points of the 12 units of the activity of design. The slopes of both lines are half a unit per day.

If the first unit of implementation is started after a one-day buffer, when the first unit of design is completed, with this production rate the implementation will cross over with the design. Thus, the last unit of implementation is started after one-day buffer when the last unit of design is completed. A one-day buffer is also set between the activity of design and the subsequent activity of implementation. The last unit of implementation starts at the end of day 27 and ends at the end of day 28. To ensure the continuous use of resources, the repetitive activity of implementation is scheduled in such a way that a continuous flow is obtained through its 12 units. Two lines are drawn joining the start and finish points of the 12 units of implementation. The slopes of both lines are one unit per day. The time duration of the project is 28 days. Figure 13.17 shows an alternate way to represent the LOB graph.

13.11 LOB Graph Development Procedure

The development of an LOB graph depends upon the start time of the first repetitive unit of the starting activity, the activities' time durations, their productivities/rates of execution, and the

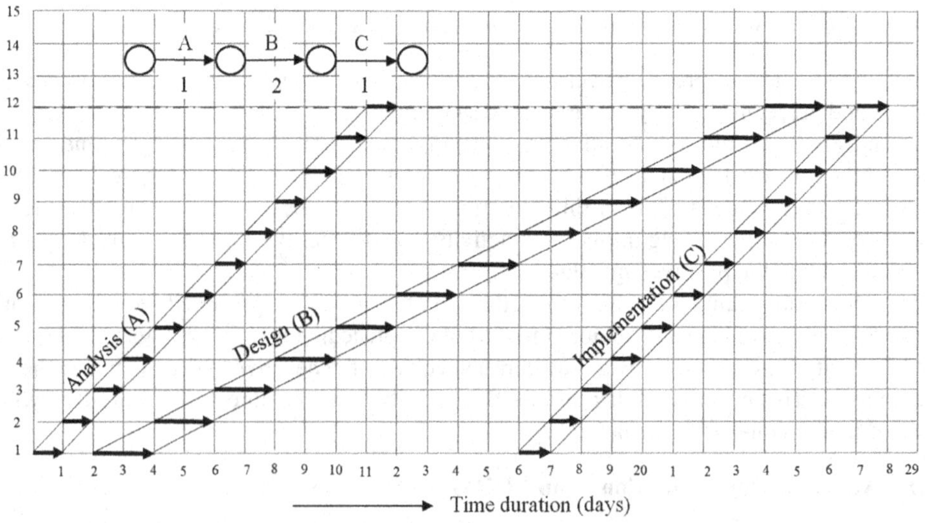

Figure 13.16 Alternate representation of an LOB graph

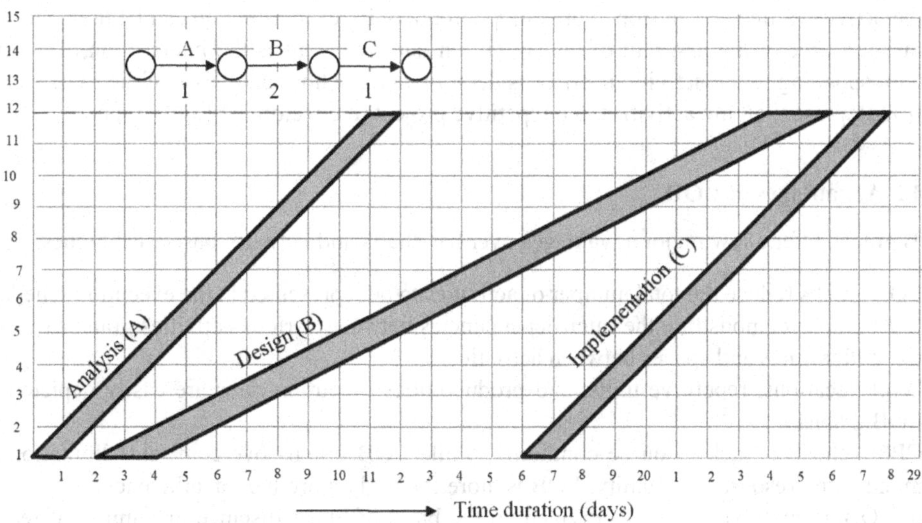

Figure 13.17 Alternate representation of the LOB graph shown in Figure 13.16

buffer times/spaces set between preceding and subsequent activities. The steps involved in the preparation of an LOB graph are as follows:

Step 1: Separate the repetitive and non-repetitive parts of the project in question. The scheduling of the repetitive part is done using LOB, and the scheduling of the non-repetitive part is done using the critical path method or PDM.

Step 2: Divide the repetitive part of the project into its repetitive units. Identify all the activities involved in a single repetitive unit. Develop a bar chart or network to represent the execution sequence of the activities involved in a single repetitive unit. The schedule of a single repetitive unit includes a group of activities in which no repetitive work is involved.

Step 3: Decide on the buffer times and spaces required between the different activities in the single repetitive unit to cater for unforeseen delays to the completion of preceding activities or the start of subsequent activities.

Step 4: Decide on the productivities/execution rates of all the activities involved in the single repetitive unit of the repetitive project. The productivities/execution rates of the various activities are used to represent the slopes of the lines that correspond to the various activities on an LOB graph. This is incidentally similar to determining the time durations of all the activities in a project.

Step 5: Decide on the start time of the first repetitive unit of the starting activity and draw a line, starting from the start time of that first repetitive unit, with a slope equal to its productivity/execution rate. The first unit of the subsequent activity is started when the first unit of the preceding activity (the starting activity) is completed, and a line is drawn with a slope equal to the productivity/execution rate of the subsequent activity. Draw the subsequent activities in the same way until an activity with a higher productivity/execution rate than the activity in question is reached.

Step 6: If an activity with a higher productivity/execution rate than the preceding activity is drawn, as discussed in the last step, it will cross over the preceding activities. Thus, the last unit of the activity with the higher productivity/execution rate is started when the last unit of the preceding activity is completed. Using the known start and finish times of the last unit of the activity, a line with a slope equal to its productivity/execution rate can be drawn to find the start time of the first unit of the activity in question. This is the intersection of the line corresponding to the activity under consideration and the horizontal axis of the LOB graph.

Step 7: Represent all the activities in a repetitive project to develop the LOB graph.

13.12 Advantages of LOB

LOB graphs offer a few major advantages over bar charts and network-based techniques.

- LOB graphs reduce the total number of activities to be represented in the execution schedule.
- LOB graphs demonstrate the inter-dependencies between various activities graphically and use buffer times and spaces between activities.
- Work locations, repetitive units, and productivities/execution rates are easily depicted on LOB graphs.
- CPM schedules lack resource continuity, while LOB graphs overcome this limitation by maintaining resource continuity. LOB is more focused on production efficiency.
- in LOB a repetitive project is divided on the basis of linear distance or number of repetitive units, rather than just activities, for the better planning, scheduling, and management of repetitive projects.
- LOB graphs take less time and effort to develop than network-based techniques. The execution sequences of various activities are easily represented and modified, and their impacts can be observed simultaneously on the graph.

13.13 Conclusion

When network-based techniques are applied to repetitive project, they suffer from the limitation of non-continuous crew engagement. For planning, scheduling, and management repetitive projects, LOB is the most widely used technique. It ensures continuous crew engagement. A repetitive activity to be carried out multiple times at multiple locations is treated as a single activity in LOB. Therefore, the entire repetitive task is represented as a single activity using a single line. Schedule representation in an LOB graph is based on two axes: time and distance/units. In general, the time duration of a project is represented along the x-axis. Distance/ units is plotted along the y-axis. Activities are represented by lines with constant or sometimes changing slopes. The slope of a line represents the progress rate of an activity. LOB graphs use buffer times and spaces between activities. LOB graphs take less time and effort to develop than network-based techniques.

Example 13.1: A fiber cable-laying project consists of the following five activities.

a. Trench excavation.
b. Laying the base course.
c. Laying fiber cable with covering.
d. Trench backfilling.
e. Soil compaction.

The length of the fiber cable is 1000 meters. The production rates of the five activities are 100, 50, 100, 100, and 200 meters per day respectively. Draw an LOB graph for the repetitive project, maintaining a one-day buffer time between each activity. Assume that a subsequent activity starts on the next day.

Solution: The total length of the fiber cable is 1000 meters. The time duration of each activity is obtained by dividing the total length of the cable by its production rate.

The time durations of the various activities are 10, 20, 10, 10, and 5 days respectively. Assume activity A starts at the start of day 1 and ends at the end of day 10.

Maintaining a one-day buffer time between each activity, activity B starts at the start of day 3 and ends at the end of day 22.

The time duration of activity C is 10 days. A one-day buffer time is required, thus activity C ends at the end of day 24. To determine the starting point of activity C, its time duration of 10 days is subtracted from its finish time at the end of day 24. Thus, activity C starts at the start of day 15 or at the end of day 14.

Maintaining a one-day buffer time between each activity, activity D starts at the start of day 17 and ends at the end of day 26.

Activity E has a time duration of 5 days, and a one-day buffer time is required. Thus, the activity finishes at the end of day 28. To determine the starting point of activity E, its time duration of 5 days is subtracted from its activity finish time – that is, the end of day 28. Thus, activity E starts at the start of day 24 or at the end of day 23.

Finally, as shown in Figure 13.18, the time duration of the project is 28 days.

Example 13.2: A project involves the construction of ten similar houses. The construction of a house involves the following five activities.

a. Preparation of the site.
b. Construction of the foundation.
c. Construction of the superstructure.
d. Electrical fittings.
e. Final finishing.

The production rates of the five activities are 5, 2, 0.5, 2, and 1 units per day respectively. Draw an LOB graph for the repetitive project. Assume that a subsequent activity starts on the next day.

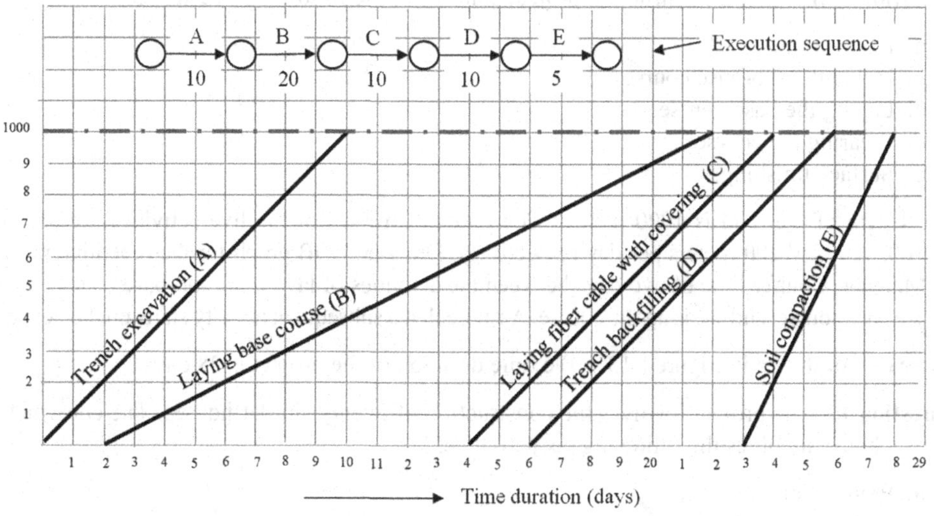

Figure 13.18 LOB graph of a project

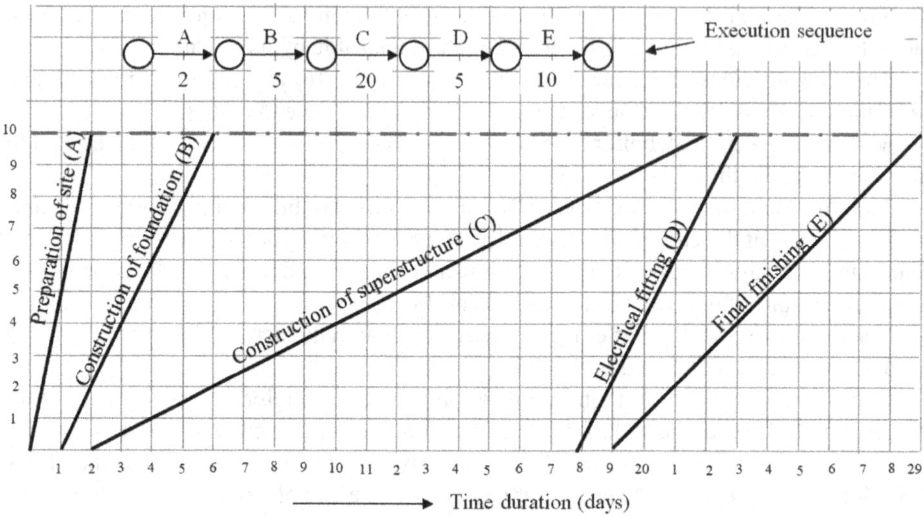

Figure 13.19 LOB graph of a project

Solution: The time duration of each activity is obtained by dividing the total number of units to be completed by its production rate. The time durations of the various activities obtained 2, 5, 20, 5, and 10 days respectively. Assume that activity A starts at the start of day 1 and ends at the end of day 2. Activity B starts at the start of day 2 and ends at the end of day 6. Activity C starts at the start of day 3 and ends at the end of day 22. Activity D finishes at the end of day 23, and its time duration is 5 days, thus it starts at the start of day 19. Activity E starts at the start of day 20 and ends at the end of day 29. Finally, as shown in Figure 13.19, the time duration of the project is 29 days.

Exercises

Question 13.1: A road construction project consists of the following five activities.

a. Excavation.
b. laying the sub-base course.
c. Laying the base course.
d. Wearing the course.
e. Surface finishing.

The length of the road is 1000 meters. The production rates of the five activities are 50, 100, 100, 100, and 200 meters per day respectively. Draw an LOB graph for the repetitive project. Maintain a buffer time of two days between the activities of laying the sub-base course, laying the base course, and wearing the course. Assume that a subsequent activity starts on the next day.

Answer: As shown in Figure E13.1, the time duration of the project is 28 days.

Question 13.2: A project involves the construction of twenty similar houses. The construction of a house involves the following six activities.

a. Preparation of the site.
b. Construction of the foundation.

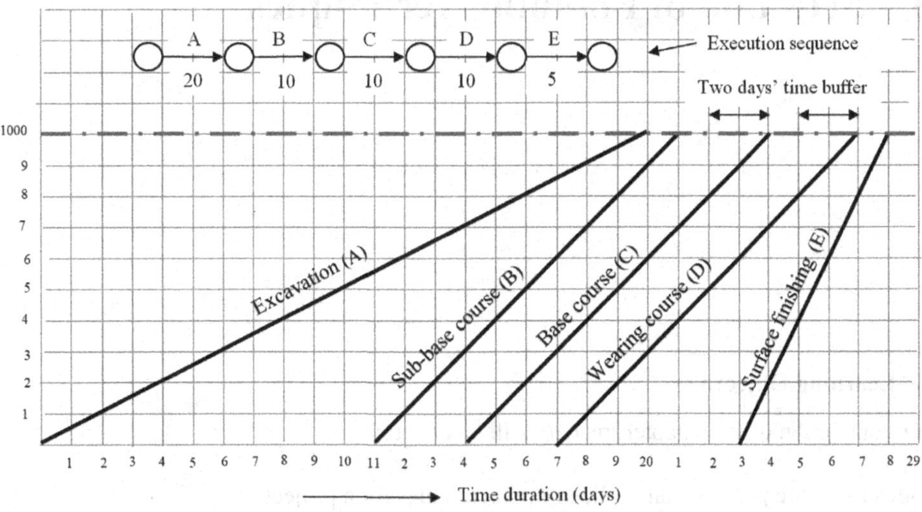

Figure E13.1 LOB graph of a project

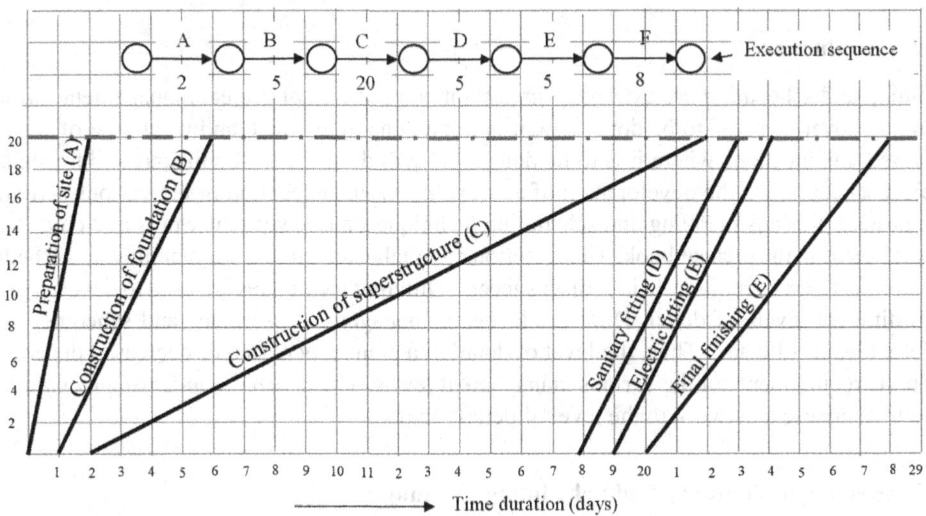

Figure E13.2 LOB graph of a project

c. Construction of the superstructure.
d. Sanitary fittings.
e. Electric fittings.
f. Final finishing.

The production rates of the five activities are 10, 4, 1, 4, 4, and 2.5 units per day respectively. Draw an LOB graph for the repetitive project. Assume that a subsequent activity starts on the next day.

Answer: As shown in Figure E13.2, the time duration of the project is 28 days.

14 Selection of Planning Techniques

14.1 Learning Objectives

After completion of this chapter, readers will be able to:

- Select suitable planning and scheduling techniques for a project,
- Determine the level of detail required in a project plan, and
- Determine a suitable technique for reducing the time duration of a project.

14.2 Introduction

Planning and scheduling are extremely important components of project management. Project planning covers what is to be done, how it is to be done, how much and what type of work is involved, the locations where it is to be done, who will do it, when it will start and finish, etc. Project scheduling is the development of a timetable for the execution of the various activities in a project. Various planning and scheduling techniques, along with their pros and cons, have been covered in the present book. The selection of suitable techniques for planning and scheduling a project plays a major role in project success. Different techniques are used to develop plans with different levels of detail. Different levels of management use plans and schedules with different levels of detail. Thus, the level of detail in a plan is increased or decreased depending upon the requirements of the various management levels. The *network condensing technique* is used to condense a network to the level of detail required.

14.3 Selection of Planning And Scheduling Techniques

The technique which should be used for planning and scheduling a given project depends upon its nature and scope. The simplest technique used for planning and scheduling routine, simple, and non-repetitive projects is the bar chart. The technique used for planning and scheduling routine, simple, and repetitive projects is LOB. Bar chart and LOB techniques are used at the project and sub-project levels in the case of simple routine projects. The technique used for planning and scheduling non-routine projects at the project and sub-project levels is PERT.

The planning and scheduling of routine, complex, and non-repetitive projects are done using network-based techniques. PDM is the most widely used technique for planning and scheduling routine, complex, and non-repetitive projects at the project level. However, CPM is the most widely used technique for planning and scheduling routine, complex, and non-repetitive projects at the sub-project level. PDM or LOB techniques are widely used for planning and scheduling routine, complex, and repetitive projects at the project level. However, the LOB technique

DOI: 10.1201/9781003428992-14

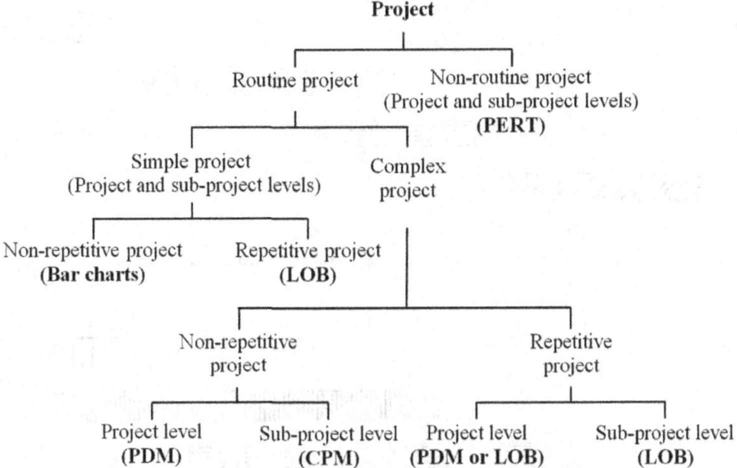

Figure 14.1 Selection of planning and scheduling techniques

is also widely used for planning and scheduling routine, complex, and repetitive projects at the sub-project level. A summary of the most commonly used planning and scheduling techniques for different types of projects is given in Figure 14.1.

PDM is the most widely used technique for planning and scheduling routine, complex, and non-repetitive projects at the project level. Sometimes, in large and complex projects, PDM is also used at the sub-project level. However, CPM is the technique used at the sub-project, task, and work package level for networks with fairly accurate time durations estimates for their project activities. At all work breakdown levels, time-scaled networks and bar charts are widely used to display and communicate schedules. Bar charts are widely used to display and convey plans and schedules to execution-level staff. However, there are no solid rules and regulations about the technique to be used when planning and scheduling a project.

14.3.1 Sample Example

The technique to be used when planning and scheduling a project varies from planner to planner depending upon their experience, level of understanding, and individual perspective. Consider a project that involves the construction of three similar parking sheds. The activities involved in the project are the construction of the foundation, the construction of the structure, roofing, and finishing. Figure 14.2(a) shows the execution sequence of the construction in the form of a bar chart in which all four activities are in a sequence. In this case, the time duration of the project is 12 days. Practically, it is not necessary to finish constructing the foundation to start putting up the structure. Similarly, it is not necessary to finish putting up the structure to start the roofing. If this is the case, the bar chart shown in Figure 14.2(a) does not reflect the correct execution sequence. Figure 14.2(b) shows the correct execution sequence for the construction of three similar parking sheds, in which these activities are given a suitable degree of overlap.

In order to make the construction plan more understandable, each activity in the project has been divided into three sub-activities. Figure 14.2(c) shows an execution sequence for the construction of three similar parking sheds in which all twelve sub-activities are shown. In the bar chart, once the foundation work of the first parking shed is finished, its structure work and the

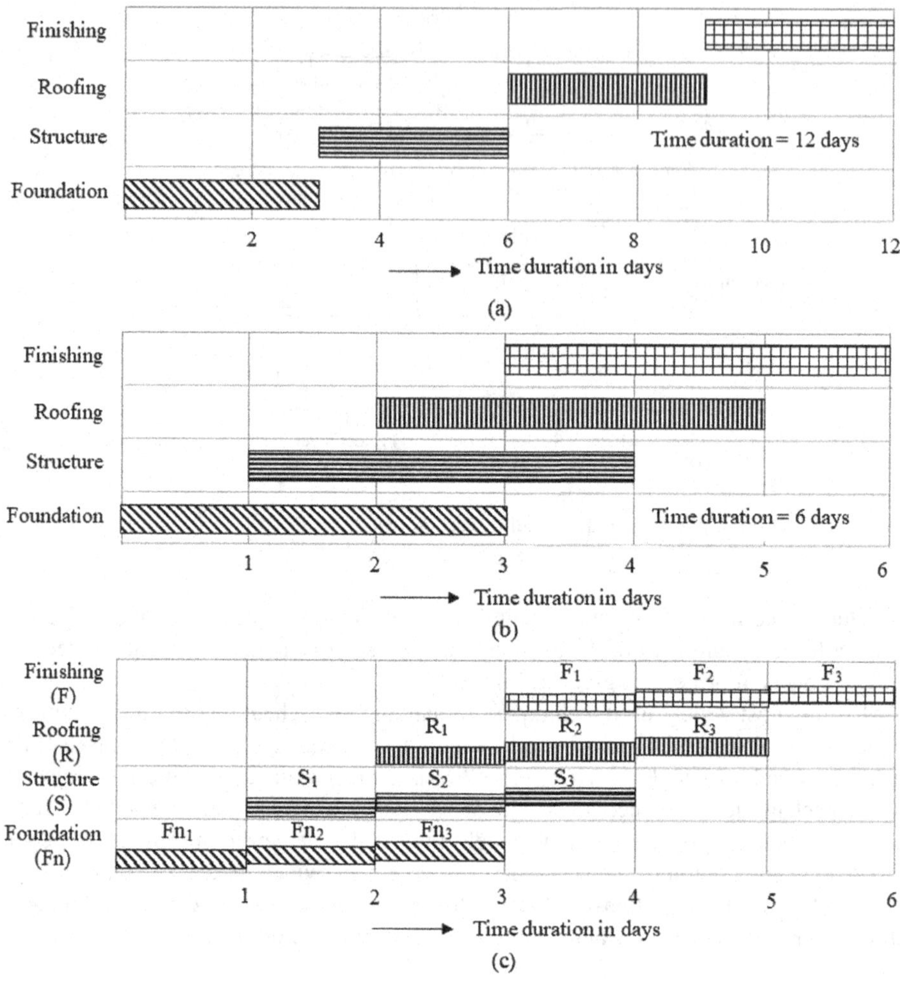

Figure 14.2 Bar charts; (a) activities in series, (b) activities with suitable degree of overlap, and (c) division of activities into sub-activities

foundation work of the second shed start at the start of day 2. Once the structure work of the first and the foundation work of the second parking sheds are finished, the roofing work of the first, the structure work of the second, and the foundation work of the third parking sheds start at the start of day 3. In this way, the construction goes on. The project plan, in the form of a bar chart, looks as shown in Figure 14.2(c).

A ladder-type network used to represent the execution sequence of the construction of three similar parking sheds is shown in Figure 14.3. This type of representation is used for projects in which activities are few in number but significant in terms of the volume of work involved, and in which the activities have long time durations. Thus, it is necessary to divide the activities of such projects into sub-activities. However, the main disadvantage of ladder-type networks is the increase in the number of activities. For the project shown in Figure 14.2(a), four activities have been sub-divided into 12 sub-activities.

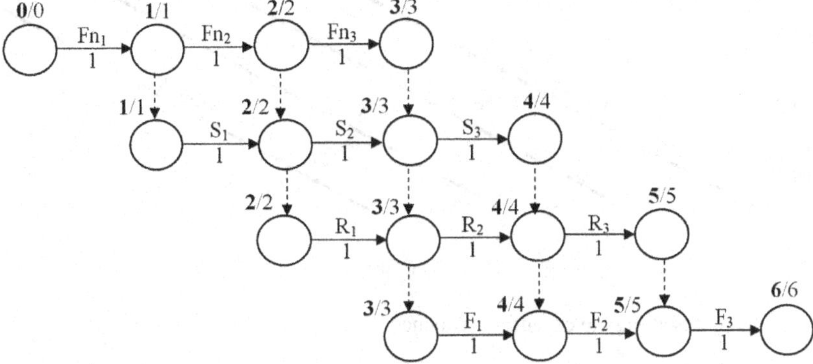

Figure 14.3 Ladder-type network representing the execution sequence

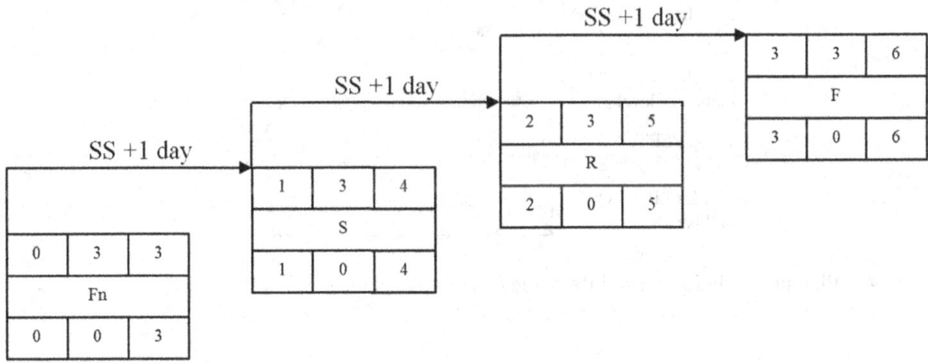

Figure 14.4 PDM representing the execution sequence

Another method to represent the plan of the project shown in Figure 14.2 is PDM. Practically, at least the foundation work of the first parking shed must be completed to start its structure work. That is, if the foundation work starts at the start of day 1, the structure work can start at the start of day 2. The roofing work can start at the start of day 3 and the finishing work at the start of day 4, and the project will finish at the end of day 6. If the time duration estimates have a fair degree of accuracy and the precedence relationships are established properly, PDM works well for such projects as shown in Figure 14.4.

LOB is another technique to represent the plan of the project shown in Figure 14.2. In Figure 14.5, the foundation work starts at the start of day 1 and the structure work starts at the start of day 2. The roofing work starts at the start of day 3 and the finishing work starts at the start of day 4 and the project ends at the end of day 6.

14.4 Techniques for Different Levels of Management

The first step in project management is breaking down a project into small parts to develop a WBS. This involves splitting up a project into small manageable divisions, sub-divisions, and further sub-divisions. The hierarchical representation of all the divisions and sub-divisions of a project in the form of a WBS facilitates its planning, scheduling, management,

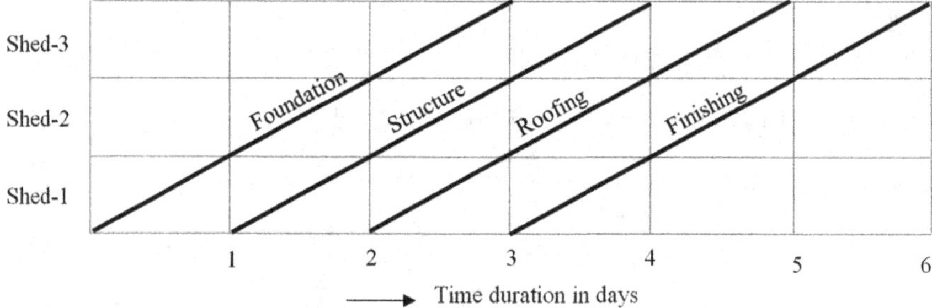

Figure 14.5 LOB representing the execution sequence

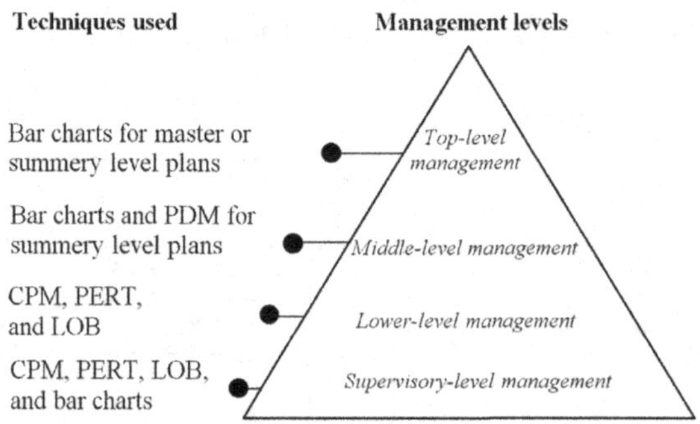

Figure 14.6 Planning techniques for different levels of management

and completion within its allocated time duration using the resources assigned. The work breakdown levels in a WBS are the project level, sub-project level, task level, work package level, and activity level. If an experienced planner is involved in the planning process, the number of levels may be lower; however, if a planner has limited experience, the number of levels may be higher. Splitting up a sub-division at the lower levels is complicated as it requires knowledge of the methods and techniques employed in project execution. However, projects are better planned, scheduled, controlled, and monitored at lower levels, like package or activity levels.

In Figure 14.6 different levels of management have been represented in a pyramid. The first level is *top-level management*; the second level is *middle-level management*. The third level is *lower-level management* and the last level is *supervisory-level management*. Different management levels need different levels of project detail. For example, top-level management is responsible for decision-making and does not require the execution-level detail of a project plan. Therefore, bar charts are used to represent master or summary-level project plans for top-level management. Bar charts and PDM are used to represent master-level project plans for middle-level management. However, lower-level management requires operational-level detail in a project plan, which is thus generally developed using CPM, PERT, or LOB techniques depending upon the type of project. Supervisory-level management needs only operational-level detail, developed using CPM, PERT, or LOB techniques, but the schedule is generally communicated using bar charts or time-scaled versions of the networks.

14.4.1 Network Condensation

Networks are condensed to reduce the level of detail in the project plan. Network condensation is used to develop networks with different levels of detail for different management levels or sometimes to reduce unnecessary detail. Top-level management generally does not require operational-level detail in a network to make the necessary decisions. Thus, a network containing operational-level detail may be condensed to develop a summary network for top-level management. Network condensation is applied throughout the network for the purpose of developing a project plan with different levels of detail. The number of activities in a network is reduced in the process of network condensation. A network containing hundreds of activities is reduced to a network of a few activities. These condensed networks are used by the different levels of management to make the necessary decisions or implement the plan.

The inter-dependencies between the various activities in a project are not disturbed in the process of network condensation. Activities which are independent of other activities are generally condensed without disturbing network logic. For example, the network shown in Figure 14.7(a) has thirteen activities. Its condensed version is shown in Figure 14.7(b). Activities C, E, and H are in a series which has been condensed into a single activity. Activities D and G have also been condensed into a single activity. Similarly, activities B and F have been condensed. The condensation of activities which are in series does not change the network logic. Activities L and M are parallel to each other and have been condensed to a single activity; this also does not change the network logic. The number of activities in the network has been reduced, through the process of condensation, to eight.

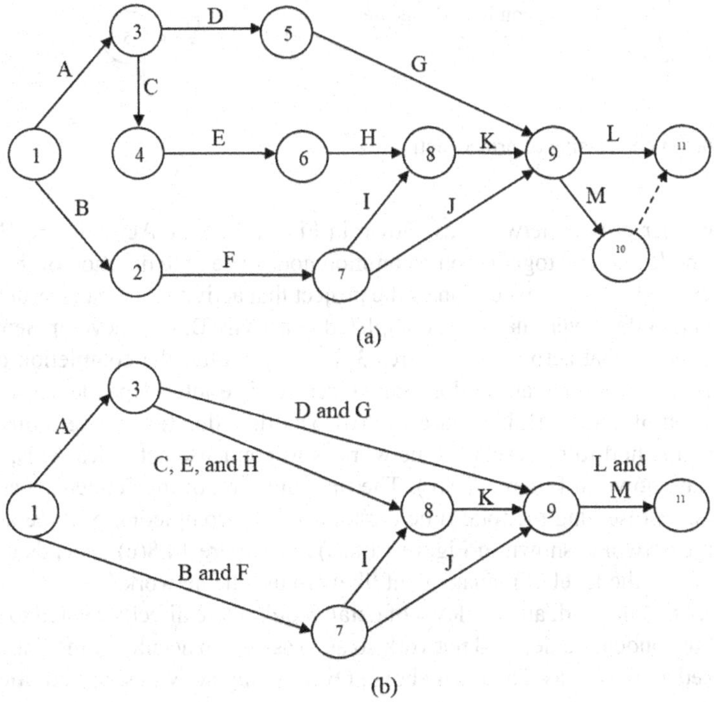

Figure 14.7 Network condensation

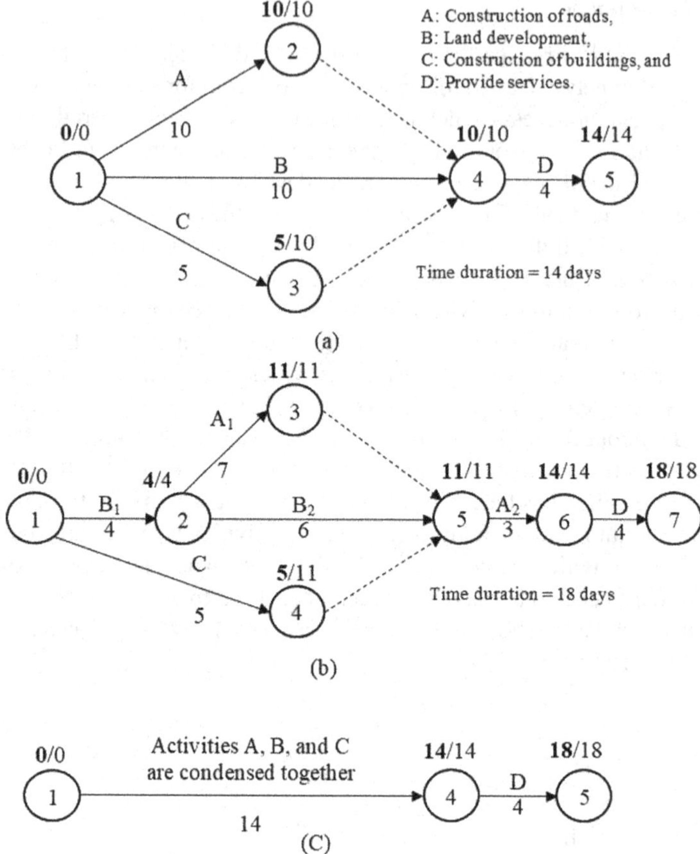

Figure 14.8 Correct condensing of project activities

Consider another project network, as shown in Figure 14.8(a). Activities A, B, and C start from the same node and end together on a common node. The time duration of the project is 14 days. It is observed during the execution of the project that activity A, the construction of roads, cannot begin until 4 days work has been completed in activity B, land development. In addition to this, it is observed that activity A requires 3 days' wait after the completion of activity B. However, activities B and C can be done concurrently. The inter-dependencies described are shown in the form of a network in Figure 14.8(b). The time duration of the project is 18 days. Another simple method to represent the network is to condense activities A, B, and C into a single activity as shown in Figure 14.8(c). The time duration of the condensed activity in this case is 14 days, representing the total time duration for the completion of three activities. The suitability of the networks shown in Figure 14.8(b) and Figure 14.8(c) depends upon the level of detail required or the level of management likely to use the network.

The accuracy of a network and the level of detail required are directly related to each other. It is often useful to condense a detailed network so as to assign an accurate time duration estimate to the condensed activities, as discussed above. Overlapping activities, or activities in parallel, occur frequently in normal or fast-tracked projects. Sometimes activities are not completely concurrent or are have some degree of overlap; in such cases, PDM is useful for representing

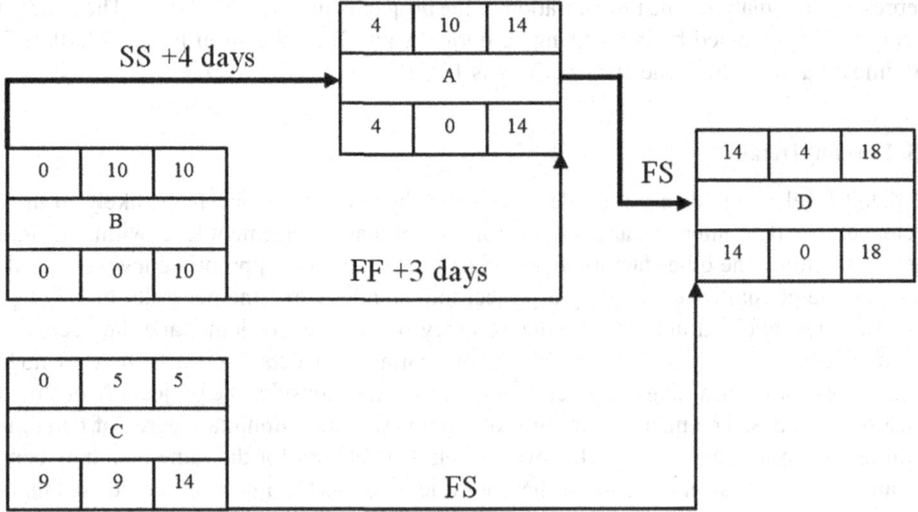

Figure 14.9 PDM representation of the activities in Figure 14.8(b)

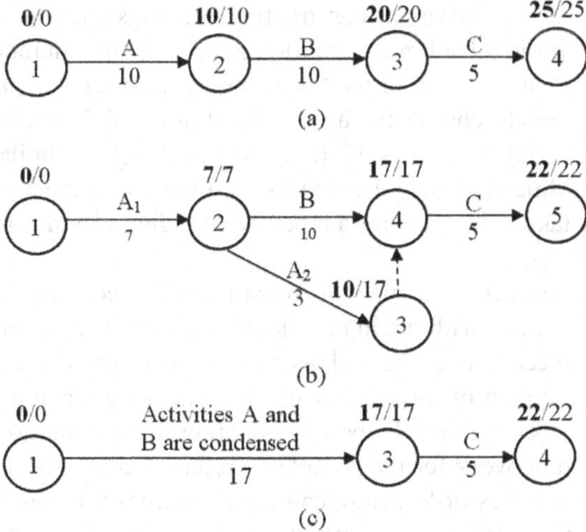

Figure 14.10 Correct condensing of project activities

concurrent activities. The network shown in Figure 14.8(b) has been redrawn using PDM, as shown in Figure 14.9. It shows activities A, B, C, and D, and the relationships between them using nodes, arrows, and time lags. The activities have not been subdivided; however, the relationships between them are correctly represented.

Another example of network condensation is shown in Figure 14.10, in which activities A, B, and C are in series. The time duration of the project shown in Figure 14.10(a) is 25 days. It is noticed during the execution of the project that activities A and B need not be completely in series; activity B may begin 3 days before the completion of activity A. The network corresponding to this is shown in Figure 14.10(b), where activity A has been divided into two parts to represent the situation. The time duration of the project in this case is 22 days. The condensed network can be obtained by condensing activities A and B as shown in Figure 14.10(c). The total time duration of the condensed activity is 17 days.

14.5 Level of Detail

The detail level required in a plan depends upon the level of management likely to use the developed plan, their interests, and the control span of that management level within its organizational structure. The other factors involved in determining the appropriate level of detail for a plan are the possibility of dividing a project into smaller parts, the necessity of sharing the responsibilities involved in a project, the accuracy of the network logic, and the necessity of providing better time duration estimates by incorporating more detail. As such, there are no hard and fast rules for determining the level of detail of a plan. Consider the project of constructing an institute campus. The plan, in the form of a network with a minimal degree of detail, may appear as shown in Figure 14.11. However, a detailed network for the same plan may contain thousands of activities. In the case of the lowest level of detail, a network could contain only a few activities, and in the case of the highest level of detail, the same network could contain thousands of activities. The network shown in Figure 14.11 may be helpful for top-level management only.

The activity of constructing buildings, in Figure 14.11, contains many buildings. An accurate time duration estimate for the activity of constructing buildings requires a list of the number of activities, of the activities involved in each building, of their time durations, and of the interdependencies between them. Further, at the lowest operational level, a plan is used to monitor day-to-day operations, which requires the highest level of detail. Thus, it is worthwhile developing a plan for each building with the highest level of detail. If similar buildings are to be constructed, it is worthwhile equipping the plan for a building with the highest level of detail. In such cases, a small mistake in the plan for a repetitive building would cause repeated problems across all similar buildings.

If a project contains several repetitive units or parts, a detailed plan of such a repetitive unit must be developed separately, with the highest level of detail. The detailed plan of the repetitive unit must contain execution-level detail, as well as a condensed version, and must use the condensed version of the plan of the repetitive unit to develop a complete project plan. The detailed network of a repetitive unit helps in estimating its time duration accurately, and in developing a condensed network for the complete project. Repetition of the detailed network of each repetitive unit unnecessarily complicates and enlarges the complete project network. The primary purpose of developing a separate detailed network of a portion of the project plan is to develop a more readable format of the complete project plan. An effective approach for improving network readability is providing details about the complex portions of a project

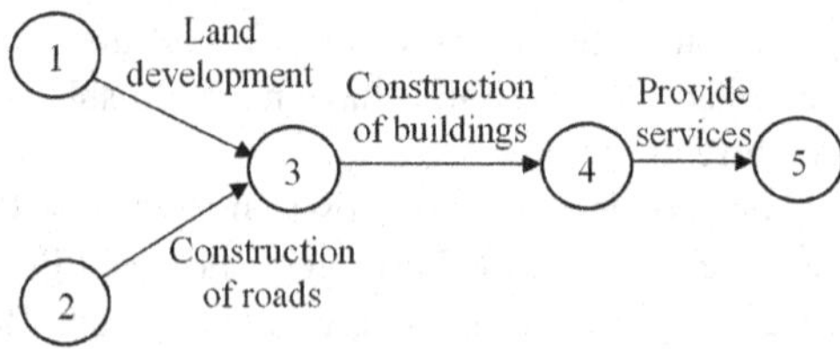

Figure 14.11 Network with low level of detail

plan separately. A WBS is used as the basis for the development of project plans, as discussed below.

Level-I: The plan at level-I is also called the *administrative summary of a project plan* or *master plan*. It is the summary plan provided along with project reports and other related documents when a detailed plan is not required. It involves the division of a project into sub-projects (or major parts) and features its key events (or milestones). Generally, plans and schedules of this level are used by top-level management, executive-level personnel, or administration, to help make necessary decisions. It is generally developed during the initial or feasibility stage of a project. Bar charts are generally used to represent such a plan and schedule.

Level-II: Plans of this level include the expansion of sub-projects into small manageable parts, sometimes called *tasks*. These sub-projects are sometimes delimited based on work locations, the nature of work involved, or the types of crew involved in the project execution. Plans of this level are sometimes provided along with detailed project reports and other related documents when detailed execution-level plans and schedules are not required. Generally, such plans and schedules are used by top-level management, project managers, or financers, to help make necessary decisions. Bar charts or PDM are generally used to represent such plans.

Level-III: Plans of this level involve the further expansion of tasks (defined at level-II) into small manageable parts, which are sometimes called work packages. Plans and schedules at this level are developed during the initial stage of project execution. Such plans are used by middle-level management. Level-II plans provide the framework and constraints used to develop this level of plans and schedules. Network-based techniques are used to develop plans and schedules at this level.

Level-IV: Plans of this level are execution-level plans. This involves the further expansion of work packages into small manageable parts called activities. These are working-level detail plans displaying the activities to be completed, the time duration of a project, and the criticalities involved. Generally, plans and schedules of this level are used by execution-level personnel, engineers, and superintendents to help attain time, cost, and quality objectives. Network-based techniques are used to develop such plans and schedules.

Level-V: This is the further sub-division of the activities defined in level-IV plans. This is the lowest-level of plans and schedules, used to monitor day-to-day operations. To make such plans executable, execution-level personnel are involved in their development. Network-based techniques are used to develop plans and schedules which are communicated to execution level staff in the form of bar charts. In general, schedules are prepared covering only short time durations, depending upon the complexity of a project. The schedules covering the remaining time durations are updated frequently. Plans and schedules of this level are used by execution-level superintendents, supervisors, team leaders, crew leaders, and foremen.

14.6 Techniques For Reducing Time Durations

Projects are often delayed, for various reasons. In such cases, to ensure a project is finished in time, its time duration is reduced. Fast-tracking and project crashing are widely used techniques for reducing the time duration of a project. Figure 14.12(a) shows three activities – A, B, and C – which are in a sequence in the normal planning situation. To complete the project in its normal time duration, the cost of the project is also normal. To reduce the time duration of the project, instead of waiting for the completion of an activity to start the subsequent activity, as many activities as possible are performed in parallel or with the required degree of overlap.

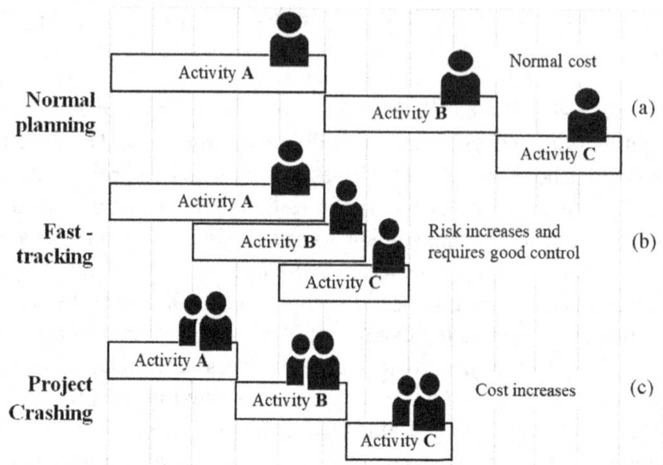

Figure 14.12 Techniques for reducing the time duration of a project

The technique of performing as many activities in parallel as possible to reduce the project time duration is called fast-tracking, as shown in Figure 14.12(b). It is important to constantly monitor project progress and take necessary corrective measures to ensure the required project progress when fast-tracking. Also, to perform multiple activities together, good project control is necessary. There is no additional cost involved in reducing the time duration of a project, however the risk increases. Fast-tracking does not apply to all activities, because only some activities are completely sequential.

If fast-tracking is not possible, crashing of a project/schedule is used. This is a technique in which additional resources are used to reduce the time durations of project activities. It involves employing additional resources to help finish a project before its normal time duration, as shown in Figure 14.12(c). The employment of these additional resources increases the project cost. To provide sufficient workspace for the additional resources employed, sometimes a second or possibly a third working shift may be planned. A project is completed in the planned sequence and in a shorter time duration than the normal one. The risk involved in the fast-tracking technique is greater than that of crashing a project/schedule. Project/schedule crashing is applied to the critical activities of a project. If the critical path of a project is not shortened, the time duration of a project will not be reduce. Initially, the fast-tracking technique is used to reduce the time duration of a project; however, if it is not possible to use this technique, project/schedule crashing is used.

14.7 Project Success or Failure

No matter how hard a planner tries, planning is not perfect every time, and sometimes a plan may fail. The main objective of project management is to complete a project within the constraints of the allocated time duration, available budget, and prescribed quality. These objectives must be made clear to the personnel involved in a project. However, the main reasons for project failure are an increase in the time duration of a project beyond the allocated time duration, or an increase in the project cost beyond the allocated budget.

The time duration of a project depends upon the time durations of its project activities and the inter-dependencies between them. The experience of a planner in handling projects like the project in question plays a major role in the accurate estimation of time durations. Limited experience and the unavailability of past information in time duration estimations result in a poor project plan. If the time duration estimates of the various activities in a project lack a fair degree of accuracy, project failure is certain. A project plan also becomes ineffective when a planner un-necessary increases or decreases the time duration of a project.

The financial success of a project is ensured by forecasting the inflow and outflow of funds required for the execution of a project. The inflow and outflow of funds jointly determine the time and the amount of net funds required to execute a project. Financial planning determines the success or failure of a project. Thus, a comprehensive examination of the estimated inflow and outflow of funds is necessary for project success.

Due to a lack of good communication systems in an organization, each management level may not understand the project objectives clearly. Thus, a good communication system must be established in an organization. Each management level must be made aware of the major activities and milestone dates, and the quality of product required. To develop an executable plan for project success, operational-level personnel must be involved in making planning decisions.

Efficient project planning also requires information about the project in question at the right time and location. The information available comes in different forms and in different formats obtained from different sources, and is available at different locations. Bringing such information together to aid effective planning for project success is sometimes a difficult task.

Other reasons for project failure may include not updating and determining plan correction measures in time, key personnel leaving the project team, resource availability problems, unavailability of skilled staff, lack of motivation, an un-reasonable management attitude, poor human relations, poor output, extreme optimism about output, not assessing risks properly, non-commitment of the personnel involved in the project, etc.

14.8 Conclusion

Various planning and scheduling techniques have been covered in the present book. The selection of suitable techniques for planning and scheduling a project depends upon the project's complexity, repetitive or non-repetitive nature, routine or non-routine nature, and the level of management likely to use the plan and schedule. The present chapter covers a sample planning example in which bar chart, network, PDM, and LOB techniques have been used to demonstrate the suitability of planning and scheduling techniques. Different levels of management use plans and schedules with different levels of detail. Thus, the network condensing process is used to develop plans and schedules with different levels of detail. Five levels of project plan detail have been discussed in the present chapter. The most widely used techniques for shortening the time duration of a project have also been covered. Finally, the factors which control the success or failure of a project have been discussed.

Exercises

Question 14.1: How are the accuracy of a network and the level of detail required in it directly related to each other?

Question 14.2: What is the network condensation process and why it is required?

Question 14.3: Define the fast-tracking of projects and briefly discuss its advantages and disadvantages.

Question 14.4: What are the different detail levels of a project plan and why are these are required?

Question 14.5: Differentiate between fast-tracking and project crashing.

Bibliography (for Additional Readings)

Ahuja, H. N., Dozzi, S. P., and Abourizk, S. M. (1994). *Project Management: Techniques in Planning and Controlling Construction Projects*. Second edition, John Wiley & Sons, Inc., New York.

Chitkara, K. K. (2014). *Construction Project Management: Planning, Scheduling, and Controlling*. Third edition, McGraw Hill Education (India) Private Limited, New Delhi.

Hajdu, M. (1997). *Network Scheduling Techniques for Construction Project Management*. Springer-Science+Business Media, B.V., Kluwer Academic Publisher, Dordrecht, The Netherlands.

Kerzner, H. (2017). *Project Management: A Systems Approach to Planning Scheduling and Controlling*. Twelfth edition, John Wiley & Sons, Inc., Hoboken, New Jersey.

Moder, J. J., Phillips, C. R., and Davis, E. W. (1983). *Project Management with CPM, PERT and Precedence Diagramming*. Third edition, Van Nostrand Reinhold Company, New York.

Mubarak, S. (2015). *Construction Project Scheduling and Control*. Third edition, John Wiley & Sons, Inc., Hoboken, New Jersey.

Naylor, H. F. W. (1995). *Construction Project Management: Planning and Scheduling*. Delmar Publishers, A division of International Thomson Publishing Inc., New York.

O'Brien, J., and Plotnick, F. L. (2006). *CPM in Construction Management*. Sixth edition, McGraw-Hill Publishing Company, New York.

PMBOK (2017). *A Guide to the Project Management Body of Knowledge (PMBOK® GUIDE)*. Sixth edition, Project Management Institute, Pennsylvania, USA.

Punmia, B. C., and Khandelwal, K. K. (2002). *Project Planning and Control with CPM and PERT*. Fourth edition, Laxmi Publication, New Delhi.

Index

Printed in the United States
by Baker & Taylor Publisher Services